Valentin L. Popov · Markus Heß · Emanuel Willert

Handbook of Contact Mechanics

Exact Solutions of Axisymmetric Contact Problems

Valentin L. Popov
Institut für Mechanik
Technische Universität Berlin
Berlin, Germany

Emanuel Willert
Institut für Mechanik
Technische Universität Berlin
Berlin, Germany

Markus Heß
Institut für Mechanik
Technische Universität Berlin
Berlin, Germany

https://doi.org/10.1007/978-3-662-58709-6

Preface

Contact mechanics deals with phenomena of critical importance for countless physical, technical, and medical applications. In classical mechanical engineering alone the scope of applications is immense, examples of which include ball bearings, gear drives, friction clutches, or brakes. The field of contact mechanics was originally driven by the desire to understand macroscopic problems such as rail-wheel contact or the calculation of stresses in building foundations. In recent decades, however, it has conquered qualitatively new areas of application at the forefront of global development trends in technology and society. The fields of micro-technology, where boundary properties play a central role, as well as biology and medicine have been particularly important additions to the vast spectrum of applications. The force-locking of screw connections, the adhesive strength of bonded joints, fretting wear of turbine blades, friction damping of aerospace structures, extraction methods of broken implants, and certain methods of materials testing shall also be named as examples of the extended scope encompassed by contact mechanics.

This expansion in scope has multiple roots, which are technological, experimental, or numerical in nature. This is owed in large part, but not limited to, the rapid development of numerical methods for the calculation of contacts. Thanks to the development of the Fast Fourier Transform(FFT)-based boundary element method in recent years, the field of contact mechanics has arguably taken on a leading role in the development of numerical calculation methods. However, this development certainly does not render analytical solutions obsolete. Due to the new numerical methods, the existing analytical solutions gained remain of immense importance. Today they are employed for the testing of numerical methods, to further the general "analytical understanding", or for empirically capturing numerical results in multi-dimensional parameter spaces. The "exact solutions" enjoy particular importance for their indisputable reliability and take on a position of great significance in contemporary science and technology.

The authors have set two goals for this book:

- The first goal is to provide a "complete" systematic catalog of all "significant" axially symmetric contact problems discovered in the last 137 years (since the classic work by Hertz 1882).
- The second goal is to provide not just the solutions of all these contact problems but also offer a detailed documentation of the solution process.

Of course these goals are not easily attained. The meanings of "complete" catalog and "significant" solutions are highly debatable already. Luckily the scientific community has done a great amount of work in the past 100 years to identify a set of characteristic problems of great practical relevance that required repeated research. This includes "generic" profile shapes such as the parabolic body, which offers a first-order approximation of nearly any curved surface. The general power law profile has also been considered repeatedly for the past 77 years, as any differentiable function can be described by a power series expansion. However, technical profiles are not necessarily differentiable shapes. Various applications employ flat cylindrical punches with a sharp edge or sharp-tipped conical indenters. In turn, absolutely sharp edges and tips cannot be realized practically. This leads to a set of profiles such as flattened spheres, flattened or rounded cones, and cylinders, etc., which offer a solid approximation of reality and have been the repeated subject of research for at least 77 years (since the work of Schubert 1942).

The second goal, which is to provide detailed documentation of the solution process, might seem overambitious at first glance, especially considering that certain historical publications dealing with a single solution of a contact mechanical problem amount to small volumes themselves. Jumping forward, this issue can be considered resolved in the year 2019. Instead of the original solution, we describe the simplest available one at present. For the normal contact problem without adhesion, the simplest known solution was found by Schubert in 1942, and later by Galin (1946) and Sneddon (1965). In this book, this solution will be used in the interpretation of the method of dimensionality reduction (Popov and Heß 2013). This approach requires zero prerequisite knowledge of contact mechanics and no great feats of mathematics except single variable calculus. Complicated contact problems such as adhesive contact, tangential contact, or contacts with viscoelastic media can be reduced to the normal contact problem. This allows a highly compact representation, with every step of solution processes being fully retraceable. The sole exceptions to this are the axially symmetric problems without a compact contact area, which so far lack a simple solution. In this case, we will mostly limit ourselves to the formal listing of the solutions.

This book also deals with contact mechanics of functionally graded materials, which are the subject of current research. Again, complete exact solutions will be provided using the method of dimensionality reduction.

Berlin
March 2019

Valentin L. Popov
Markus Heß
Emanuel Willert

Acknowledgments

In addition to the worldwide, century-long development of contact mechanics, this book builds on a great number of scientific results within the department of "System Dynamics and Friction Physics" from the years 2005–2017. We would like to express a heartfelt thanks to all of our colleagues who have been a part of these developments.

We would also like to offer our sincere gratitude to Dr.-Ing. J. Starčević for her complete support in the writing of this book, and to Ms. J. Wallendorf for her help in creating the countless illustrations contained within it.

English translation from the first German Edition was performed by J. Meissner and J. Wallendorf under editorship of the authors.

We would like to acknowledge support from the German Research Foundation and the Open Access Publication Funds of TU Berlin.

Contents

Introduction 1

1.1 On the Goal of this Book

The works of Hertz (1882) and Boussinesq (1885) are generally considered the beginning of classical contact mechanics. The solutions for the pressure distribution under a cylindrical flat punch and a sphere that are featured in those works certainly enjoy a high level of prominence. Yet multiple exact solutions exist which are of similar technical relevance to the Hertzian contact problems, but only a limited number of specialists know about these. Among other reasons, this is due to the fact that many individual contact mechanical solutions were published in relevant journals, however, a generalized representation in any complete monograph is lacking. Exceptions to this can be found in the books by Galin (2008) and Gladwell (1980), yet even these were written with scientific usage in mind rather than as a handbook for technical applications. This book aims to provide a compendium of exact solutions for rotationally symmetric contact problems which are suitable for practical applications.

Mathematically the terms "rotationally symmetric" and "contact problem" are quite straightforward to define. But what is an "exact solution"? The answer to this question is dual-faceted and involves an aspect of modeling; consideration must also be given to the final structure of what one accepts as a "solution". The first aspect is unproblematic: any model represents a certain degree of abstraction of the world, and makes assumptions and simplifications. Any solution derived from this model can, of course, only be as exact as the model itself. For example, all solutions in this book operate under the assumption that the resulting deformations and gradients of contacting surfaces within the contact area are small.

The second aspect is tougher to define. A "naïve" approach would be that an exact solution can be derived and evaluated without the aid of a computer. However, even the evaluation of trigonometric functions requires computation devices. Does a solution in the form of a numerically evaluated integral or a generalized, perhaps hypergeometric function count as "exact"? Or is it a solution in the form of a differential or integral equation? In exaggerated terms, assuming the valid-

V.L. Popov, M. Heß, E. Willert, *Handbook of Contact Mechanics*,
https://doi.org/10.1007/978-3-662-58709-6_1

ity of a respective existence and uniqueness theorem, simply stating the complete mathematical description of a problem already represents the implicit formulation of its solution. Recursive solutions are also exact but not in closed form. Therefore, distinguishing between solutions to be included in this compendium and those not "exact enough" remains, for better or worse, a question of personal estimation and taste. This is one of the reasons why any encyclopedic work cannot ever—even at the time of release—make a claim of comprehensiveness.

The selection of the problems to be included in this book were guided by two main premises: the first one being the technical relevance of the particular problem, and secondly, their place in the logical structure of this book, which will be explained in greater detail in the next section.

1.2 On Using this Book

Mechanical contact problems can be cataloged according to very different aspects. For instance, the type of the foundational material law (elastic/viscoelastic, homogeneous/inhomogeneous, degree of isotropy, etc.), the geometry of the applied load (normal contact, tangential contact, etc.), the contact configuration (complete/incomplete, simply connected/ring-shaped, etc.), the friction and adhesion regime (frictionless, no-slip, etc.), or the shape of contacting bodies are all possible categories for systematization. To implement such a multi-dimensional structure while retaining legibility and avoiding excessive repetition is a tough task within the constraints of a book.

We decided to dedicate the first five chapters to the most commonly used material model: the linear-elastic, homogeneous, isotropic half-space. Chapters 7 through to 9 are devoted to other material models. Chapter 10 deals with ring-shaped contact areas. The chapters are further broken down into sub-chapters and sections, and are hierarchically structured according to load geometry, the friction regime, and the indenter shape (in that order). While each section aims to be understandable on its own for ease of reference, it is usually necessary to pay attention to the introductory sentences of e.g. Chap. 4 and Sect. 4.6 prior to Sect. 4.6.5.

Furthermore, many contact problems are equivalent to each other, even though it may not be obvious at first glance. For example, Ciavarella (1998) and Jäger (1998) proved that the tangential contact problem for an axially symmetric body can be reduced under the Hertz–Mindlin assumptions to the respective normal contact problem. In order to avoid these duplicate cases cross-references are provided to previously documented solutions in the book which can be looked up. Where they occur, these references are presented and explained as clearly and unambiguously as possible.

References

Boussinesq, J.: Application des Potentiels a L'etude de L'Equilibre et du Mouvement des Solides Elastiques. Gauthier-Villars, Paris (1885)

Ciavarella, M.: Tangential loading of general three-dimensional contacts. J. Appl. Mech. **65**(4), 998–1003 (1998)

Galin, L.A., Gladwell, G.M.L.: Contact problems—the legacy of L.A. Galin. Springer, the Netherlands (2008). ISBN 978-1-4020-9042-4

Gladwell, G.M.L.: Contact problems in the classical theory of elasticity. Sijthoff & Noordhoff International Publishers B.V., Alphen aan den Rijn (1980). ISBN 90-286-0440-5

Hertz, H.: Über die Berührung fester elastischer Körper. J. Rein. Angew. Math. **92**, 156–171 (1882)

Jäger, J.: A new principle in contact mechanics. J. Tribol. **120**(4), 677–684 (1998)

Normal Contact Without Adhesion

2

2.1 Introduction

We begin our consideration of contact phenomena with the *normal contact problem*. Consider two bodies pressed together by forces *perpendicular to their surfaces*. A prominent example is the wheel of a train on a rail. The two most important relationships that the theory of normal contact should deliver are:

1. The relationship between the normal force and the normal displacement of the body, which determines the stiffness of the contact and, therefore, the dynamic properties of the entire system.
2. The stresses occurring in the contact area, which (for example) are required for component strength analysis.

Without physical contact, there are no other contact phenomena, no friction, and no wear. Therefore, normal contact can be regarded as a basic prerequisite for all other tribological phenomena. The solution to the adhesive contact problem, the tangential contact problem, and contact between elastomers can also be reduced to the non-adhesive normal contact problem. In this sense, the non-adhesive contact problem forms a fundamental basis of contact mechanics. It should be noted that even during normal contact, a relative *tangential* movement between contacting surfaces can occur due to different transverse contraction of contacting bodies. As a result, friction forces between the surfaces can play a role, even for normal contact problems, and it must be specified how these tangential stresses are to be treated. The two most well-known and sudied limiting cases are, firstly, the frictionless normal contact problem and, secondly, the contact problem with complete stick. All frictionless contact problems will be referred to as *"Boussinesq problems"* since the famous Boussinesq solution for a cylindrical flat punch belongs to this category. The other limiting case of complete stick will be referred to as *"Mossakovskii problems"*.

V.L. Popov, M. Heß, E. Willert, *Handbook of Contact Mechanics*,
https://doi.org/10.1007/978-3-662-58709-6_2

2.2 Boussinesq Problems (Frictionless)

We consider the frictionless normal contact between two elastic bodies with the elasticity moduli E_1 and E_2, and Poisson's ratios ν_1 and ν_2, as well as shear moduli G_1 and G_2. The axisymmetric difference between the profiles will be written as $\tilde{z} = f(r)$, where r is the polar radius in the contact plane. This contact problem is equivalent to the contact of a rigid indenter with the profile $\tilde{z} = f(r)$ and an elastic half-space with the effective elasticity modulus E^* (Hertz 1882):

$$\frac{1}{E^*} = \frac{1-\nu_1^2}{E_1} + \frac{1-\nu_2^2}{E_2}. \tag{2.1}$$

The positive direction of $\tilde{z}$ is defined by the outward-surface normal of the elastic half-space. The normal component of the displacement of the medium w, under the influence of a concentrated normal force F_z in the coordinate origin, is given by the fundamental solution (Boussinesq 1885):

$$w(r) = \frac{1}{\pi E^*}\frac{F_z}{r}. \tag{2.2}$$

Applying the superposition principle to an arbitrary pressure distribution $p(x, y) = -\sigma_{zz}(x, y)$ yields the displacement field:

$$w(x, y) = \frac{1}{\pi E^*}\iint p(x', y')\frac{\mathrm{d}x'\mathrm{d}y'}{r}, \quad r = \sqrt{(x-x')^2 + (y-y')^2}. \tag{2.3}$$

The positive direction of the normal force and normal displacement are defined by the inward-surface normal of the elastic half-space. If we call the indentation depth of the contact d and the contact radius a, the mixed boundary conditions for the displacement w and the stresses σ at the half-space surface (i.e., $z = 0$) are as follows:

$$\begin{aligned} &w(r) = d - f(r), \quad r \le a,\\ &\sigma_{zz}(r) = 0, \quad r > a,\\ &\sigma_{zr}(r) = 0. \end{aligned} \tag{2.4}$$

Usually, a is not known *a priori*, but has to be determined in the solution process. The solution of the contact problem is found by determining the pressure distribution, which satisfies (2.3) and the boundary conditions (2.4). It should be noted that the application of both the superposition principle and the boundary conditions in the form (2.4) require linearity of the material behavior as well as the half-space approximation to be met; i.e., the surface gradient must be small in the relevant area of the given contact problem in the non-deformed and deformed state. If we call the gradient θ then the condition is $\theta \ll 1$. The relative error resulting from the application of the half-space approximation is of the order of θ^2.

For ordinarily connected contacts the non-adhesive normal contact problem was solved in its general form by Schubert (1942) (based on the paper by Föppl (1941)),

Galin (1946), Shtaerman (1949), and Sneddon (1965). In Sect. 2.3 we will describe these solutions using the interpretation given by the method of dimensionality reduction (MDR) (Popov and Heß 2013). Naturally, it is fully equivalent to the classical solutions.

2.3 Solution Algorithm Using MDR

The contact of any given axially symmetric bodies can be solved very easily and elegantly with the so-called MDR. The MDR maps three-dimensional contacts to contacts with a one-dimensional array of independent springs (Winkler foundation). Despite its simplicity, all results are exact for axially symmetrical contacts. The MDR allows the study of non-adhesive and adhesive contacts, tangential contacts with friction, as well as contacts with viscoelastic media. In this section we will describe the application of the MDR for non-adhesive normal contact problems. Generalizations for other problems will be presented where appropriate in later chapters. Complete derivations can be found in works by Popov and Heß (2013, 2015), as well as in Chap. 11 in this book (Appendix).

2.3.1 Preparatory Steps

Solving the contact problem by way of the MDR requires two preparatory steps.

1. First, the three-dimensional elastic (or viscoelastic) bodies are replaced by a Winkler foundation. This is a linear arrangement of elements with independent degrees of freedom, with a sufficiently small distance Δx between the elements.

In the case of elastic bodies, the foundation consists of linear-elastic springs with a normal stiffness (Fig. 2.1):

$$\Delta k_z = E^* \Delta x, \tag{2.5}$$

whereby E^* is given by (2.1).

2. Next, the three-dimensional profile $\tilde{z} = f(r)$ (left in Fig. 2.2) is transformed to a plane profile $g(x)$ (right in Fig. 2.2) according to:

$$g(x) = |x| \int_0^{|x|} \frac{f'(r)}{\sqrt{x^2 - r^2}} \mathrm{d}r. \tag{2.6}$$

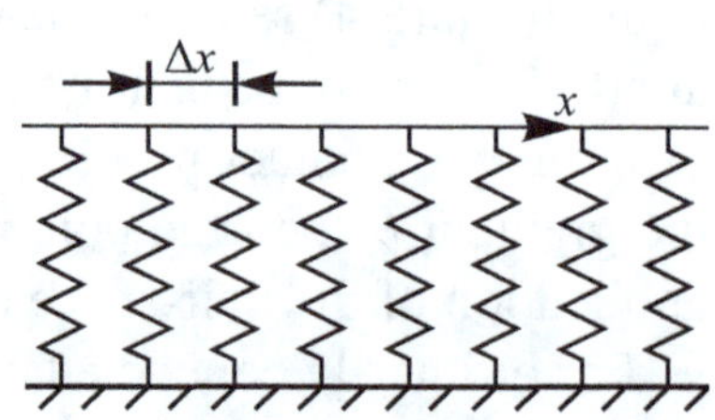

Fig. 2.1 One-dimensional elastic foundation

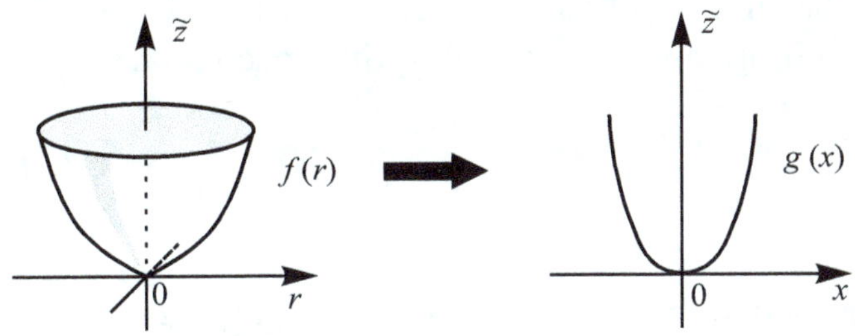

Fig. 2.2 Within the MDR the three-dimensional profile is transformed to a plane profile

The inverse transform is:

$$f(r) = \frac{2}{\pi}\int_0^r \frac{g(x)}{\sqrt{r^2 - x^2}}\mathrm{d}x. \tag{2.7}$$

2.3.2 Calculation Procedure of the MDR

The plane profile $g(x)$ of (2.6) is now pressed into the elastic foundation with the normal force F_N (see Fig. 2.3).

The normal surface displacement at position x within the contact area is equal to the difference of the indentation depth d and the profile shape g:

$$w_{1D}(x) = d - g(x). \tag{2.8}$$

At the boundary of the non-adhesive contact, $x = \pm a$, the surface displacement must be zero:

$$w_{1D}(\pm a) = 0 \quad \Rightarrow \quad d = g(a). \tag{2.9}$$

This equation determines the relationship between the indentation depth and the contact radius a. Note that this relationship does not depend upon the elastic properties of the medium.

The force of a spring at position x is proportional to the displacement at this position:

$$\Delta F_N(x) = \Delta k_z w_{1D}(x) = E^* w_{1D}(x)\Delta x. \tag{2.10}$$

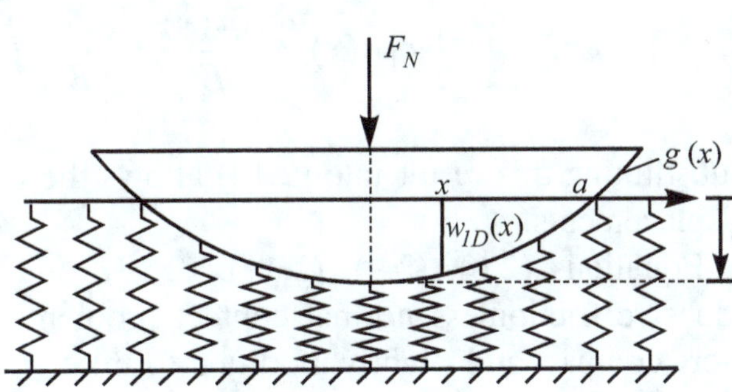

Fig. 2.3 MDR substitute model for the normal contact problem

The sum of all spring forces must balance out the external normal force. In the limiting case of very small spring spacing, $\Delta x \to \mathrm{d}x$, the sum turns into an integral:

$$F_N = E^* \int_{-a}^{a} w_{1D}(x)\mathrm{d}x = 2E^* \int_{0}^{a} [d - g(x)]\mathrm{d}x. \tag{2.11}$$

Equation (2.11) provides the normal force as a function of the contact radius and, under consideration of (2.9), of the indentation depth. Let us now define the linear force density $q_z(x)$:

$$q_z(x) = \frac{\Delta F_N(x)}{\Delta x} = E^* w_{1D}(x) = E^* \begin{cases} d - g(x), & |x| < a \\ 0, & |x| > a \end{cases}. \tag{2.12}$$

As shown in the appendix to this book, the stress distribution of the original three-dimensional system can be determined from the one-dimensional linear force density via the integral transform:

$$\sigma_{zz}(r) = -p(r) = \frac{1}{\pi} \int_{r}^{\infty} \frac{q_z'(x)}{\sqrt{x^2 - r^2}}\mathrm{d}x. \tag{2.13}$$

The normal surface displacement $w(r)$ (inside as well as outside the contact area) is given by the transform:

$$w(r) = \frac{2}{\pi} \int_{0}^{r} \frac{w_{1D}(x)}{\sqrt{r^2 - x^2}}\mathrm{d}x. \tag{2.14}$$

For the sake of completeness, we will provide the inverse transform to (2.13):

$$q_z(x) = 2 \int_{x}^{\infty} \frac{rp(r)}{\sqrt{r^2 - x^2}}\mathrm{d}r. \tag{2.15}$$

With the MDR it is also possible to determine the displacements for a prescribed stress distribution at the surface of the half-space. First, the displacement of the Winkler foundation w_{1D} must be calculated from the stresses according to:

$$w_{1D}(x) = \frac{q_z(x)}{E^*} = \frac{2}{E^*} \int_{x}^{\infty} \frac{rp(r)}{\sqrt{r^2 - x^2}}\mathrm{d}r. \tag{2.16}$$

Substituting this result into (2.14) allows the calculation of the three-dimensional displacements.

Equations (2.6), (2.9), (2.11), (2.13), and (2.14) completely solve the non-adhesive frictionless normal contact problem, so we state them once again in a more compact and slightly modified form:

$$
\begin{aligned}
g(x) &= |x| \int_0^{|x|} \frac{f'(r)\mathrm{d}r}{\sqrt{x^2 - r^2}} \\
d &= g(a), \quad \text{if } g \text{ continous at } x = a, \\
F_N &= 2E^* \int_0^a [d - g(x)]\mathrm{d}x, \\
\sigma_{zz}(r) &= -\frac{E^*}{\pi}\left[\int_r^a \frac{g'(x)\mathrm{d}x}{\sqrt{x^2 - r^2}} + \frac{d - g(a)}{\sqrt{a^2 - r^2}}\right], \quad \text{for } r \leq a, \\
w(r) &= \frac{2}{\pi}\begin{cases} \displaystyle\int_0^r \frac{[d - g(x)]\mathrm{d}x}{\sqrt{r^2 - x^2}}, & \text{for } r < a, \\ \displaystyle\int_0^a \frac{[d - g(x)]\mathrm{d}x}{\sqrt{r^2 - x^2}}, & \text{for } r > a. \end{cases}
\end{aligned} \tag{2.17}
$$

In the following we will present the relationships between the normal force F_N, indentation depth d, and contact radius a, as well as the stresses and displacements outside the contact area for various technically relevant profiles $f(r)$.

2.4 Areas of Application

The most well-known normal contact problem is likely the Hertzian contact (see Sects. 2.5.3 and 2.5.4). While Hertz (1882) examined the contact of two parabolic bodies with different radii of curvature around the x-axis and y-axis in his work, we will consider the more specific axisymmetric case of the contact of two elastic spheres or, equivalently, of a rigid sphere and an elastic half-space. This problem occurs ubiquitously in technical applications; for example, in roller bearings, joints, or the contact between wheel and rail. Hertz also proposed using this contact for measuring material hardness in Hertzian contact, however, the stress maximum lies underneath the surface of the half-space. Therefore cones (see Sect. 2.5.2) are more suitable for this task. For punching, flat indenters (see Sect. 2.5.1) or even flat rings (see Sect. 2.5.7) are very commonly used because of the stress singularity at the edge of the contact.

These three shapes—flat, cone-shaped, and spherical indenters—essentially form the three ideal base shapes for most contacts in technical applications. Additionally, it is also of great value to examine the effects of imperfections on these base shapes, for example through manufacturing or wear. Such imperfect indenters may be truncated cones or spheres (see Sects. 2.5.9 and 2.5.10), bodies with rounded tips (see Sects. 2.5.11 to 2.5.13) or rounded edges (see Sect. 2.5.14), as well as ellipsoid profiles (see Sect. 2.5.5).

Furthermore, any infinitely often continuously differentiable profile can be expanded in a Taylor series. By utilizing a profile defined by a power-law (see Sect. 2.5.8), the solution for a more complex profile—assuming it satisfies the aforementioned differentiability criterion—can be constructed to arbitrary precision with the Taylor series.

Furthermore, this chapter contains profiles relevant for applications where adhesive normal contact comes into play. This includes a profile which generates a constant pressure distribution (see Sect. 2.5.5), concave bodies (see Sects. 2.5.15 and 2.5.16), or bodies with a periodic roughness (see Sect. 2.5.17). Since the next chapter dealing with adhesive normal contact reveals that the frictionless normal contact problem with adhesion can, under certain circumstances, be reduced to the non-adhesive one, we will provide the corresponding non-adhesive solutions already in this chapter, even though the practical significance of the respective problems will only become apparent later.

The contact problem is fully defined by the profile $f(r)$ and one of the global contact quantities F_N, d or a. Generally, we will assume that, of these three, the contact radius is given, consequentially yielding the solution as a function of this contact radius. Should, instead of the contact radius, the normal force or the indentation depth be given, the given equations must be substituted as necessary.

2.5 Explicit Solutions for Axially Symmetric Profiles

2.5.1 The Cylindrical Flat Punch

The solution of the normal contact problem for the flat cylindrical punch of radius a, which can be described by the profile:

$$f(r) = \begin{cases} 0, & r \leq a, \\ \infty, & r > a, \end{cases} \tag{2.18}$$

goes back to Boussinesq (1885). The utilized notation is illustrated in Fig. 2.4. The original solution by Boussinesq is based on the methods of potential theory. The solution using the MDR is significantly simpler. The equivalent flat profile for the purposes of the MDR, $g(x)$, is given by:

$$g(x) = \begin{cases} 0, & |x| \leq a, \\ \infty, & |x| > a. \end{cases} \tag{2.19}$$

The contact radius corresponds to the radius of the indenter. The only remaining global contact quantities to be determined are the indentation depth d and the

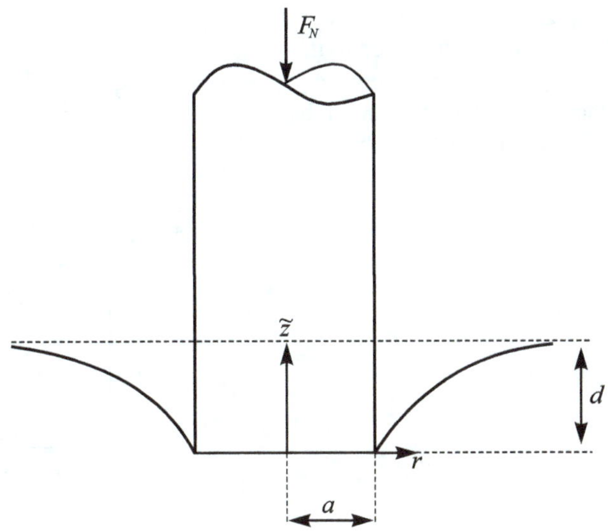

Fig. 2.4 Normal indentation by a cylindrical flat punch

normal force F_N. For the latter we obtain:

$$F_N(d) = 2E^* da. \tag{2.20}$$

Hence, the contact stiffness equals:

$$k_z := \frac{\mathrm{d}F_N}{\mathrm{d}d} = 2E^* a. \tag{2.21}$$

The stress distribution in the contact area and the displacements of the half-space outside the contact are, due to (2.17):

$$\begin{aligned} \sigma_{zz}(r;d) &= -\frac{E^* d}{\pi \sqrt{a^2 - r^2}}, \quad r \le a, \\ w(r;d) &= \frac{2d}{\pi} \arcsin\left(\frac{a}{r}\right), \quad r > a. \end{aligned} \tag{2.22}$$

The average pressure in the contact is:

$$p_0 = \frac{F_N}{\pi a^2} = \frac{2E^* d}{\pi a}. \tag{2.23}$$

The stress distribution and displacements within and outside the contact area are shown in Figs. 2.5 and 2.6.

Finally, it should be noted that, although the notation E^* is used for the cylindrical indenter, implying a possible elasticity of both contact bodies, the aforementioned solution described previously is solely valid for *rigid* cylindrical indenters. While the deformation of the half-space can satisfy the conditions of the *half-space approximation*, this is generally not the case for the cylindrical indenter. The discrepancies which occur for elastic indenters are discussed in Sect. 2.5.19.

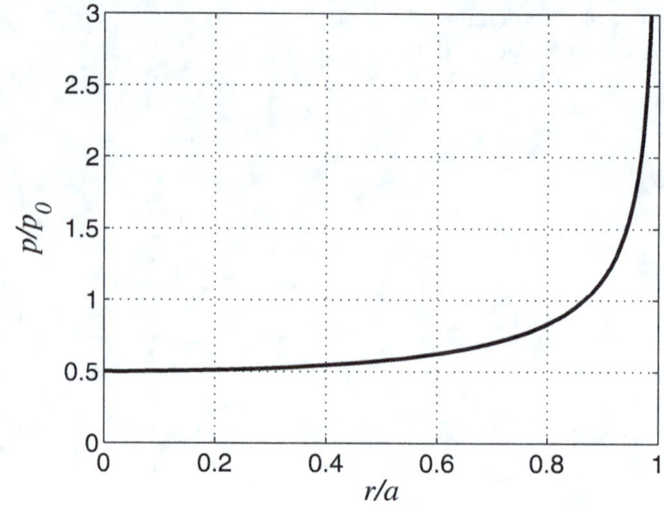

Fig. 2.5 Normal pressure $p = -\sigma_{zz}$, normalized to the average pressure in the contact p_0, for the indentation by a flat cylindrical punch

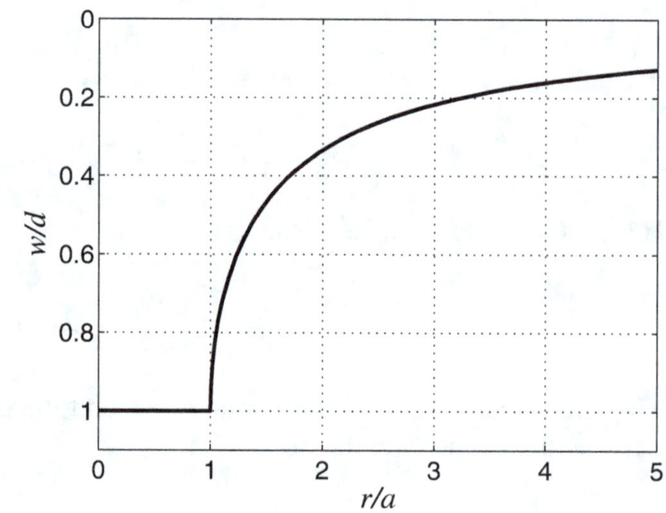

Fig. 2.6 Displacement of the half-space, normalized to the indentation depth d, for the indentation by a flat cylindrical punch

2.5.2 The Cone

The case of the conical indenter (see Fig. 2.7),

$$f(r) = r \tan\theta, \tag{2.24}$$

with a small inclination angle θ, was first solved by Love (1939). He also made use of potential theory and used several tricky series expansions to obtain the solution. Again, we describe the easier way of using the MDR. The equivalent profile is given by:

$$g(x) = |x| \tan\theta \int_0^{|x|} \frac{\mathrm{d}r}{\sqrt{x^2 - r^2}} = \frac{\pi}{2} |x| \tan\theta. \tag{2.25}$$

For the relationships between contact radius a, indentation depth d, normal force F_N, average pressure p_0, and for the stresses σ_{zz} and displacements w, according

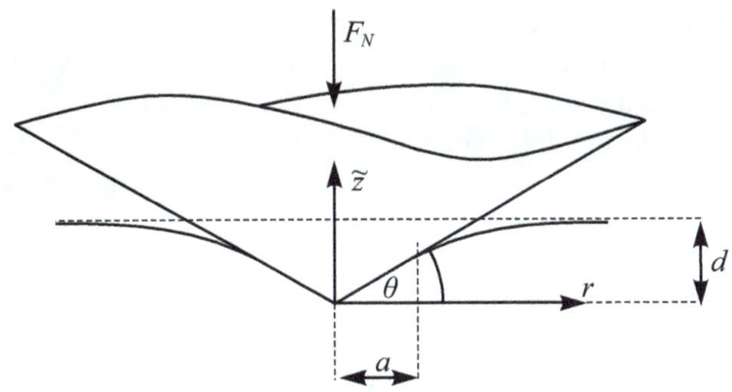

Fig. 2.7 Normal indentation by a conical indenter

to (2.17) we obtain:

$$
\begin{aligned}
d(a) &= \frac{\pi}{2} a \tan\theta, \\
F_N(a) &= \pi E^* \tan\theta \int_0^a (a-x)\,\mathrm{d}x = \frac{\pi a^2}{2} E^* \tan\theta, \\
p_0 &= \frac{1}{2} E^* \tan\theta, \\
\sigma_{zz}(r;a) &= -\frac{E^* \tan\theta}{2} \int_r^a \frac{\mathrm{d}x}{\sqrt{x^2-r^2}} = -p_0 \operatorname{arcosh}\left(\frac{a}{r}\right), \quad r \le a, \\
w(r;a) &= \tan\theta \int_0^a \frac{(a-x)\mathrm{d}x}{\sqrt{r^2-x^2}} \\
&= a \tan\theta \left[\arcsin\left(\frac{a}{r}\right) + \frac{\sqrt{r^2-a^2}-r}{a}\right], \quad r > a.
\end{aligned}
\tag{2.26}
$$

Here arcosh(·) denotes the area hyperbolic cosine function, which can also be represented explicitly by the natural logarithm:

$$
\operatorname{arcosh}\left(\frac{a}{r}\right) = \ln\left(\frac{a+\sqrt{a^2-r^2}}{r}\right). \tag{2.27}
$$

The stress distribution, normalized to the average pressure in the contact, is shown in Fig. 2.8. One recognizes the logarithmic singularity at the apex of the cone. In Fig. 2.9, the displacement of the half-space normalized to the indentation depth d is shown.

Finally, it should be noted—in analogy to the previous section—that although the notation E^* is used for the conical punch as well (implying that both contacting bodies are allowed to be elastic), the previously described solution is correct without restrictions only for *rigid* conical indenters. While for the half-space the requirements of the *half-space approximation* can still be fulfilled, for the conical

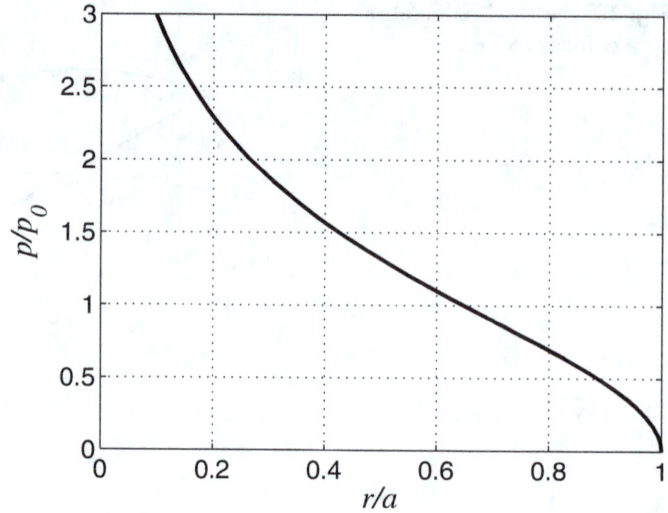

Fig. 2.8 Course of normal pressure $p = -\sigma_{zz}$, normalized to the average pressure, for indentation by a cone

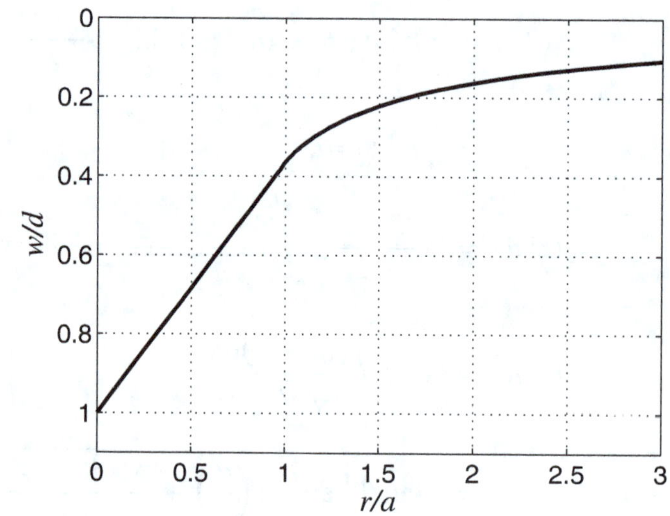

Fig. 2.9 Displacement of the half-space, normalized to the indentation depth d, for indentation by a cone

punch this is the case only for small angles θ. The deviations that occur in the case of elastic indenters are addressed in Sect. 2.5.19.

2.5.3 The Paraboloid

The solution to the problem illustrated in Fig. 2.10 goes back to the classical work of Hertz (1882), although he studied the generalized problem of an elliptic contact area. Hertz made use of potential theory too. For a small contact radius a compared to the radius of the sphere R, the profile shape in the contact is characterized by the parabola:

$$f(r) = \frac{r^2}{2R}. \tag{2.28}$$

Fig. 2.10 Normal indentation by a parabolic indenter

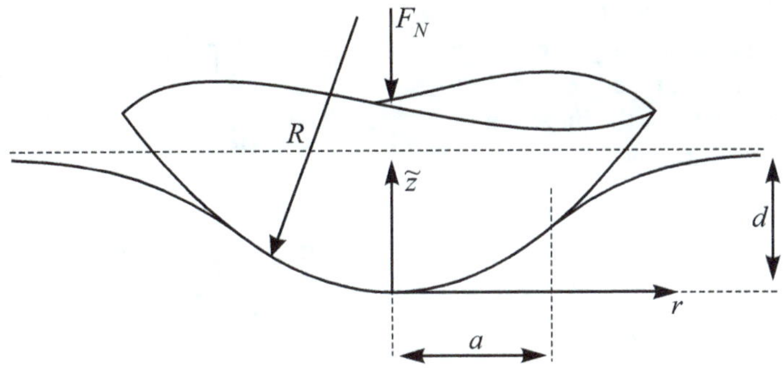

The solution of the contact problem as per (2.17) is given by:

$$g(x) = \frac{|x|}{R} \int_0^{|x|} \frac{r\,\mathrm{d}r}{\sqrt{x^2 - r^2}} = \frac{x^2}{R},$$

$$d(a) = \frac{a^2}{R},$$

$$F_N(a) = \frac{2E^*}{R} \int_0^a \left(a^2 - x^2\right) \mathrm{d}x = \frac{4}{3} \frac{E^* a^3}{R}. \tag{2.29}$$

And by:

$$\sigma_{zz}(r;a) = -\frac{2E^*}{\pi R} \int_r^a \frac{x\,\mathrm{d}x}{\sqrt{x^2 - r^2}} = -\frac{2E^*}{\pi R} \sqrt{a^2 - r^2}, \quad r \le a,$$

$$w(r;a) = \frac{2}{\pi R} \int_0^a \frac{\left(a^2 - x^2\right) \mathrm{d}x}{\sqrt{r^2 - x^2}}$$

$$= \frac{a^2}{\pi R} \left[\left(2 - \frac{r^2}{a^2} \right) \arcsin\left(\frac{a}{r} \right) + \frac{\sqrt{r^2 - a^2}}{a} \right], \quad r > a. \tag{2.30}$$

The average pressure in the contact is equal to:

$$p_0 = \frac{4E^* a}{3\pi R}. \tag{2.31}$$

The stress curve normalized to this pressure and the displacement curve normalized to the indentation depth d are shown in Figs. 2.11 and 2.12, respectively. In this normalized representation, the curves of the contact quantities are independent of the curvature radius R.

Stresses within the Half-Space

As Huber (1904) demonstrated, the stresses inside the half-space can also be calculated for this contact problem. After a lengthy calculation, he provided the

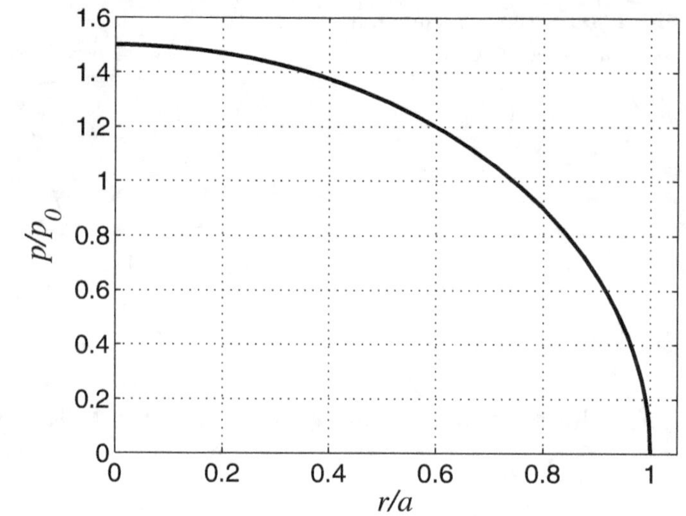

Fig. 2.11 Normal pressure curve $p = -\sigma_{zz}$, normalized to the average pressure in the contact, for the indentation by a paraboloid

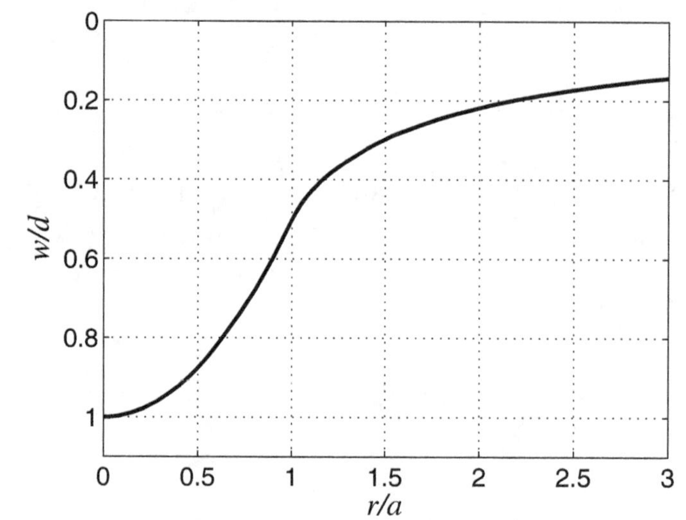

Fig. 2.12 Displacement of the half-space, normalized to the indentation depth d, for the indentation by a paraboloid

following solution:

$$\frac{\sigma_{rr}(r,z;a)}{p_0} = \frac{3}{2}\left\{\frac{1-2\nu}{3}\frac{a^2}{r^2}\left[1-\left(\frac{z}{\sqrt{u}}\right)^3\right]+\left(\frac{z}{\sqrt{u}}\right)^3\frac{a^2u}{u^2+a^2z^2}\right.$$
$$\left.+\frac{z}{\sqrt{u}}\left[\frac{(1-\nu)u}{a^2+u}+(1+\nu)\frac{\sqrt{u}}{a}\arctan\left(\frac{a}{\sqrt{u}}\right)-2\right]\right\},$$
$$\frac{\sigma_{\varphi\varphi}(r,z;a)}{p_0} = -\frac{3}{2}\left\{\frac{1-2\nu}{3}\frac{a^2}{r^2}\left[1-\left(\frac{z}{\sqrt{u}}\right)^3\right]\right.$$
$$\left.+\frac{z}{\sqrt{u}}\left[2\nu+\frac{(1-\nu)u}{a^2+u}-(1+\nu)\frac{\sqrt{u}}{a}\arctan\left(\frac{a}{\sqrt{u}}\right)\right]\right\},$$

$$\frac{\sigma_{zz}(r,z;a)}{p_0} = -\frac{3}{2}\left(\frac{z}{\sqrt{u}}\right)^3 \frac{a^2 u}{u^2 + a^2 z^2},$$
$$\frac{\sigma_{rz}(r,z;a)}{p_0} = -\frac{3}{2}\frac{rz^2}{u^2 + a^2 z^2}\frac{a^2\sqrt{u}}{u + a^2}, \tag{2.32}$$

with the Poisson's ratio ν and the expression:

$$u(r,z;a) = \frac{1}{2}\left(r^2 + z^2 - a^2 + \sqrt{(r^2 + z^2 - a^2)^2 + 4a^2 z^2}\right). \tag{2.33}$$

The stresses $\sigma_{r\varphi}$ and $\sigma_{\varphi z}$ vanish due to rotational symmetry. Figure 2.13 displays the von-Mises equivalent stress—normalized to the average pressure in the contact—resulting from this stress tensor. It is apparent that the equivalent stress reaches its greatest value in the middle of contact, yet underneath the half-space surface. Therefore, the parabolic indenter is not suitable for measuring hardness as Hertz (1882) had originally assumed. Figure 2.14 displays the greatest resulting principle stresses in the half-space.

An alternative, yet naturally fully equivalent formulation of the stresses in the half-space given by (2.32), can be referenced within work by Hamilton and Goodman (1966).

The Hertzian Impact Problem

Hertz (1882) studied the impact problem for this contact also. Consider the parabolic indenter of mass m normally impacting the initially non-deformed half-space with the initial speed v_0. Let the impact be quasi-static, i.e., $v_0 \ll c$, where c represents the characteristic propagation speed of elastic waves in the half-space. The energy radiation in the form of elastic waves in the half-space can then be neglected, as demonstrated by Hunter (1957). In (2.29) the relationship between

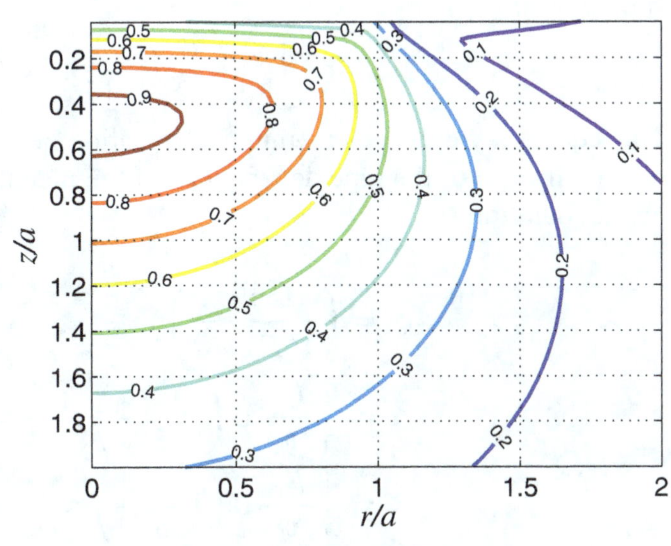

Fig. 2.13 Von-Mises equivalent stress curve, normalized to the average pressure in the half-space for the indentation by a paraboloid assuming $\nu = 0.3$

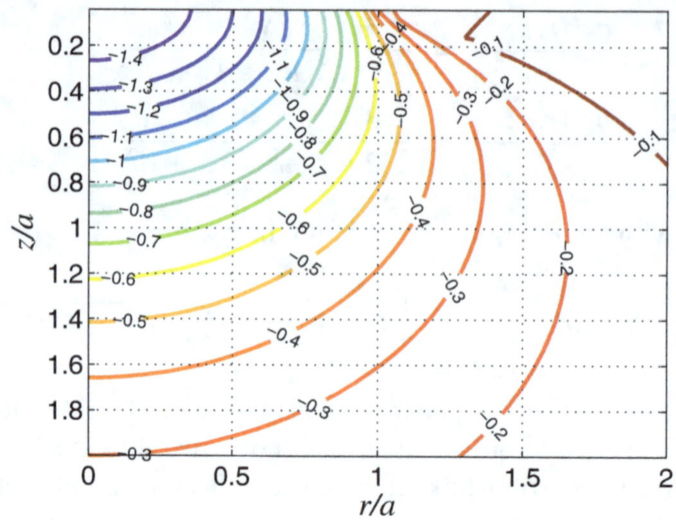

Fig. 2.14 Greatest principle stress curve, normalized to the average pressure in the half-space for the indentation by a paraboloid assuming $\nu = 0.3$

normal force F_N and indentation depth d is implicitly given:

$$F_N(d) = \frac{4}{3} E^* \sqrt{R d^3}. \tag{2.34}$$

Therefore, the potential energy stored in the elastic deformation of the elastic half-space, U, is given by:

$$U(d) = \int_0^d F\left(\tilde{d}\right) \mathrm{d}\tilde{d} = \frac{8}{15} E^* \sqrt{R d^5}. \tag{2.35}$$

The indentation depth during the impact is a function of time, $d = d(t)$, and for the quasi-static case the energy conservation during impact takes on the following simple form:

$$m\frac{v_0^2}{2} = m\frac{\dot{d}^2}{2} + \frac{8}{15} E^* \sqrt{R d^5}. \tag{2.36}$$

This equation yields the maximum indentation depth $d_{\max}$, the function $t = t(d)$, i.e., the inverse of the time dependence of the indentation depth $d = d(t)$, and the impact duration t_S:

$$\begin{aligned} d_{\max} &= \left(\frac{15 m v_0^2}{16 E^* \sqrt{R}}\right)^{2/5}, \\ t &= \frac{2}{5}\frac{d_{\max}}{v_0} \mathrm{B}\left(\xi; \frac{2}{5}, \frac{1}{2}\right), \quad \xi = \left(\frac{d}{d_{\max}}\right)^{5/2} \\ t_S &= \frac{4}{5}\frac{d_{\max}}{v_0} \mathrm{B}\left(1; \frac{2}{5}, \frac{1}{2}\right) \approx 2.94 \frac{d_{\max}}{v_0}. \end{aligned} \tag{2.37}$$

Here, $\mathrm{B}(\cdot;\cdot,\cdot)$ is the incomplete beta function

$$\mathrm{B}(z;a,b) := \int_0^z t^{a-1}(1-t)^{b-1}\,\mathrm{d}t. \tag{2.38}$$

2.5.4 The Sphere

The problem of the spherical indenter is very closely related to the problem of the parabolic indenter described in the previous section. The profile of a sphere of radius R is:

$$f(r) = R - \sqrt{R^2 - r^2}. \tag{2.39}$$

For the case of $r \ll R$ this can be approximated as:

$$f(r) \approx \frac{r^2}{2R}, \tag{2.40}$$

which obviously coincides with (2.28) from Sect. 2.5.3. It may now be necessary to use the exact spherical shape instead of the parabolic approximation. However, the assumption of small deformations that underlies the whole theory used in this book requires the validity of the half-space hypothesis, which in our case can be written as $a \ll R$. If the latter is fulfilled, one can still work with the parabolic approximation. Nevertheless, we want to present the solution of the contact problem with a spherical indenter, which was first published by Segedin (1957). Applying (2.17) to (2.39) yields:

$$\begin{aligned}
g(x) &= |x| \int_0^{|x|} \frac{r\,\mathrm{d}r}{\sqrt{R^2 - r^2}\sqrt{x^2 - r^2}} = |x| \operatorname{artanh}\left(\frac{|x|}{R}\right), \\
d(a) &= a \operatorname{artanh}\left(\frac{a}{R}\right), \\
F_N(a) &= 2E^* \int_0^a \left[a \operatorname{artanh}\left(\frac{a}{R}\right) - x \operatorname{artanh}\left(\frac{x}{R}\right)\right] \mathrm{d}x \\
&= E^* R^2 \left[\left(1 + \frac{a^2}{R^2}\right) \operatorname{artanh}\left(\frac{a}{R}\right) - \frac{a}{R}\right].
\end{aligned} \tag{2.41}$$

Here $\operatorname{artanh}(\cdot)$ refers to the area hyperbolic tangent function, which can also be represented explicitly by the natural logarithm:

$$\operatorname{artanh}\left(\frac{a}{R}\right) = \frac{1}{2} \ln\left(\frac{R + a}{R - a}\right). \tag{2.42}$$

The average pressure in the contact is:

$$p_0 = \frac{E^* R^2}{\pi a^2}\left[\left(1 + \frac{a^2}{R^2}\right)\operatorname{artanh}\left(\frac{a}{R}\right) - \frac{a}{R}\right]. \tag{2.43}$$

The stresses and displacements were not determined by Segedin (1957), and can only be partially expressed by elementary functions. With the help of (2.17) we obtain the relationships:

$$\begin{aligned}
\sigma_{zz}(r;a) &= -\frac{E^*}{\pi}\int\limits_r^a \left[\frac{xR}{R^2 - x^2} + \operatorname{artanh}\left(\frac{x}{R}\right)\right]\frac{\mathrm{d}x}{\sqrt{x^2 - r^2}}, \quad r \le a, \\
&= -\frac{E^*}{\pi}\left[\frac{R}{\sqrt{R^2 - r^2}}\operatorname{artanh}\left(\frac{\sqrt{a^2 - r^2}}{\sqrt{R^2 - r^2}}\right)\right. \\
&\quad \left. + \int\limits_r^a \operatorname{artanh}\left(\frac{x}{R}\right)\frac{\mathrm{d}x}{\sqrt{x^2 - r^2}}\right], \\
w(r;a) &= \frac{2}{\pi}\left[a \operatorname{artanh}\left(\frac{a}{R}\right)\arcsin\left(\frac{a}{r}\right) - \int\limits_0^a x \operatorname{artanh}\left(\frac{x}{R}\right)\frac{\mathrm{d}x}{\sqrt{r^2 - x^2}}\right], \\
&\quad r > a, \\
&= \frac{2}{\pi}\left\{\operatorname{artanh}\left(\frac{a}{R}\right)\left[a \arcsin\left(\frac{a}{r}\right) + \sqrt{r^2 - a^2}\right] - R \arcsin\left(\frac{a}{r}\right)\right. \\
&\quad \left. + \sqrt{R^2 - r^2}\arctan\left(\frac{a\sqrt{R^2 - r^2}}{R\sqrt{r^2 - a^2}}\right)\right\}.
\end{aligned} \tag{2.44}$$

These functions are shown in normalized form in Figs. 2.15 and 2.16 for different values of R/a. It can be seen that, even for small values of this ratio such as 1.5 (which already strongly violates the half-space approximation), the stresses are only slightly different and the displacements almost indistinguishable from the parabolic approximation in Sect. 2.5.3.

2.5.5 The Ellipsoid

The solution for an indenter in the form of an ellipsoid of rotation also originated from Segedin (1957). The profile is given by:

$$f(r) = R\left(1 - \sqrt{1 - k^2 r^2}\right), \tag{2.45}$$

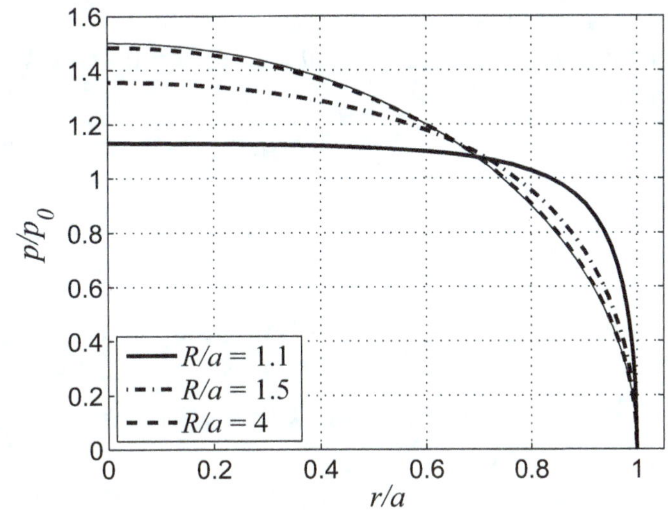

Fig. 2.15 Normal pressure $p = -\sigma_{zz}$, normalized to the average pressure in the contact, p_0, for indentation by a sphere for different ratios R/a. The thin solid line indicates the parabolic approximation from (2.30)

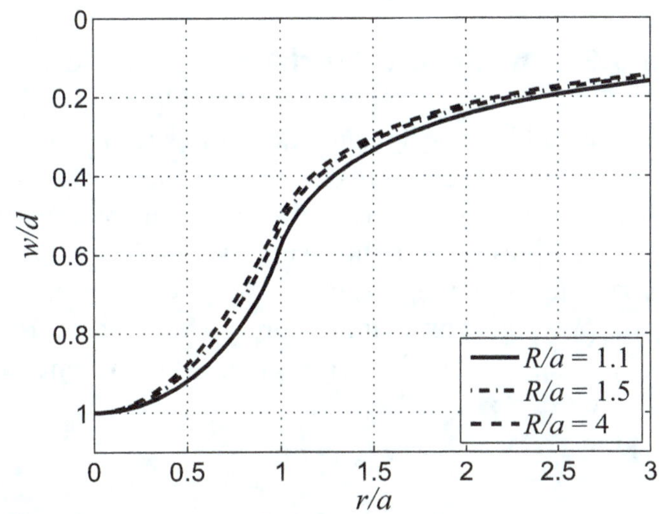

Fig. 2.16 Displacement of the half-space, normalized to the indentation depth d for indentation by a sphere for different ratios R/a. Since the curves are approximately on top of each other, the line of the parabolic approximation, according to (2.30), has been omitted

with the two parameters, R and k. $kR = 1$, resulting in the spherical indenter of the previous section. In general cases, the equivalent profile is as follows:

$$\begin{aligned} g(x) &= |x| R \int_0^{|x|} \frac{k^2 r \, \mathrm{d}r}{\sqrt{1 - k^2 r^2}\sqrt{x^2 - r^2}} = |x| k R \operatorname{artanh}(k|x|) \\ &= k R g_{\text{sphere}}\left(x; R = \frac{1}{k}\right). \end{aligned} \tag{2.46}$$

Here, $g_{\text{sphere}}(x; R)$ denotes the solution:

$$g_{\text{sphere}}(x; R) := |x| \operatorname{artanh}\left(\frac{|x|}{R}\right) \tag{2.47}$$

derived in the previous section for a sphere with the radius R. Because of the superposition principle, all expressions for the stresses and displacements—and, correspondingly of course, also for the macroscopic quantities—are linear in g. Therefore, it is clear without calculation that the solution of the contact problem is given by:

$$\begin{aligned} d(a) &= kRd_{\text{sphere}}\left(a; R = \frac{1}{k}\right), \\ F_N(a) &= kRF_{N,\text{sphere}}\left(a; R = \frac{1}{k}\right), \\ \sigma_{zz}(r;a) &= kR\sigma_{zz,\text{sphere}}\left(r;a; R = \frac{1}{k}\right), \quad r \leq a, \\ w(r;a) &= kRw_{\text{sphere}}\left(r;a; R = \frac{1}{k}\right), \quad r > a. \end{aligned} \tag{2.48}$$

The index "sphere" denotes the respective solution from Sect. 2.5.4.

2.5.6 The Profile Which Generates Constant Pressure

It is possible to design an indenter in such a way that the generated pressure in the contact is constant. This contact problem was initially solved by Lamb (1902) in the form of hypergeometric functions and by utilizing the potentials of Boussinesq. We will present a slightly simplified solution based on elliptical integrals, which goes back to Föppl (1941).

Applying a constant pressure p_0 to a circular region with the radius a yields the following vertical displacements $w_{1D}(x)$ in a one-dimensional MDR model according to (2.16):

$$w_{1D}(x) = \frac{2}{E^*}\int_x^a \frac{rp_0}{\sqrt{r^2 - x^2}}\mathrm{d}r = \frac{2p_0}{E^*}\sqrt{a^2 - x^2}. \tag{2.49}$$

The displacement in the real three-dimensional space is given by:

$$\begin{aligned} w(r) &= \frac{2}{\pi}\int_0^r \frac{w_{1D}(x)}{\sqrt{r^2 - x^2}}\mathrm{d}x = \frac{4p_0}{\pi E^*}\int_0^r \frac{\sqrt{a^2 - x^2}}{\sqrt{r^2 - x^2}}\mathrm{d}x \\ &= \frac{4p_0 a}{\pi E^*}\mathrm{E}\left(\frac{r}{a}\right), \quad r \leq a. \end{aligned} \tag{2.50}$$

Here, $E(\cdot)$ denotes the complete elliptical integral of the second kind:

$$\mathrm{E}(k) := \int_0^{\pi/2} \sqrt{1 - k^2 \sin^2 \varphi}\, \mathrm{d}\varphi. \tag{2.51}$$

The indentation depth d is therefore:

$$d = w(0) = \frac{2p_0 a}{E^*}, \tag{2.52}$$

and the shape of the profile is given by:

$$f(r) = d - w(r) = \frac{2p_0 a}{E^*}\left[1 - \frac{2}{\pi}\mathrm{E}\left(\frac{r}{a}\right)\right]. \tag{2.53}$$

It is apparent that this is not a classical indenter with a constant shape: varying p_0 causes the profile to be scaled. In other words, different pairings $\{a, p_0\}$ require different indenter profiles $f(r)$. Concrete applications, usually in biological systems, are discovered upon considering the adhesive normal contact. We will examine them at a later point. For completeness, we will calculate the displacement outside the contact area:

$$\begin{aligned} w(r; a, p_0) &= \frac{4p_0}{\pi E^*}\int_0^a \frac{\sqrt{a^2 - x^2}\,\mathrm{d}x}{\sqrt{r^2 - x^2}}, \qquad r > a, \\ &= \frac{4p_0 r}{\pi E^*}\left[\mathrm{E}\left(\frac{a}{r}\right) - \left(1 - \frac{a^2}{r^2}\right)\mathrm{K}\left(\frac{a}{r}\right)\right], \end{aligned} \tag{2.54}$$

with the complete elliptical integral of the first kind:

$$\mathrm{K}(k) := \int_0^{\pi/2} \frac{\mathrm{d}\varphi}{\sqrt{1 - k^2 \sin^2\varphi}}. \tag{2.55}$$

The displacement w of the half-space is shown in Fig. 2.17.

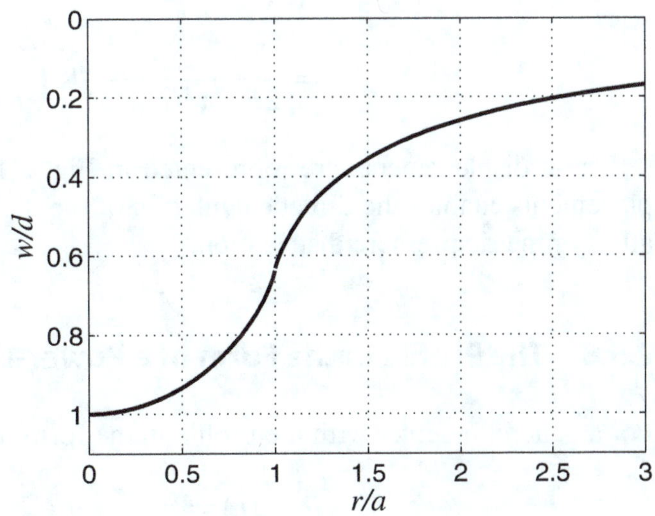

Fig. 2.17 Displacements within and outside the contact area, normalized to the indentation depth, for an indenter generating constant pressure

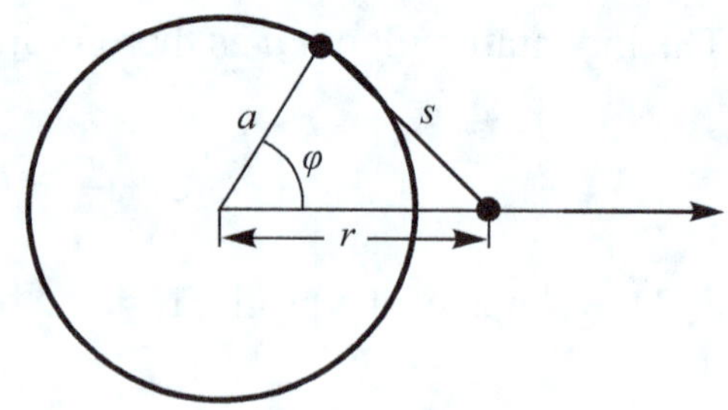

Fig. 2.18 Thin circular ring. Visualized derivation of the integral (2.58)

2.5.7 Displacement from Indentation by a Thin Circular Ring

We now examine the indentation of the elastic half-space by a thin circular ring of radius a. Let the ring be sufficiently thin so that the pressure distribution can be regarded as a Dirac distribution:

$$\sigma_{zz}(r;a) = -\frac{F_N}{2\pi a}\delta(r-a), \tag{2.56}$$

where F_N denotes the normal force loading the ring. The resulting displacement of the half-space can be determined from the superposition of fundamental solutions of elasticity theory. The half-space normal displacement resulting from the point force acting on the origin in the z-direction is, according to (2.2), given by:

$$w(s) = \frac{F_z}{\pi E^* s}, \tag{2.57}$$

with the distance s to the acting point of the force.

The displacements (see notations used in the diagram in Fig. 2.18) produced by this pressure distribution (2.56) are given by:

$$\begin{aligned} w(r;a) &= \frac{1}{\pi E^*}\int_0^{2\pi} \frac{F_N}{2\pi}\frac{\mathrm{d}\varphi}{\sqrt{a^2+r^2-2ar\cos\varphi}} \\ &= \frac{F_N}{2E^*}\frac{4}{\pi^2(r+a)}\mathrm{K}\left(\frac{2\sqrt{ra}}{r+a}\right). \end{aligned} \tag{2.58}$$

These displacements are represented in Fig. 2.19. A superposition of the displacements enables the direct calculation of the displacements from any given axially symmetric pressure distribution.

2.5.8 The Profile in the Form of a Power-Law

For a general indenter with the profile in the form of a power-law (see Fig. 2.20),

$$f(r) = cr^n, \quad n \in \mathbb{R}^+, \tag{2.59}$$

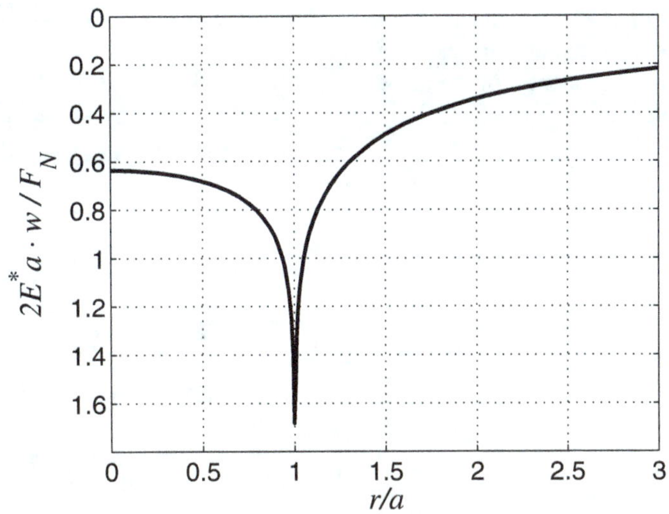

Fig. 2.19 Normalized surface displacement from indentation by a thin circular ring

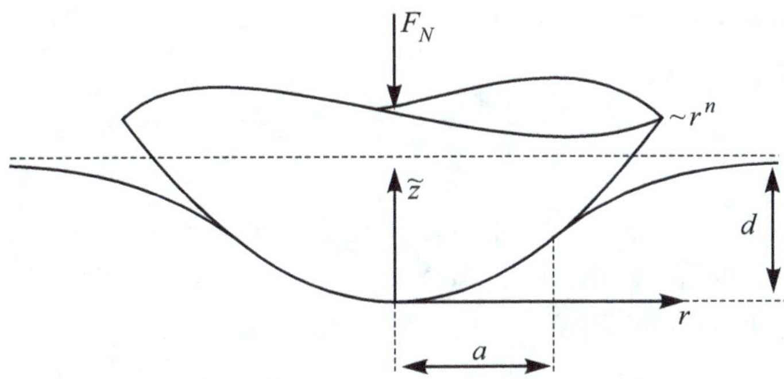

Fig. 2.20 Normal indentation by a mnemonic indenter

the solution of the contact problem can also be given in explicit form. Here, c is a constant and n is a positive real number. For example, for $n = 1$ and $n = 2$ arise the already considered cases of a conical or parabolic indenter. The general solution was first found by Galin (1946). Shtaerman (1939) gave a solution in faculty expressions for even integers n.

The equivalent plane profile $g(x)$ is, in this case, also a power function with the exponent n:

$$g(x) = |x|nc \int_0^{|x|} \frac{r^{n-1}\mathrm{d}r}{\sqrt{x^2 - r^2}} = \kappa(n)c|x|^n, \tag{2.60}$$

with the scaling factor

$$\kappa(n) := \sqrt{\pi}\frac{\Gamma(n/2+1)}{\Gamma\left[(n+1)/2\right]}. \tag{2.61}$$

Here, $\Gamma(\cdot)$ denotes the gamma function

$$\Gamma(z) := \int_0^{\infty} t^{z-1} \exp(-t)\,\mathrm{d}t. \tag{2.62}$$

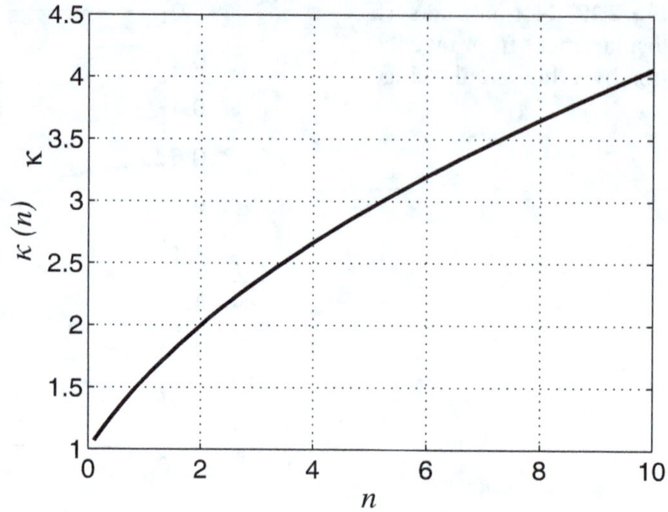

Fig. 2.21 Dependence of the stretch factor κ in (2.61) on the exponent n of the power profile

Table 2.1 Scaling factor $\kappa(n)$ for selected exponents of the shape function

n	0.5	1	2	3	4	5	6	7	8	9	10
$\kappa(n)$	1.311	1.571	2	2.356	2.667	2.945	3.2	3.436	3.657	3.866	4.063

The dependence of the scaling factor on the exponent n is shown in Fig. 2.21 and in Table 2.1. For the relationships between the normal force F_N, indentation depth d, and contact radius a we get:

$$\begin{aligned} d(a) &= \kappa(n)ca^n, \\ F_N(a) &= E^* \frac{2n}{n+1}\kappa(n)ca^{n+1}. \end{aligned} \tag{2.63}$$

The mean pressure in contact is:

$$p_0 = \frac{E^*}{\pi}\frac{2n}{n+1}\kappa(n)ca^{n-1}. \tag{2.64}$$

For the stress and displacement distributions, the expressions given in (2.17) will result in:

$$\begin{aligned} \sigma_{zz}(r;a) &= -\frac{E^*}{\pi}n\kappa(n)c\int_r^a x^{n-1}\frac{\mathrm{d}x}{\sqrt{x^2-r^2}}, \quad r \le a, \\ &= -\frac{E^*}{2\pi}n\kappa(n)cr^{n-1}\left[\mathrm{B}\left(1;\frac{1-n}{2},\frac{1}{2}\right) - \mathrm{B}\left(\frac{r^2}{a^2};\frac{1-n}{2},\frac{1}{2}\right)\right], \\ w(r;a) &= \frac{2}{\pi}\kappa(n)c\left[a^n\arcsin\left(\frac{a}{r}\right) - \int_0^a x^n\frac{\mathrm{d}x}{\sqrt{r^2-x^2}}\right], \quad r > a, \\ &= \frac{2}{\pi}\kappa(n)ca^n\left[\arcsin\left(\frac{a}{r}\right) - \frac{1}{n+1}\frac{a}{r}{}_2\mathrm{F}_1\left(\frac{1}{2},\frac{n+1}{2};\frac{n+3}{2};\frac{a^2}{r^2}\right)\right]. \end{aligned} \tag{2.65}$$

Here, $\mathrm{B}(\cdot;\cdot,\cdot)$ denotes the incomplete beta function

$$\mathrm{B}(z;a,b) := \frac{z^a}{a}{}_2\mathrm{F}_1(a, 1-b; a+1; z) \tag{2.66}$$

and ${}_2\mathrm{F}_1(\cdot,\cdot;\cdot;\cdot)$ the hypergeometric function

$${}_2\mathrm{F}_1(a,b;c;z) := \sum_{n=0}^{\infty} \frac{\Gamma(a+n)\Gamma(b+n)\Gamma(c)}{\Gamma(a)\Gamma(b)\Gamma(c+n)} \frac{z^n}{n!}. \tag{2.67}$$

However, the evaluation of these functions is somewhat cumbersome, which is why we give two recursion formulas with which all stresses and displacements for natural values of n can be recursively determined from the known solutions for a cylindrical flat punch and a cone:

$$\begin{aligned} \int_r^a \frac{x^{n-1}\mathrm{d}x}{\sqrt{x^2-r^2}} &= \frac{a^{n-2}\sqrt{a^2-r^2}}{n-1} + r^2\frac{n-2}{n-1}\int_r^a \frac{x^{n-3}\mathrm{d}x}{\sqrt{x^2-r^2}}, \quad r \le a, \\ -\!\!\!\!\!\!\int_0^a \frac{x^n\mathrm{d}x}{\sqrt{r^2-x^2}} &= \frac{a^{n-1}\sqrt{r^2-a^2}}{n} - r^2\frac{n-1}{n}\int_0^a \frac{x^{n-2}\mathrm{d}x}{\sqrt{r^2-x^2}}, \quad r > a. \end{aligned} \tag{2.68}$$

In the case of the cone ($n = 1$), the normal stress has a logarithmic singularity at the apex of the cone. For all $n > 1$ the pressure distribution is not singular; the pressure maximum remains at the center of the contact until $n = 2$ and then begins to shift to the edge of the contact at higher n. In the limiting case $n \to \infty$ corresponding to a cylindrical flat punch, the pressure distribution is singular at the contact edge.

The Impact Problem for the Indenter with Power-Law Profile

Since the relationships between normal force, contact radius, and indentation depth are known to be power functions, the normal impact problem can also be solved easily for this indenter profile. Consider a rotationally symmetric rigid body of mass m with the power-law profile just described, which impacts on the elastic half-space with the initial velocity v_0. The collision should be quasi-static, i.e., this impact velocity would be much smaller than the propagation speed of elastic waves.

The potential energy as a function of the current indentation depth can be derived from (2.63) and is given by:

$$U(d) = \int_0^d F(\tilde{d})\,\mathrm{d}\tilde{d} = \frac{E^*}{[c\kappa(n)]^{1/n}} \frac{2n^2}{(2n+1)(n+1)} d^{\frac{2n+1}{n}}. \tag{2.69}$$

Due to energy conservation, the following expressions for the maximum indentation depth $d_{\max}$, the function $t = t(d)$, and the impact duration t_S are valid:

$$d_{\max} = \left\{ \frac{[c\kappa(n)]^{1/n} (2n+1)(n+1) m v_0^2}{4n^2 E^*} \right\}^{\frac{n}{2n+1}},$$

$$t = \frac{n}{2n+1} \frac{d_{\max}}{v_0} \mathrm{B}\left(\xi; \frac{n}{2n+1}, \frac{1}{2}\right), \quad \xi = \left(\frac{d}{d_{\max}}\right)^{\frac{2n+1}{n}},$$

$$t_S = \frac{2n}{2n+1} \frac{d_{\max}}{v_0} \mathrm{B}\left(1; \frac{n}{2n+1}, \frac{1}{2}\right), \tag{2.70}$$

which, for $n = 2$, coincides with the solution of the Hertzian impact problem described in Sect. 2.5.3.

2.5.9 The Truncated Cone

For certain technical applications it can be of interest to consider an indenter with a flattened tip, perhaps worn down by wear. The solution for the indentation by a truncated cone was first published by Ejike (1969). This solution—as well as the general solution by Sneddon (1965)—is based on appropriate integral transforms. A simple solution in the framework of MDR has been provided in this section.

The axisymmetric profile (see Fig. 2.22) can be written as:

$$f(r) = \begin{cases} 0, & r \le b, \\ (r-b)\tan\theta, & r > b. \end{cases} \tag{2.71}$$

Here, θ denotes the slope angle of the cone and b the radius of the blunt end. As in previous cases, the problem can be dealt with elementarily by utilizing (2.17),

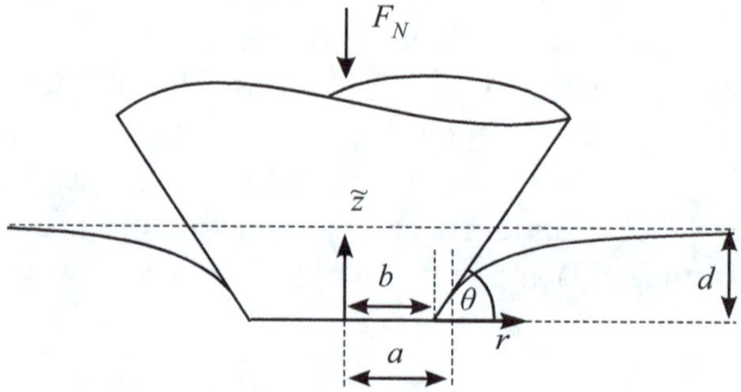

Fig. 2.22 Normal indentation by a truncated cone

which provide the solution:

$$g(x) = \begin{cases} 0, & |x| \leq b, \\ |x| \tan\theta \arccos\left(\dfrac{b}{|x|}\right), & |x| > b, \end{cases}$$

$$d(a) = a \tan\theta \arccos\left(\frac{b}{a}\right) = \varphi_0 a \tan\theta,$$

$$F_N(a) = E^* \tan\theta a^2 \left[\arccos\left(\frac{b}{a}\right) + \frac{b}{a}\sqrt{1 - \frac{b^2}{a^2}}\right]$$
$$= E^* \tan\theta a^2 \left(\varphi_0 + \cos\varphi_0 \sin\varphi_0\right), \tag{2.72}$$

where we introduced the angle

$$\varphi_0 := \arccos\left(\frac{b}{a}\right). \tag{2.73}$$

The average pressure in the contact is:

$$p_0 = \frac{E^*}{\pi} \tan\theta(\varphi_0 + \cos\varphi_0 \sin\varphi_0). \tag{2.74}$$

Setting $b = 0$, i.e., $\varphi_0 = \pi/2$, yields the solution of the ideal conical indenter from Sect. 2.5.2. For the stresses in the contact area we obtain:

$$\sigma_{zz}(r; a) =$$
$$-\frac{E^* \tan\theta}{\pi} \begin{cases} \displaystyle\int_b^a \left\{\frac{b}{\sqrt{x^2 - b^2}} + \arccos\left(\frac{b}{x}\right)\right\} \frac{\mathrm{d}x}{\sqrt{x^2 - r^2}}, & r \leq b, \\ \displaystyle\int_r^a \left\{\frac{b}{\sqrt{x^2 - b^2}} + \arccos\left(\frac{b}{x}\right)\right\} \frac{\mathrm{d}x}{\sqrt{x^2 - r^2}}, & b < r \leq a. \end{cases} \tag{2.75}$$

The stresses are shown in Fig. 2.23. The singularity at the sharp edge of the blunt end, $r = b$, is clearly visible. For small values of b/a, the curves approach the solution for the complete cone, and for $b/a \to 1$ the limiting case of the flat cylindrical punch. Both limiting cases are represented as thin solid lines in Figs. 2.23 and 2.24. The displacements outside of the contact area are:

$$w(r; a) = \frac{2\tan\theta}{\pi} \left\{\varphi_0 a \arcsin\left(\frac{a}{r}\right) - \int_b^a x \arccos\left(\frac{b}{x}\right) \frac{\mathrm{d}x}{\sqrt{r^2 - x^2}}\right\},$$
$$r > a. \tag{2.76}$$

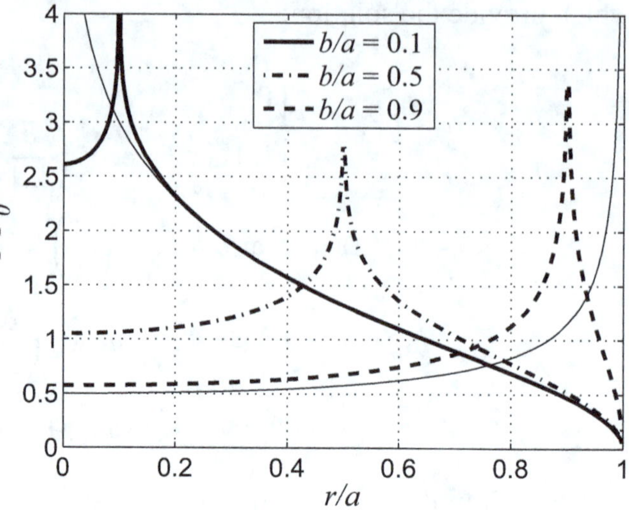

Fig. 2.23 Stress distribution, normalized to the average pressure in the contact, for the indentation by a truncated cone with different values b/a. The thin solid lines represent the solutions for the complete cone and the flat cylindrical punch

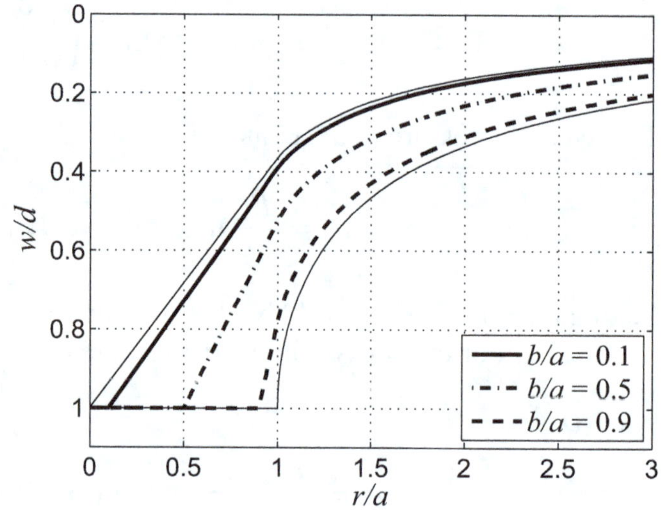

Fig. 2.24 Displacements of the half-space, normalized to the indentation depth, for the indentation by a truncated cone with different values b/a. The thin solid lines represent the solutions for the complete cone and the flat cylindrical punch

Theoretically, this integral can be solved analytically using tabulated functions, but due to its complexity, a numerical solution seems more feasible. These displacements are displayed in Fig. 2.24.

2.5.10 The Truncated Paraboloid

The indentation problem for a truncated paraboloid (see Fig. 2.25) was also first solved by Ejike (1981). The profile is described by:

$$f(r) = \begin{cases} 0, & r \le b, \\ \dfrac{r^2 - b^2}{2R}, & r > b, \end{cases} \tag{2.77}$$

Fig. 2.25 Normal indentation by a truncated paraboloid

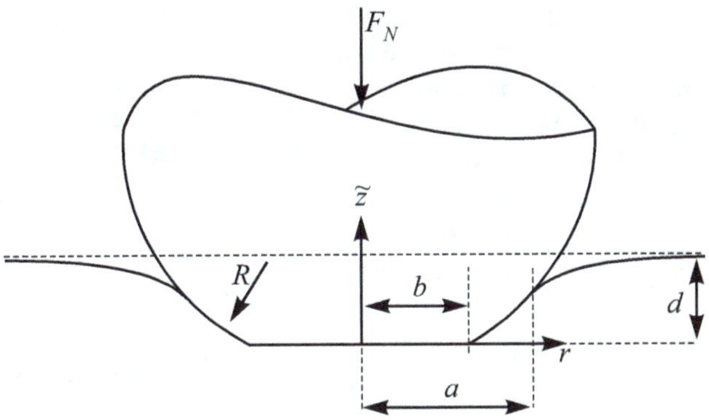

with the radius of the blunt end, b, and the radius of curvature of the paraboloid R. The solution within the MDR (see (2.17)) gives:

$$
\begin{aligned}
g(x) &= \begin{cases} 0, & |x| \le b, \\ \dfrac{|x|}{R}\sqrt{x^2 - b^2}, & |x| > b, \end{cases} \\
d(a) &= \frac{a}{R}\sqrt{a^2 - b^2}, \\
F_N(a) &= \frac{2E^*}{3R}(2a^2 + b^2)\sqrt{a^2 - b^2}.
\end{aligned}
\tag{2.78}
$$

For the stresses within the contact area, we obtain:

$$
\sigma_{zz}(r;a) = -\frac{E^*}{\pi R}\begin{cases} \displaystyle\int_b^a \frac{(2x^2 - b^2)\mathrm{d}x}{\sqrt{x^2 - b^2}\sqrt{x^2 - r^2}}, & r \le b, \\ \displaystyle\int_r^a \frac{(2x^2 - b^2)\mathrm{d}x}{\sqrt{x^2 - b^2}\sqrt{x^2 - r^2}}, & b < r \le a. \end{cases}
\tag{2.79}
$$

These can theoretically be expressed in closed form via elliptic integrals. However, the resulting expressions are very extensive and cumbersome to evaluate, which is why a numerical evaluation will be preferable. The stresses are shown in Fig. 2.26. One recognizes the singularities at the sharp edge of the blunt end at $r = b$, which are increasingly localized for small values of b, and the limiting cases of the paraboloid and the cylindrical flat punch (indicated by thin solid lines). The displacements outside the contact area are as follows:

$$
\begin{aligned}
w(r;a) = {} & \frac{2a}{\pi R}\sqrt{a^2 - b^2}\arcsin\left(\frac{a}{r}\right) \\
& - \frac{1}{\pi R}\left[(r^2 - b^2)\arcsin\left(\frac{\sqrt{a^2 - b^2}}{\sqrt{r^2 - b^2}}\right) - \sqrt{a^2 - b^2}\sqrt{r^2 - a^2}\right], \\
& r > a.
\end{aligned}
\tag{2.80}
$$

These are shown in Fig. 2.27. For $b = 0$ we obtain the Hertzian solutions for the parabolic indenter given in Sect. 2.5.3.

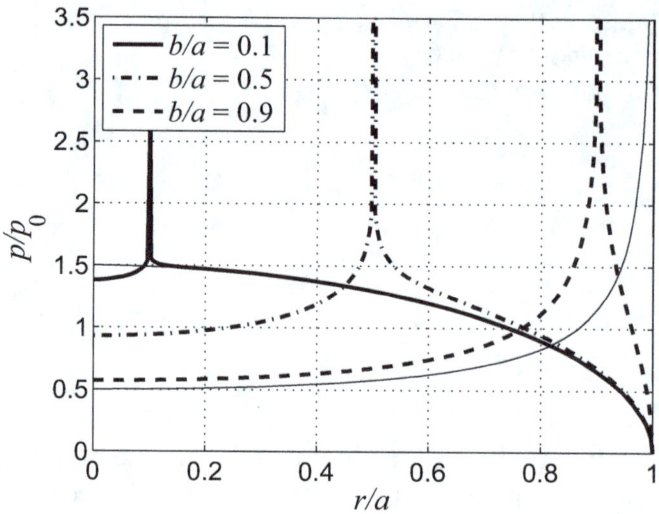

Fig. 2.26 Stress distribution, normalized to the mean pressure in contact, for indentation by a flattened paraboloid at different values b/a. The thin solid lines indicate the solutions for the complete paraboloid and the flat punch

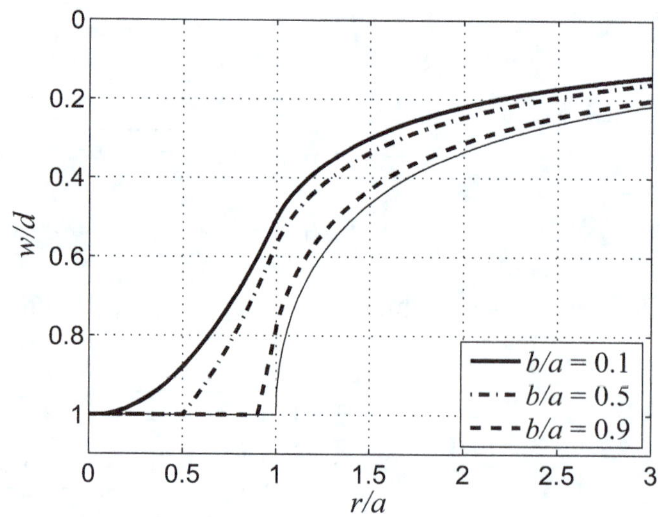

Fig. 2.27 Displacements of the half-space, normalized to the indentation depth, for indentation by a flattened paraboloid at different values b/a. The thin solid lines indicate the solutions for the complete paraboloid and the flat punch

2.5.11 The Cylindrical Flat Punch with Parabolic Cap

For mechanical microscopy and other metrological applications, indenters with a parabolic cap are important. In the following, we examine such indenters, assuming that the indentation depth is large enough to ensure that the face comes into complete contact (otherwise we would deal with the simple parabolic contact studied in Sect. 2.5.3).

The solution for a flat cylindrical punch (as a base body) stems from Abramian et al. (1964). In their solution the authors used series expansions to fulfill the field equations. We again show the easier solution within the framework of MDR. The indenter has the profile (see Fig. 2.28)

$$f(r) = \begin{cases} \dfrac{r^2}{2R}, & r \le a, \\ \infty, & r > a, \end{cases} \tag{2.81}$$

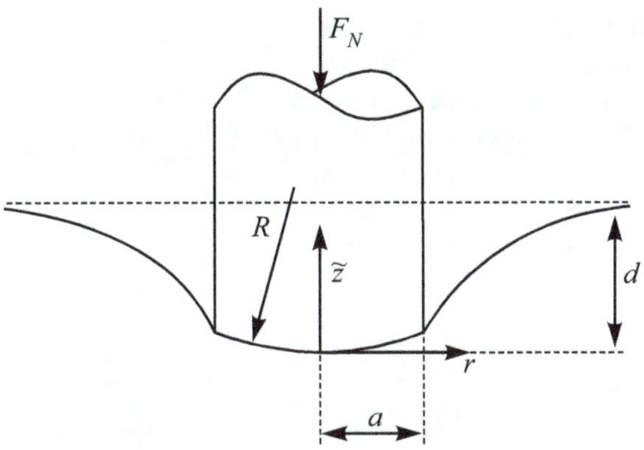

Fig. 2.28 Normal indentation by a cylindrical punch with parabolic cap

with the radius of curvature of the cap R, and the radius of the base punch a; the solution of the contact problem is, according to (2.17), given by:

$$
\begin{aligned}
g(x) &= \begin{cases} \dfrac{x^2}{R}, & |x| \leq a \\ \infty, & |x| > a \end{cases}, \\
F_N(d) &= 2E^* \left(da - \frac{a^3}{3R} \right), \quad dR \geq a^2, \\
\sigma_{zz}(r; d) &= -\frac{E^*}{\pi R} \frac{a^2 - 2r^2 + dR}{\sqrt{a^2 - r^2}}, \quad r \leq a,\ dR \geq a^2, \\
w(r; d) &= \frac{1}{\pi R} \left[\left(2dR - r^2\right) \arcsin\left(\frac{a}{r}\right) + a\sqrt{r^2 - a^2} \right], \\
& \quad r > a,\ dR \geq a^2.
\end{aligned}
\tag{2.82}
$$

For $dR = a^2$ (of course), we obtain the Hertzian solution from Sect. 2.5.3, and for $R \to \infty$ that for the cylindrical flat punch from Sect. 2.5.1. The stresses and

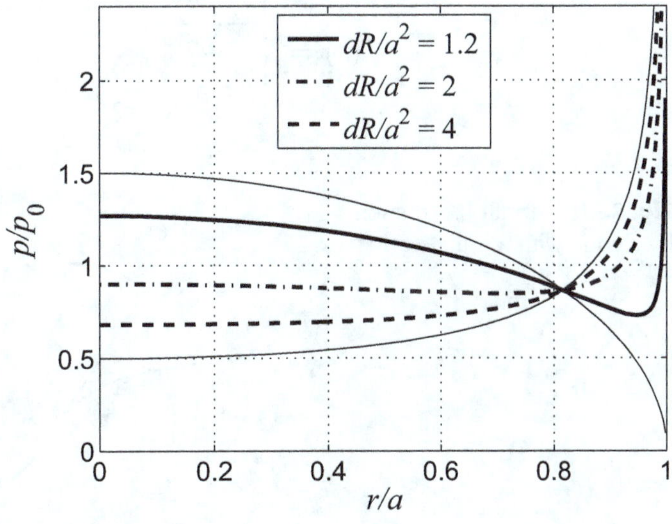

Fig. 2.29 Stress distribution, normalized to the mean pressure in contact, for indentation by a flat punch with parabolic cap at different values dR/a^2. The thin solid lines indicate the solutions for the paraboloid and the flat punch

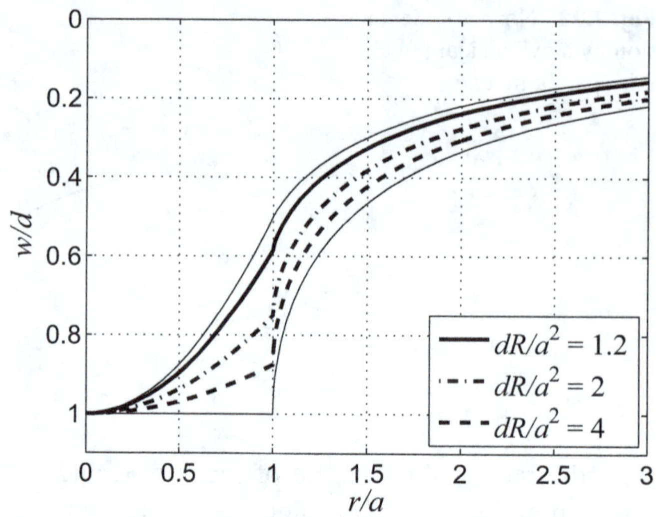

Fig. 2.30 Displacements normalized to the indentation depth, for indentation by a flat punch with parabolic cap at different values dR/a^2. The thin solid lines indicate the solutions for the paraboloid and the flat punch

displacements for various values of dR/a^2 are shown in Figs. 2.29 and 2.30. One recognizes the limiting cases of the paraboloid and the flat punch marked by solid lines. The intersection of all normalized pressure curves is at $r/a = \sqrt{2/3}$ and $p/p_0 = \sqrt{3}/2$.

2.5.12 The Cone with Parabolic Cap

The solution for a cone with a rounded tip, despite its metrological significance, was found first by Ciavarella (1999). However, Maugis and Barquins (1983) already knew the relations between the global contact quantities—normal force, contact radius, and indentation depth—by solving the respective problem with adhesion. Ciavarella's solution is based on the general solution by Shtaerman (1949).

The profile (see Fig. 2.31) is described by:

$$f(r) = \begin{cases} \dfrac{r^2 \tan\theta}{2b}, & r \leq b, \\ r\tan\theta - \dfrac{b}{2}\tan\theta, & r > b, \end{cases} \tag{2.83}$$

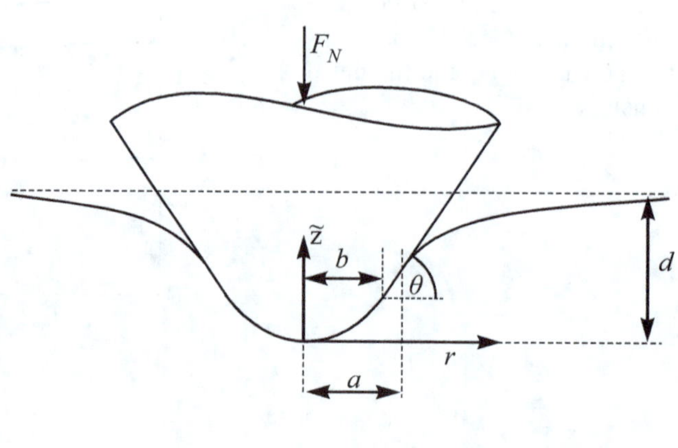

Fig. 2.31 Normal indentation by a cone with a rounded tip

with the radius b at the base of the parabolic cap and the slope angle θ of the cone. Note that the radius of curvature of the cap is given by $R := b/\tan\theta$ to guarantee continuous differentiability. For the solution to the contact problem, according to (2.17) one obtains:

$$
\begin{aligned}
g(x) &= \begin{cases} \dfrac{x^2 \tan\theta}{b}, & |x| \le b, \\ \dfrac{|x|\tan\theta}{b}\left\{|x| - \sqrt{x^2-b^2} + b\arccos\left(\dfrac{b}{|x|}\right)\right\}, & |x| > b, \end{cases} \\
d(a) &= \frac{a}{b}\tan\theta\left(a - \sqrt{a^2-b^2} + b\arccos\left(\frac{b}{a}\right)\right) \\
&= a\tan\theta\left(\frac{1-\sin\varphi_0}{\cos\varphi_0} + \varphi_0\right), \\
F_N(a) &= \frac{E^* a^2 \tan\theta}{b}\left[b\arccos\left(\frac{b}{a}\right) + \frac{4}{3}\left(a - \sqrt{a^2-b^2}\right) + \frac{1}{3}\frac{b^2\sqrt{a^2-b^2}}{a^2}\right] \\
&= E^* a^2 \tan\theta\left(\varphi_0 + \frac{4}{3}\frac{1-\sin\varphi_0}{\cos\varphi_0} + \frac{1}{3}\sin\varphi_0\cos\varphi_0\right),
\end{aligned}
\tag{2.84}
$$

where we introduced the angle:

$$
\varphi_0 := \arccos\left(\frac{b}{a}\right). \tag{2.85}
$$

In the limit $b \to 0$ we obtain the expressions for the conical indenter from Sect. 2.5.2, because

$$
\lim_{\varphi_0 \to \pi/2} \frac{1-\sin\varphi_0}{\cos\varphi_0} = 0. \tag{2.86}
$$

For $b \to a$, i.e., $\varphi_0 = 0$, we obtain solutions of the Hertzian contact problem from Sect. 2.5.3. For the stresses in the contact area the following expressions result:

$$
\begin{aligned}
\sigma_{zz}(r;a) = & -\frac{E^*\tan\theta}{\pi b} \\
& \cdot \begin{cases} 2\sqrt{a^2-r^2} + b\displaystyle\int_0^{\varphi_0} \frac{(\varphi - 2\tan\varphi)\tan\varphi \,\mathrm{d}\varphi}{\sqrt{1-k^2\cos^2\varphi}}, & r \le b, \\ 2\sqrt{a^2-r^2} + b\displaystyle\int_0^{\operatorname{arcosh}(a/r)} \left[\arccos\left(\frac{b}{r\cosh\varphi}\right) - 2\sqrt{k^2\cosh^2\varphi - 1}\right]\mathrm{d}\varphi, \\ b < r \le a, \end{cases}
\end{aligned}
\tag{2.87}
$$

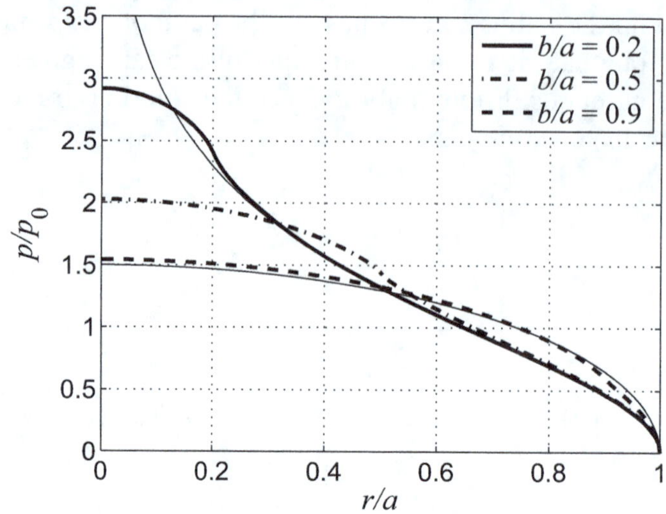

Fig. 2.32 Stress distribution, normalized to the mean pressure in contact, for the indentation by a cone with a rounded tip at different values b/a. The thin solid lines indicate the solutions for the paraboloid and the cone

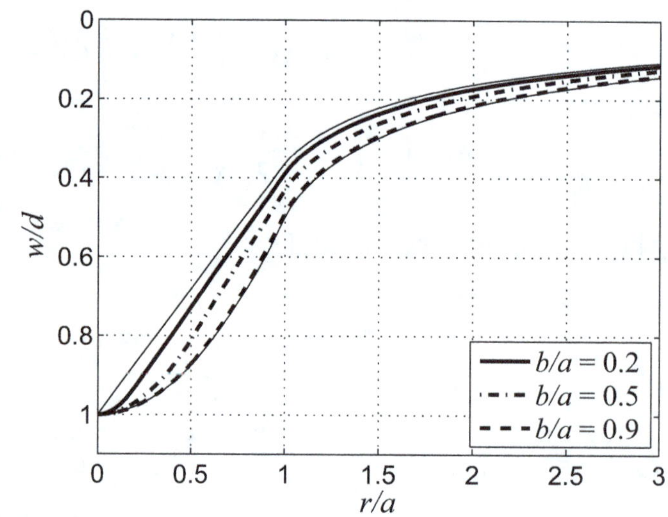

Fig. 2.33 Displacements of the half-space, normalized to the indentation depth, for indentation by a cone with a rounded tip at different values b/a. The thin solid lines indicate the solutions for the paraboloid and the cone

with the parameter $k := r/b$. For the displacement outside the contact area we get:

$$w(r;a) = \frac{2d(a)}{\pi} \arcsin\left(\frac{a}{r}\right) - \frac{\tan\theta}{\pi b}\left[r^2 \arcsin\left(\frac{a}{r}\right) - a\sqrt{r^2-a^2} + 2b^2 \int_0^{\varphi_0} \frac{(\varphi - \tan\varphi)\tan\varphi \mathrm{d}\varphi}{\cos\varphi\sqrt{k^2\cos^2\varphi - 1}}\right], \quad r > a. \tag{2.88}$$

In Figs. 2.32 and 2.33, the normalized stresses and displacements are shown for some values of the ratio b/a. It is easy to see the limiting cases of the conical and parabolic indenters marked by the thin lines. The stress distribution is not singular in the origin due to the rounded tip; this is in contrast to the indentation by an ideal cone.

From the equivalent plane profile $g(x)$ in (2.84) it can be seen that this contact problem is actually a superposition of three previously solved problems, and the solution can therefore be obtained from the sum of the corresponding solutions given in this section. It is indeed

$$g(x) = g_P\left(x; R = \frac{b}{\tan\theta}\right) + g_{KS}(x) - g_{PS}\left(x; R = \frac{b}{\tan\theta}\right). \tag{2.89}$$

Here, g_P denotes the equivalent profile of a paraboloid of radius R (see Sect. 2.5.3), $g_{KS}(x)$ the equivalent profile of a truncated cone with the inclination angle θ and the radius b of the flat "tip" (see Sect. 2.5.8), and g_{PS} the equivalent profile of a truncated paraboloid with the radius R and the radius b of the flat "tip" (see Sect. 2.5.10).

2.5.13 The Paraboloid with Parabolic Cap

The contact problem of a paraboloid with parabolic cap indenting an elastic half-space was first solved by Maugis and Barquins (1983) for the adhesive contact. However, the case without adhesion is of course included in this solution. As will be seen, the problem can be considered as a superposition of the indentations by a paraboloid (see Sect. 2.5.3) and a truncated paraboloid (see Sect. 2.5.10), where the weighting factors depend on the two radii of the cap and the paraboloid R_1 and R_2. The profile has the shape (see Fig. 2.34)

$$f(r) = \begin{cases} \dfrac{r^2}{2R_1}, & r \leq b, \\ \dfrac{r^2 - h^2}{2R_2}, & r > b. \end{cases} \tag{2.90}$$

The radius of curvature of the cap, R_1, must be greater than that of the base paraboloid, R_2, to ensure a compact contact area. The continuity of the profile also requires

$$h^2 = b^2\left(1 - \frac{R_2}{R_1}\right), \tag{2.91}$$

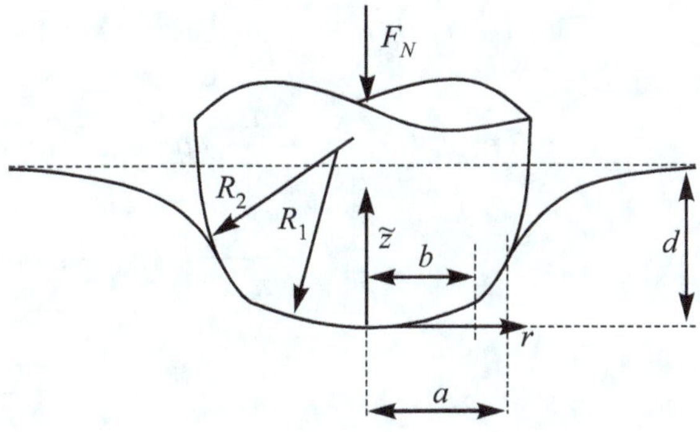

Fig. 2.34 Normal indentation by a sphere with spherical cap

for the length h. The equivalent profile within the MDR, seen in (2.17), is given by:

$$g(x) = \begin{cases} \dfrac{x^2}{R_1}, & |x| \leq b, \\ \dfrac{x^2}{R_1} + \dfrac{|x|}{R^*}\sqrt{x^2 - b^2}, & |x| > b, \end{cases} \tag{2.92}$$

with the effective radius of curvature

$$R^* := \frac{R_1 R_2}{R_1 - R_2}. \tag{2.93}$$

It is easy to see that (2.92) can be thought of as a sum:

$$g(x) = g_P(x; R = R_1) + g_{PS}(x; R = R^*). \tag{2.94}$$

Here, g_P and g_{PS} denote the equivalent profiles of the paraboloid (see (2.29)) and the truncated paraboloid (see (2.78)). The solution of the contact problem is, therefore, given by:

$$\begin{aligned} d(a) &= \frac{a^2}{R_1} + \frac{a}{R^*}\sqrt{a^2 - b^2}, \\ F_N(a) &= \frac{2E^*}{3}\left\{\frac{2a^3}{R_1} + \frac{1}{R^*}(2a^2 + b^2)\sqrt{a^2 - b^2}\right\}, \\ \sigma_{zz}(r; a) &= -\frac{E^*}{\pi}\begin{cases} \dfrac{2\sqrt{a^2 - r^2}}{R_1} + \displaystyle\int_b^a \frac{(2x^2 - b^2)\mathrm{d}x}{R^*\sqrt{x^2 - b^2}\sqrt{x^2 - r^2}}, & r \leq b, \\ \dfrac{2\sqrt{a^2 - r^2}}{R_1} + \displaystyle\int_r^a \frac{(2x^2 - b^2)\mathrm{d}x}{R^*\sqrt{x^2 - b^2}\sqrt{x^2 - r^2}}, & b < r \leq a, \end{cases} \\ w(r; a) &= w_P(r; a; R = R_1) + w_{PS}(r; a; R = R^*), \quad r > a. \end{aligned} \tag{2.95}$$

Here, w_P and w_{PS} denote the displacements in the indentation by a complete and cut-off paraboloid:

$$\begin{aligned} w_P(r; a; R_1) &= \frac{a^2}{\pi R_1}\left[\left(2 - \frac{r^2}{a^2}\right)\arcsin\left(\frac{a}{r}\right) + \frac{\sqrt{r^2 - a^2}}{a}\right], \\ w_{PS}(r; a; R^*) &= \frac{2a}{\pi R^*}\sqrt{a^2 - b^2}\arcsin\left(\frac{a}{r}\right) \\ &\quad - \frac{1}{\pi R^*}\left[(r^2 - b^2)\arcsin\left(\frac{\sqrt{a^2 - b^2}}{\sqrt{r^2 - b^2}}\right)\right. \\ &\quad \left. - \sqrt{a^2 - b^2}\sqrt{r^2 - a^2}\right]. \end{aligned} \tag{2.96}$$

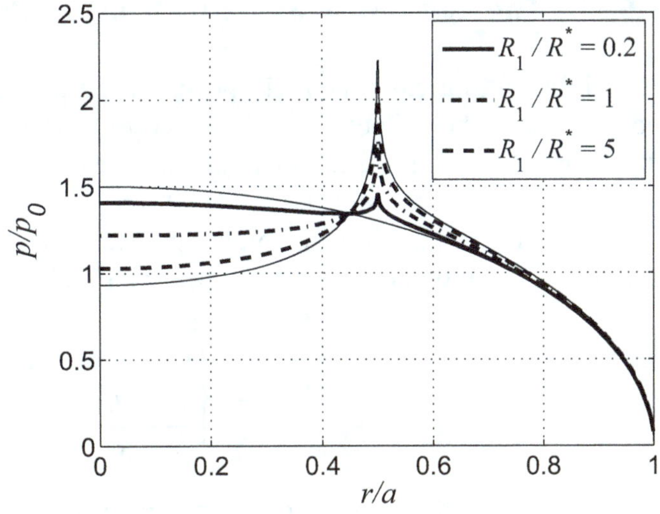

Fig. 2.35 Stress distribution, normalized to the mean pressure in contact, for indentation by a paraboloid with round cap for $b/a = 0.5$ at different values R_1/R^*. The thin solid lines indicate the solutions for the complete paraboloid and the truncated paraboloid

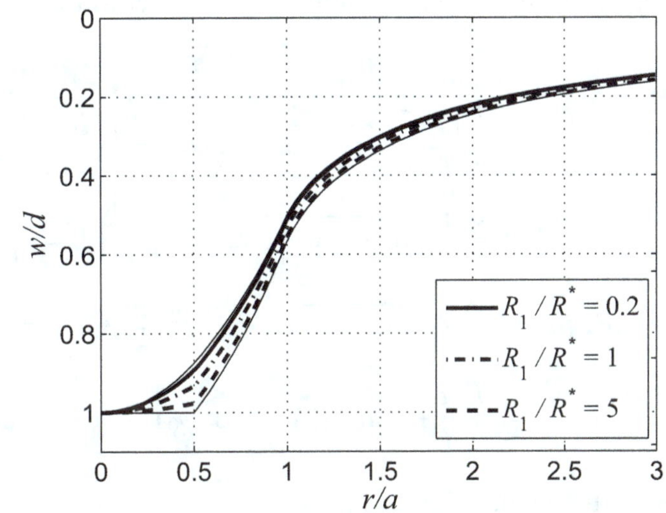

Fig. 2.36 Displacements of the half-space, normalized to the indentation depth, for indentation by a paraboloid with a round cap for $b/a = 0.5$ at different values R_1/R^*. The thin solid lines indicate the solutions for the complete paraboloid and the truncated paraboloid

For $R_1 = R_2$, respectively $R^* \to \infty$, we obtain the Hertzian solution from Sect. 2.5.3, and for $R_1 \to \infty$ the solution of the truncated paraboloid from Sect. 2.5.10.

Since the solution is determined by two parameters, b/a and R_1/R^*, it would be very ponderous to try to show all solutions for the stresses and displacements in a single diagram. Moreover, the superposition (2.94) already clarifies the structure of the solution. Therefore, the dependencies are shown only for $b/a = 0.5$ in Figs. 2.35 and 2.36. One recognizes the limiting cases given by the thin solid lines and the stress singularity at the sharp edge at $r = b$, which is more localized when approaching the parabolic solution.

2.5.14 The Cylindrical Flat Punch with a Rounded Edge

A real flat cylindrical punch will never have a perfectly sharp edge but will always be rounded. The influence of this curvature on the normal contact problem was investigated for both plain and axisymmetric contacts by Schubert (1942) (see also later papers by Ciavarella et al. (1998) and Ciavarella (1999), who presented a solution to the rotationally symmetric problem based on Shtaerman's (1949) general solution). The indenter has the profile (see Fig. 2.37):

$$f(r) = \begin{cases} 0, & r \leq b, \\ \dfrac{(r-b)^2}{2R}, & r > b, \end{cases} \tag{2.97}$$

with the radius of curvature of the rounded corner, R, and the radius of the blunt end, b. The contact problem is solved according to (2.17) by the relations:

$$\begin{aligned} g(x) &= \begin{cases} 0, & |x| \leq b, \\ \dfrac{|x|}{R}\left[\sqrt{x^2-b^2} - b\arccos\left(\dfrac{b}{|x|}\right)\right], & |x| > b, \end{cases} \\ d(a) &= \frac{a}{R}\left[\sqrt{a^2-b^2} - b\arccos\left(\frac{b}{a}\right)\right] = \frac{a^2}{R}\left(\sin\varphi_0 - \varphi_0\cos\varphi_0\right), \\ F_N(a) &= \frac{E^*}{3R}\left[\sqrt{a^2-b^2}(4a^2-b^2) - 3ba^2\arccos\left(\frac{b}{a}\right)\right] \\ &= \frac{E^* a^3}{3R}\left[\sin\varphi_0(4-\cos^2\varphi_0) - 3\varphi_0\cos\varphi_0\right], \end{aligned} \tag{2.98}$$

with the angle:

$$\varphi_0 := \arccos\left(\frac{b}{a}\right). \tag{2.99}$$

For $b = 0$, the Hertzian solution from Sect. 2.5.3 is recovered, and for $R = 0$ and $b = a$, i.e., $\varphi_0 = 0$, the solution for the cylindrical flat punch from Sect. 2.5.1.

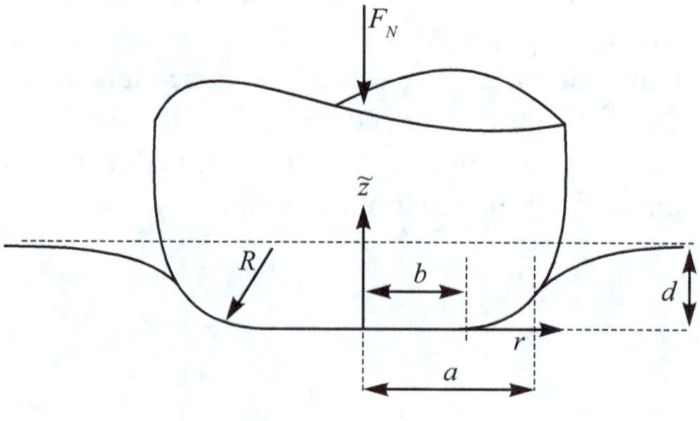

Fig. 2.37 Normal indentation by a flat cylindrical punch with a rounded edge

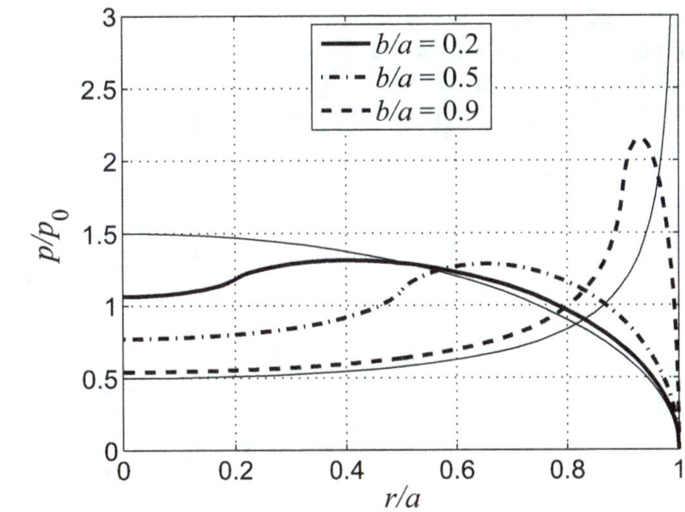

Fig. 2.38 Stress distribution, normalized to the mean pressure in contact, for indentation by a flat punch with rounded corners at different values b/a. The thin solid lines indicate the solutions for the paraboloid and the flat punch

It can be seen from the form of the function $g(x)$ that this contact problem actually represents a sum of two already solved problems and, therefore, the solution can be obtained via an appropriate superposition. It is:

$$g(x) = g_{PS}(x) - g_{KS}\left(x; \tan\theta = \frac{b}{R}\right). \tag{2.100}$$

Here $g_{PS}(x)$ denotes the equivalent profile of a truncated paraboloid (see (2.78)) and g_{KS} that of a truncated cone (see (2.72)) whose conical angle of inclination θ is determined by the relationship $\tan\theta = b/R$. For the stresses and displacements we get:

$$\sigma_{zz}(r;a) = -\frac{E^*}{\pi R}\begin{cases}\displaystyle\int_b^a \left(2\sqrt{x^2-b^2} - b\arccos\left(\frac{b}{x}\right)\right)\frac{\mathrm{d}x}{\sqrt{x^2-r^2}}, & r \le b,\\ \displaystyle\int_r^a \left(2\sqrt{x^2-b^2} - b\arccos\left(\frac{b}{x}\right)\right)\frac{\mathrm{d}x}{\sqrt{x^2-r^2}}, & b < r \le a,\end{cases}$$

$$\begin{aligned} w(r;a) = {} & \frac{2d(a)}{\pi}\arcsin\left(\frac{a}{r}\right) \\ & - \frac{2}{\pi}\left[\int_b^a \frac{x}{R}\left(\sqrt{x^2-b^2} - b\arccos\left(\frac{b}{x}\right)\right)\frac{\mathrm{d}x}{\sqrt{r^2-x^2}}\right], \quad r > a. \end{aligned} \tag{2.101}$$

These are shown in Figs. 2.38 and 2.39. Due to the rounded edge, the stress at the edge of the contact—in contrast to the indentation by a flat cylindrical punch—is not singular.

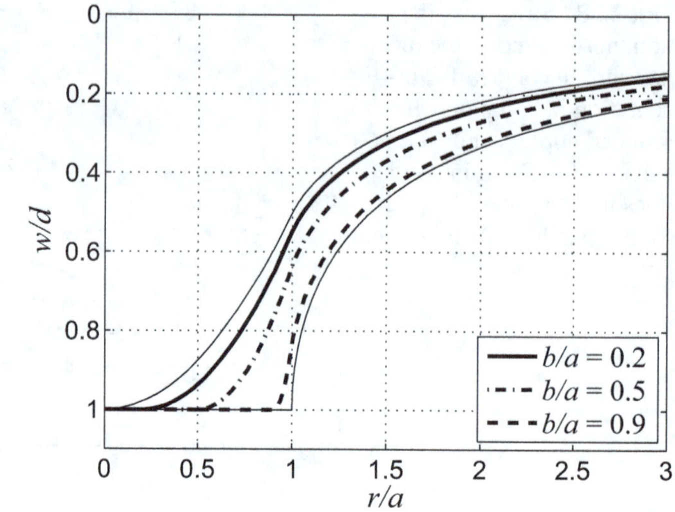

Fig. 2.39 Displacements, normalized to the indentation depth, for the indentation by a flat punch with rounded corners at different values b/a. The thin solid lines indicate the solutions for the paraboloid and the flat punch

2.5.15 The Concave Paraboloid (Complete Contact)

For concave indenters of certain profile geometries, the contact problem can be solved analytically and in closed form using the previously applied methods, assuming the contact area remains compact. Sharp concave corners or edges—such as the cases of a cylinder with a central recess or a concave cone—render this impossible. For these cases of annular contact areas, there sometimes exist semi-analytical solutions in the form of series expansions, which will be detailed in Chap. 10 of this book. The interested reader can also refer to the respective publications of Collins (1963) and Barber (1976, 1983). Complete contact of the concave paraboloid can be ensured if the normal force is large enough. The solution to this problem was discovered by Schubert (1942) (see also Barber 1976). The profile can be characterized by:

$$f(r) = \begin{cases} -\dfrac{hr^2}{a^2}, & r \leq a, \\ \infty, & r > a. \end{cases} \tag{2.102}$$

For the notation, see Fig. 2.40.

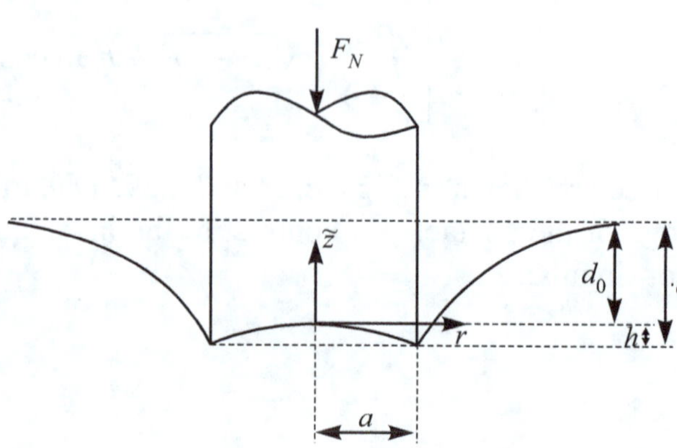

Fig. 2.40 Normal indentation by a parabolic concave indenter

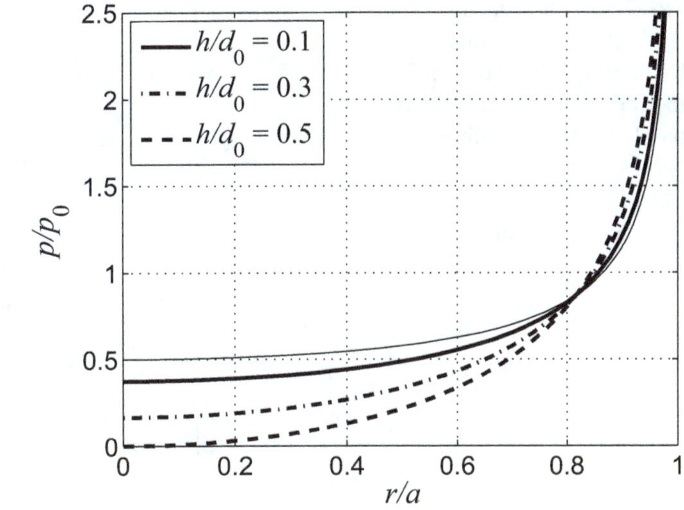

Fig. 2.41 Stress distribution, normalized to the average pressure, for the indentation by concave paraboloid with different values h/d_0. The thin solid line represents the solution for the flat cylindrical punch

The contact problem is solved according to (2.17) by the relationships:

$$g(x) = \begin{cases} -\dfrac{2h|x|}{a^2} \displaystyle\int_0^{|x|} \dfrac{r\,\mathrm{d}r}{\sqrt{x^2 - r^2}} = -\dfrac{2hx^2}{a^2}, & |x| \le a, \\ \infty, & |x| > a, \end{cases}$$

$$F_N(d_0) = 2E^* a\left(d_0 + \frac{2h}{3}\right),$$

$$\sigma_{zz}(r; d_0) = -\frac{E^*\left[d_0 a^2 - 2h(a^2 - 2r^2)\right]}{\pi a^2 \sqrt{a^2 - r^2}}, \quad r \le a,$$

$$w(r; d_0) = \frac{2}{\pi} \int_0^a \left(d_0 + \frac{2hx^2}{a^2}\right) \frac{\mathrm{d}x}{\sqrt{r^2 - x^2}}$$

$$= \frac{2d_0}{\pi} \arcsin\left(\frac{a}{r}\right) + \frac{2h}{\pi a^2}\left(r^2 \arcsin\left(\frac{a}{r}\right) - a\sqrt{r^2 - a^2}\right), \quad r > a. \tag{2.103}$$

Setting $h = 0$ yields the solution for the flat cylindrical punch. For complete contact, condition $\sigma_{zz}(r = 0) < 0$ must be satisfied, which leads to the conditions $d_0 > 2h$, or equivalently, $F_N > 16E^* ah/3$. The normalized stresses and displacements are shown in Figs. 2.41 and 2.42.

2.5.16 The Concave Profile in the Form of a Power-Law (Complete Contact)

We will briefly focus on the indentation through a concave power profile in the form of:

$$f(r) = \begin{cases} -\dfrac{hr^n}{a^n}, & r \le a, \\ \infty, & r > a \end{cases} \quad n \in \mathbb{R}^+. \tag{2.104}$$

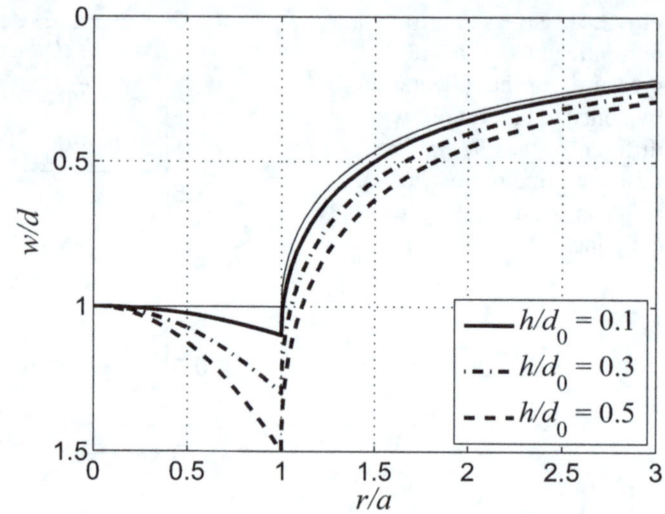

Fig. 2.42 Displacements of the half-space, normalized to the indentation depth, for the indentation by a concave paraboloid with different values h/d_0. The thin solid line represents the solution for the flat cylindrical punch

The notations can again be taken from Fig. 2.40. If the normal force is large enough to ensure complete contact, the solution to the contact problem is, following (2.17), given by:

$$
\begin{aligned}
g(x) &= \begin{cases} -\kappa(n)\dfrac{h|x|^n}{a^n}, & |x| \leq a, \\ \infty, & |x| > a \end{cases} \\
F_N(d_0) &= 2E^* a\left(d_0 + \frac{\kappa(n)h}{n+1}\right), \\
\sigma_{zz}(r; d_0) &= -\frac{E^*}{\pi}\left\{\int_r^a -\frac{\kappa(n)nhx^{n-1}}{a^n}\frac{\mathrm{d}x}{\sqrt{x^2-r^2}} + \frac{d_0+\kappa(n)h}{\sqrt{a^2-r^2}}\right\}, \quad r \leq a, \\
w(r; d_0) &= \frac{2}{\pi}\int_0^a \left(d_0 + \frac{\kappa(n)hx^n}{a^n}\right)\frac{\mathrm{d}x}{\sqrt{r^2-x^2}}, \quad r > a,
\end{aligned} \tag{2.105}
$$

with the indentation depth in the middle of the contact, d_0, and the scaling factor

$$
\kappa(n) := \sqrt{\pi}\frac{\Gamma(n/2+1)}{\Gamma\left[(n+1)/2\right]}, \tag{2.106}
$$

with the gamma function $\Gamma(\cdot)$

$$
\Gamma(z) := \int_0^\infty t^{z-1}\exp(-t)\,\mathrm{d}t. \tag{2.107}
$$

For details relating to integrals occurring in the stresses and displacements, see Sect. 2.5.8.

One can ask how large the indentation depth must be in order to ensure complete contact. For $n \leq 2$, the minimum of the contact pressure will be in the center of the contact (and the center point will be the last one to make contact during indentation). In this case the condition of complete contact can be determined easily and without evaluating the aforementioned integrals. One obtains that $\sigma_{zz}(r = 0) < 0$ if

$$d_0 > h\frac{\kappa(n)}{n-1}, \quad 1 < n \leq 2. \tag{2.108}$$

From this it can be seen that, for a concave cone (and hence $n = 1$), and in general for all concave profiles with $n \leq 1$, no complete contact can be realized. This was already known to Barber (1976). Therefore, for the relations occurring in (2.105), it must be $n > 1$. For exponents $n > 2$ the minimum contact pressure in complete contact lies away from the contact midpoint. During indentation, first the contact annulus propagates inside. After a first critical point, contact is established in the contact center and an additional circular contact area grows from the inside. Complete contact is established if the inner contact circle and the outer contact annulus overlap. The determination of the criterion of complete contact for $n > 2$ is a non-trivial task (and usually only possible numerically by finding the minimum of the contact pressure and setting the minimal pressure to zero), which was solved by Popov et al. (2018). For a simple example which is possible to analyze in closed form, we can examine the results for the case $n = 4$. The pressure distribution for complete contact is given by:

$$p(r) = -\sigma_{zz}(r; d) = \frac{E^*}{\pi}\left[\frac{3d + 5h}{3\sqrt{a^2 - r^2}} - \frac{32h}{9a}\sqrt{1 - \frac{r^2}{a^2}}\left(1 + \frac{2r^2}{a^2}\right)\right]. \tag{2.109}$$

Conditions $p'(r_c) = 0$, and simultaneously $p(r_c) = 0$, lead to $3d = 7h$, i.e., full contact is established for:

$$d_0 > \frac{4}{3}h. \tag{2.110}$$

2.5.17 The Paraboloid with Small Periodic Roughness (Complete Contact)

Finally, a simple analytical model of a parabolic indenter with periodic surface roughness is presented. The influence of roughness gains great importance in the treatment of the adhesive normal contact (see Sect. 3.5.14 in Chap. 3). Nonetheless, it can also be of interest for the non-adhesive contact. The contact problem was first solved by Guduru (2007). His solution uses a superposition method, which amounts to the same algorithm as the MDR, i.e., determining the auxiliary function $g(x)$, which basically solves the contact problem.

Let the examined three-dimensional profile have the shape:

$$f(r) = \frac{r^2}{2R} + h\left(1 - \cos\left(\frac{2\pi}{\lambda}r\right)\right), \tag{2.111}$$

with the amplitude h and the wavelength λ of the roughness, as well as the usual radius of the paraboloid R. Once more we assume a compact contact area. The chapter on adhesive normal contact imposes even stricter limitations requiring $f'(r) \geq 0$ (see Sect. 3.5.14). For the equivalent profile we obtain:

$$
\begin{aligned}
g(x) &= |x| \int_0^{|x|} \left(\frac{r}{R} + \frac{2\pi}{\lambda} h \sin\left(\frac{2\pi}{\lambda} r\right) \right) \frac{dr}{\sqrt{x^2 - r^2}} \\
&= \frac{x^2}{R} + \frac{\pi^2 h}{\lambda} |x| \mathrm{H}_0\left(\frac{2\pi}{\lambda}|x|\right).
\end{aligned}
\tag{2.112}
$$

Here, $H_n(\cdot)$ denotes the n-th order Struve function, which can be expanded as a power series:

$$
\mathrm{H}_n(u) := \sum_{k=0}^{\infty} \frac{(-1)^k}{\Gamma\left(k + \frac{3}{2}\right) \Gamma\left(k + n + \frac{3}{2}\right)} \left(\frac{u}{2}\right)^{2k+n+1}, \tag{2.113}
$$

with the Gamma function $\Gamma(\cdot)$:

$$
\Gamma(z) := \int_0^{\infty} t^{z-1} \exp(-t)\, dt. \tag{2.114}
$$

From (2.113) the following differentiation property of the Struve function can be proven:

$$
\frac{d}{du}[\mathrm{H}_n(u)] = \mathrm{H}_{n-1}(u) - \frac{n}{u} \mathrm{H}_n(u). \tag{2.115}
$$

The relationships between the global contact quantities F_N, d and a using (2.17) are given by equations:

$$
\begin{aligned}
d(a) &= \frac{a^2}{R} + \frac{\pi^2 h a}{\lambda} \mathrm{H}_0\left(\frac{2\pi}{\lambda} a\right), \\
F_N(a) &= \frac{4E^* a^3}{3R} + E^* \pi a h \left[\frac{2\pi}{\lambda} a \mathrm{H}_0\left(\frac{2\pi}{\lambda} a\right) - \mathrm{H}_1\left(\frac{2\pi}{\lambda} a\right) \right].
\end{aligned}
\tag{2.116}
$$

For the stresses in the contact area we obtain:

$$
\begin{aligned}
&\sigma_{zz}(r; a) = \\
&- \frac{E^*}{\pi} \left\{ \frac{2\sqrt{a^2 - r^2}}{R} + \frac{\pi^2 h}{\lambda} \int_r^a \left[\mathrm{H}_0\left(\frac{2\pi}{\lambda} x\right) + \frac{2\pi}{\lambda} x \mathrm{H}_{-1}\left(\frac{2\pi}{\lambda} x\right) \right] \frac{dx}{\sqrt{x^2 - r^2}} \right\}, \\
&r \leq a
\end{aligned}
\tag{2.117}
$$

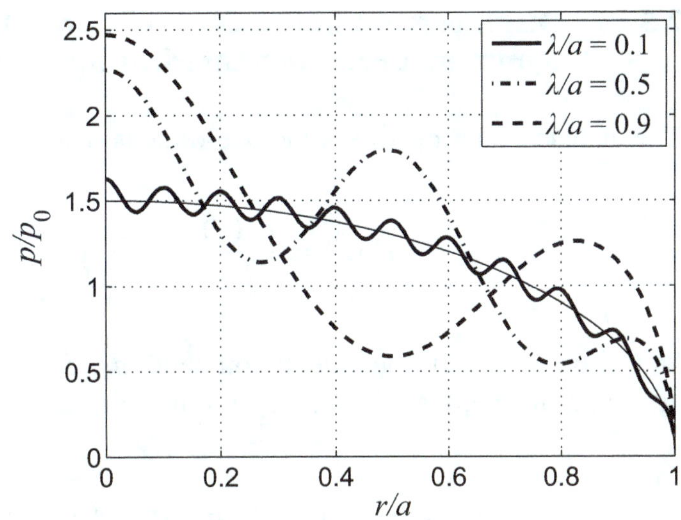

Fig. 2.43 Pressure distribution, normalized to the average pressure in the contact, for the indentation by a paraboloid with small periodic roughness at $hR/\lambda^2 = 0.1$ and different values λ/a. The thin solid line represents the solution for the smooth paraboloid

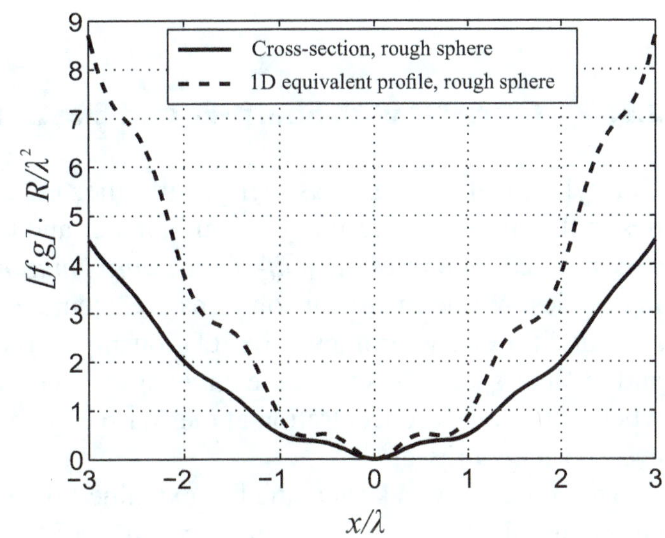

Fig. 2.44 Profile cross-section of the three-dimensional rotationally symmetric profile, and the equivalent profile within the framework of the MDR for a paraboloid with small periodic roughness

and the displacements outside the contact area are given by:

$$w(r;a) = \frac{2d(a)}{\pi}\arcsin\left(\frac{a}{r}\right) - \frac{1}{\pi R}\left(r^2\arcsin\left(\frac{a}{r}\right) - a\sqrt{r^2-a^2}\right)$$
$$-\frac{2\pi}{\lambda}h\int_0^a x\mathrm{H}_0\left(\frac{2\pi}{\lambda}x\right)\frac{\mathrm{d}x}{\sqrt{r^2-x^2}}, \quad r > a. \tag{2.118}$$

The stresses normalized to the average pressure in the contact are displayed in Fig. 2.43. The influence of the periodic roughness on the pressure distribution is quite apparent. Figure 2.44 shows the three-dimensional profile and also the equivalent one-dimensional profile.

2.5.18 Displacement in the Center of an Arbitrary Axially Symmetric Pressure Distribution

The displacement $w_{1D}(x)$ of the one-dimensional MDR-model is given by (2.16):

$$w_{1D}(x) = \frac{q_z(x)}{E^*} = \frac{2}{E^*}\int_x^\infty \frac{rp(r)}{\sqrt{r^2 - x^2}} \mathrm{d}r. \tag{2.119}$$

Setting $x = 0$ yields the indentation depth d_0, which in the context of the MDR is defined as the vertical displacement of the coordinate origin:

$$d_0 = w_{1D}(x = 0) = w(r = 0) = \frac{2}{E^*}\int_0^\infty p(r)\mathrm{d}r. \tag{2.120}$$

2.5.19 Contacts with Sharp-Edged Elastic Indenters

Flat cylindrical indenters, as well as all other indenters which generate singularities in the stress distribution (conical indenter and truncated cone, etc.), violate the necessary conditions of the *half-space approximation* at least for one of the contacting bodies. While the given solutions are accurate for *rigid* profiles, they must be modified for elastic profiles. The relationship of the elastic moduli of the indenter and the half-space, as well as the angle at the sharp edge of the indenter, determine whether the stress concentration at the sharp edge result in different singularities or even no singularity at all.

This problem was systematically examined by Rao (1971). Rao considered the stress and displacement fields in the vicinity of the sharp edge for widely varying classes of problems. Here, we will limit our consideration to the frictionless normal contact between an axisymmetric elastic indenter and an elastic half-space. The used notation is displayed in Fig. 2.45.

The stress state in the region of the sharp edge is approximately two-dimensional, and the normal stress at distance s from the sharp edge is given by:

$$\sigma_{zz} \sim s^{\lambda-1}, \tag{2.121}$$

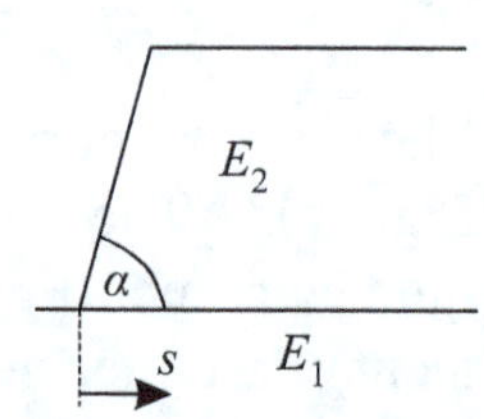

Fig. 2.45 Edge of an elastic indenter (diagram visualizing the notations used)

where λ indicates the smallest eigenvalue of the respective boundary value problem. The eigenvalue equation of the considered problem can be written as:

$$\begin{aligned} &\tan(\pi\lambda)\,[\lambda\sin(2\alpha)+\sin(2\lambda\alpha)] \\ &+e\left[1-\cos(2\lambda\alpha)-\lambda^2(1-\cos(2\alpha))\right]=0. \end{aligned} \tag{2.122}$$

Here, α is the angle at the sharp edge (measurement taken within the indenting body) and $e = E_2/E_1$ is the relationship of the elasticity moduli of the indenter (index "2") and the half-space (index "1"). Interestingly, the respective Poisson's ratios have no effect on this characteristic equation.

A singularity in the stress distribution occurs at the sharp edge if the smallest non-trivial solution (2.122) is smaller than one. The limiting case, $\lambda = 1$, arises precisely when the relationship of the moduli is set to:

$$e_{\text{crit}} = \frac{\pi\cos\alpha}{\sin\alpha - \cos\alpha}. \tag{2.123}$$

Greater values of e will result in singularities. However, (2.123) only has a positive solution for e if $\alpha \le \pi/2$. Therefore, in principle, greater angles always result in singularities, whose concrete shape is determined by the solution of the eigenvalue equation (2.122). As an example, for the rigid flat cylindrical punch with $\alpha = \pi/2$ and $e \to \infty$, the result is the familiar singularity of the Boussinesq solution, $\lambda = 1/2$.

How can the stresses be expressed in the entire contact area? For this, Jordan and Urban (1999) proposed the following expression for rectangular indenters in a plane (which can be directly applied to rotationally symmetric problems with cylindrical flat punches):

$$\sigma_{zz}(r;a,\lambda) = -F_N\cdot M(\lambda,a)(a^2-r^2)^{\lambda-1}. \tag{2.124}$$

The case of the rigid cylindrical punch coincides with the known Boussinesq solution, and the stresses exhibit the required singularity property (2.121) since for $s = a - r \ll a$:

$$a^2-r^2=(a-r)(a+r)=s(a+r)\approx 2as. \tag{2.125}$$

The function $M(\lambda, a)$ is derived from the normalization:

$$F_N = -2\pi\int_0^a \sigma_{zz}(r)r\,\mathrm{d}r, \tag{2.126}$$

and therefore:

$$M(\lambda,a)=\frac{\lambda}{\pi a^{2\lambda}}. \tag{2.127}$$

2.6 Mossakovskii Problems (No-Slip)

In the preceding sections, we considered *frictionless* normal contact problems. Assuming different elastic materials, the material points of the contacting surfaces will experience differing radial displacements. If we set aside the unrealistic assumption of ideally frictionless surfaces ($\mu = 0$), this slip will always cause radial tangential stresses. Consequently, the frictionless normal contact can only be considered the theoretical limiting case. The contact area generally consists of an inner stick zone and an outer slip zone. This normal contact problem with friction was examined by Spence (1975). It is supremely complicated, and thus only permits numerical solutions. Instead, we will just consider the other *theoretical limiting case* which describes complete stick within the contact area ($\mu \to \infty$). Figure 2.46 illustrates the indentation of an elastic half-space by a rigid curved indenter with complete stick. Also displayed are the displacement paths of individual surface points. Surface points that contact the indenter during the indentation process stick completely, eliminating the possibility of any further radial displacement. This condition is expressed through the boundary condition:

$$\frac{\partial u_r(r, a)}{\partial a} = 0, \quad r \leq a, \tag{2.128}$$

and it is used instead of the boundary condition for the frictionless normal contact according to (2.4) which assumes a frictionless contact. Here, it should be noted that for the contact of *two* elastic bodies, the radial displacement from (2.128) simply represents the relative radial displacement of the surface points.

The solution of the normal contact problem with complete stick for arbitrary axially symmetric normal contacts goes back to Mossakovskii (1963). In 1954, he had already developed the solution for the contact of the flat cylindrical punch. Therefore, normal contacts with complete stick are also referred to as *Mossakovskii problems*. Their solutions are significantly more complex than frictionless contacts, and the simplest approach to solving them is with the MDR. This requires redefining the spring stiffness compared to (2.5) and deriving the equivalent one-dimensional

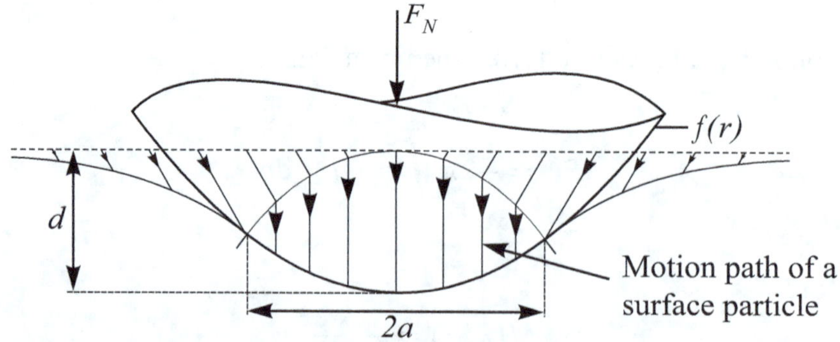

Fig. 2.46 Illustrating the boundary condition of full stick for the indentation of an elastic half-space by a rigid indenter (modeled after Spence 1968)

profile from an equation significantly more complicated than (2.6). The calculation of the global relationships between normal force, indentation depth, and contact radius follows the exact same template as described in Sect. 2.3.2. To describe the normal contact between *two* completely sticking elastic bodies, the spring stiffness is set to:

$$\Delta k_z = \frac{E^*}{2\beta_D} \ln\left(\frac{1+\beta_D}{1-\beta_D}\right) \Delta x. \tag{2.129}$$

Here, E^* denotes the effective elasticity modulus defined in (2.1):

$$\frac{1}{E^*} = \frac{1-\nu_1^2}{E_1} + \frac{1-\nu_2^2}{E_2}, \tag{2.130}$$

and β_D Dundur's second constant:

$$\beta_D := \frac{(1-2\nu_1)G_2 - (1-2\nu_2)G_1}{2(1-\nu_1)G_2 + 2(1-\nu_2)G_1}. \tag{2.131}$$

Many contact problems with complete stick featured in the literature make the simplifying assumption of one body being rigid. In this case, the spring stiffness from (2.129) is simplified to:

$$\Delta k_z = \frac{2G \ln(3-4\nu)}{1-2\nu} \Delta x. \tag{2.132}$$

The relationship between the three-dimensional profile $f(r)$ and the equivalent one-dimensional profile $g(x)$ is given by equation:

$$f(r) = \frac{2}{\pi} \int_0^r \frac{1}{\sqrt{r^2-x^2}} \int_0^x g'(t) \cos\left(\vartheta \ln\left(\frac{x-t}{x+t}\right)\right) \mathrm{d}t\,\mathrm{d}x$$
$$\text{with} \quad \vartheta := \frac{1}{2\pi} \ln\left(\frac{1+\beta_D}{1-\beta_D}\right). \tag{2.133}$$

For $\beta_D = 0$, it coincides with the inverse transform of the profile for a frictionless contact according to (2.7). While explicitly solving for $g(x)$ is possible in principle, it leads to either a very unwieldy, extremely complicated calculation which can be referenced in Fabrikant's (1986) work, or a notation using the Mellin inverse transform (Spence 1968). We will forgo providing an explicit expression since an analytical calculation of the equivalent one-dimensional profile $g(x)$ for a given axially symmetric profile $f(r)$ is generally only possible using numerical methods anyway. The sole exception seems to be the power-law profile, which we will examine in greater detail in Sect. 2.6.2. To calculate the respective one-dimensional equivalent profile, the implicit formulation in (2.133) will prove sufficient.

The elasticity parameter ϑ in (2.133) illustrates a major difference to the frictionless contact: the equivalent one-dimensional profile is no longer exclusively dependent on the geometry of the axially symmetric contact; it is also affected by the elastic properties of the contacting bodies.

As previously mentioned, the calculation of the relationships between indentation depth, contact radius, and normal force follows the same template as the frictionless normal contact. Only the modified spring stiffness from (2.129) and the equivalent profile determined by (2.133) must be accounted for. We will summarize the essential equations. First, the surface displacement of the Winkler foundation is determined to be:

$$w_{1D}(x) = d - g(x). \tag{2.134}$$

At the edge of all non-adhesive contacts, the displacement must be zero, thus formulating a condition for calculating the indentation depth:

$$w_{1D}(\pm a) = 0 \quad \Rightarrow \quad d = g(a). \tag{2.135}$$

Additionally, the sum of all spring forces must balance out the applied normal force. Consideration of the modified stiffness in accordance with (2.129) yields:

$$F_N = \frac{E^*}{\beta_D} \ln\left(\frac{1+\beta_D}{1-\beta_D}\right) \int\limits_0^a w_{1D}(x)\mathrm{d}x. \tag{2.136}$$

Naturally, the local quantities obey different calculation formulas compared to the frictionless contact. Apart from the normal surface displacement and the pressure distribution, the tangential stresses within the contact area are also of importance. Said quantities can be determined solely from the known normal displacement of the Winkler foundation. Here, we will state them for the special case of a rigid indenter pressed into an elastic half-space:

$$\begin{aligned}
p(r,a) &= -\frac{8G(1-\nu)\ln(3-4\nu)}{\pi^2(1-2\nu)\sqrt{3-4\nu}} \int\limits_r^a x w_{1d}'(x) \int\limits_0^r \frac{\cos\left[\vartheta \ln\left(\frac{x+t}{x-t}\right)\right]}{\sqrt{r^2-t^2}(x^2-t^2)} \mathrm{d}t\,\mathrm{d}x \\
\tau_{zr}(r,a) &= \frac{8G(1-\nu)\ln(3-4\nu)}{\pi^2(1-2\nu)\sqrt{3-4\nu}} \int\limits_r^a \frac{x}{r} w_{1d}'(x) \int\limits_0^r \frac{\sin\left[\vartheta \ln\left(\frac{x+t}{x-t}\right)\right] t}{\sqrt{r^2-t^2}(x^2-t^2)} \mathrm{d}t\,\mathrm{d}x \\
w(r,a) &= -\frac{4(1-\nu)}{\pi\sqrt{3-4\nu}} \int\limits_0^a w_{1d}'(x) \int\limits_0^x \frac{\cos\left[\vartheta \ln\left(\frac{x+t}{x-t}\right)\right]}{\sqrt{r^2-t^2}} \mathrm{d}t\,\mathrm{d}x.
\end{aligned} \tag{2.137}$$

2.6.1 The Cylindrical Flat Punch

The normal contact of a rigid flat cylindrical punch with an elastic half-space with complete stick was initially solved by Mossakovskii (1954). Within the framework of the MDR, the equivalent plane profile is derived from the cross-section of the flat indenter in the x–z-plane—no change in the geometry is required. The displacement of the Winkler foundation is then given by:

$$w_{1D}(x) = d\,[\mathrm{H}(x+a) - \mathrm{H}(x-a)]\,, \tag{2.138}$$

where H(·) represents the Heaviside step function. Substituting this into (2.136) under consideration of the rigidity of the indenter yields the normal force:

$$F_N = \frac{4G\ln(3-4\nu)}{1-2\nu}da. \tag{2.139}$$

The contact stiffness then takes on the form:

$$k_z^{\mathrm{M}} := \frac{\mathrm{d}F_N}{\mathrm{d}d} = \frac{4G\ln(3-4\nu)}{1-2\nu}a. \tag{2.140}$$

The validity of this contact stiffness for arbitrary axially symmetric contacts follows immediately from Mossakovskii's (1963) work. Accordingly, the incremental difference of two-contact configurations with the contact radii a and $a+\mathrm{d}a$ is equivalent to the infinitesimal indentation of an elastic half-space by a cylindrical flat punch with radius a. This applies regardless of whether a normal contact with complete stick or a frictionless contact is being considered. Nevertheless, the works of Borodich and Keer (2004) and also Pharr et al. (1992) are still frequently cited, which prove the universal validity of the normal contact stiffness using a different, more complex approach. For the frictionless normal contact, the contact stiffness is given by (2.21). A comparison of the two values of contact stiffness reveals that the contact stiffness for complete stick is generally greater than for frictionless contact. This is a direct result of the suppressed relative displacement of the contacting surfaces. The contact stiffness only coincides for incompressible materials since, in this case, no tangential forces arise in the contact area due to the radial displacement of the material being restricted. For the ratio between the contact stiffness for complete stick and frictionless contact, it follows that:

$$\frac{k_z^{\mathrm{M}}}{k_z} = \frac{(1-\nu)\ln(3-4\nu)}{1-2\nu}. \tag{2.141}$$

For common materials with Poisson's ratios in the range of $0 \le \nu \le 0.5$, the contact stiffness for complete stick is at most 10% greater than for frictionless normal contacts. This maximum is reached at $\nu = 0$. For synthetic materials characterized by negative Poisson's ratios, the relative discrepancy can reach values of up to 30%, which is documented in Fig. 2.47. Viewed over the entire physical domain, the relative discrepancy decreases monotonically with a rising Poisson's ratio. The relative discrepancy reaches its maximum for the limiting case $\nu \to -1$. It should be noted that both values of contact stiffness approach infinity for $\nu \to -1$. At this point, the

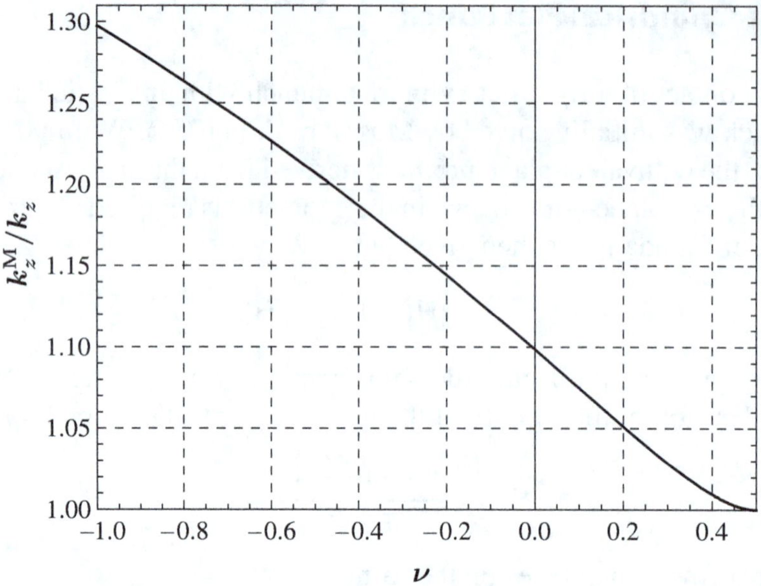

Fig. 2.47 Relationship between the normal contact stiffness for complete stick and normal contact stiffness for frictionless contacts as a function of the Poisson's ratio ν

contact compliance Π, which represents the inverse of the stiffness, is zero. Argatov et al. (2012) examined the impact of negative Poisson's ratios on on the stress distribution, thereby discovering that the location of the greatest shear stress moves towards the surface for smaller Poisson's ratios. This is because of the increasing tangential stresses in the contact area for decreasing Poisson's ratios. Figure 2.48 puts in contrast the differing contact compliances for Boussinesq and Mossakovskii

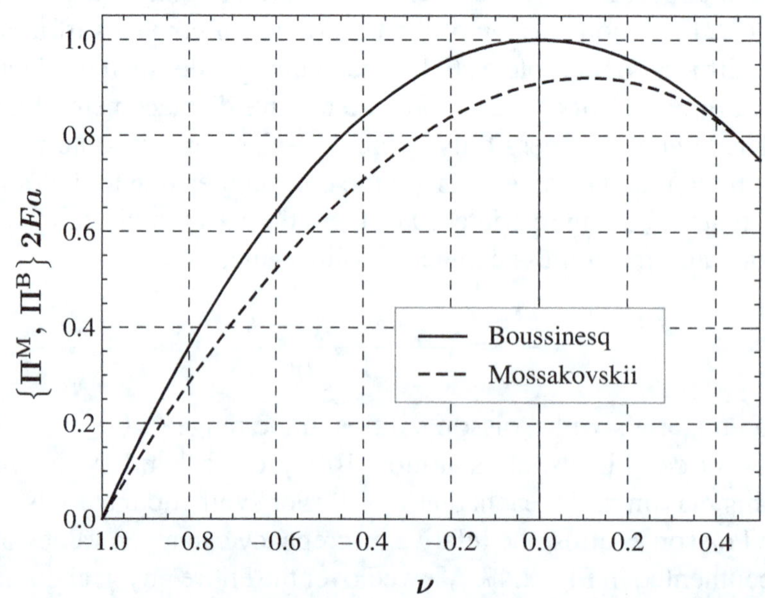

Fig. 2.48 Comparison of the normalized normal contact compliances as functions of the Poisson's ratio ν for Boussinesq and Mossakovkii problems

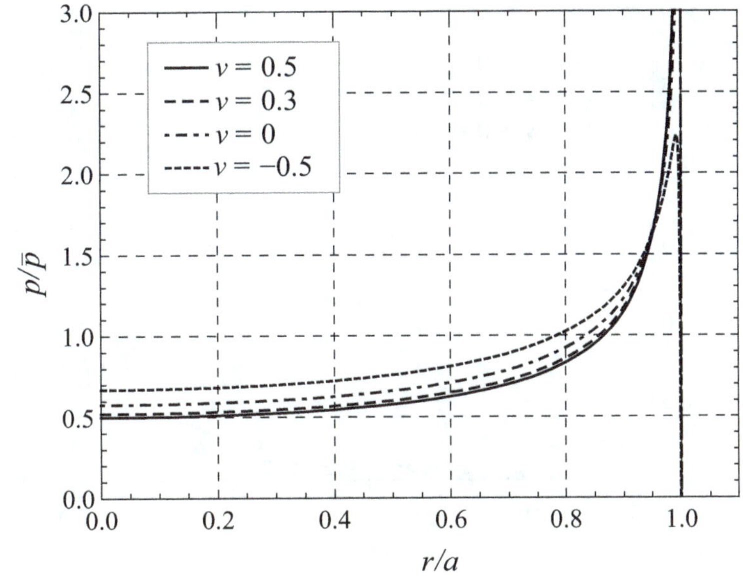

Fig. 2.49 Pressure distribution for the indentation by a flat cylindrical indenter for different Poisson's ratios ν, normalized to the average pressure $\overline{p}$

problems. It is apparent that the compliance maximum of Mossakovskii problems are not located at $\nu = 0$ but instead at $\nu \approx 0.11$, as stated by Argatov et al. (2012).

The derivative of the profile (2.138) with respect to the coordinate x is required for calculating the local quantities according to (2.137), and is:

$$w'_{1D}(x) = d\,[\delta(x+a) - \delta(x-a)]\,, \tag{2.142}$$

with the Delta function $\delta(\cdot)$. Taking into account its filtering properties, it follows from (2.137) that the solutions for the local quantities are:

$$\begin{aligned}
p(r,a) &= \frac{8Gad(1-\nu)\ln(3-4\nu)}{\pi^2(1-2\nu)\sqrt{3-4\nu}} \int_0^r \frac{\cos\left[\vartheta \ln\left(\frac{a+t}{a-t}\right)\right]}{\sqrt{r^2-t^2}(a^2-t^2)}\mathrm{d}t\\
\tau_{zr}(r,a) &= -\frac{8Gad(1-\nu)\ln(3-4\nu)}{\pi^2(1-2\nu)\sqrt{3-4\nu}}\frac{1}{r} \int_0^r \frac{\sin\left[\vartheta \ln\left(\frac{a+t}{a-t}\right)\right]t}{\sqrt{r^2-t^2}(a^2-t^2)}\mathrm{d}t\\
w(r,a) &= \frac{4d(1-\nu)}{\pi\sqrt{3-4\nu}} \int_0^a \frac{\cos\left[\vartheta \ln\left(\frac{a+t}{a-t}\right)\right]}{\sqrt{r^2-t^2}}\mathrm{d}t.
\end{aligned} \tag{2.143}$$

They correspond exactly to the quantities calculated by Mossakovskii (1963) and Spence (1968). Regrettably, the remaining integrals are only solvable using numerical methods. The pressure distribution normalized to the average pressure $\overline{p}$ within the contact area is visualized for different Poisson's ratios in Fig. 2.49. For incompressible materials, i.e., $\nu = 1/2$, the curve exactly matches the one for the

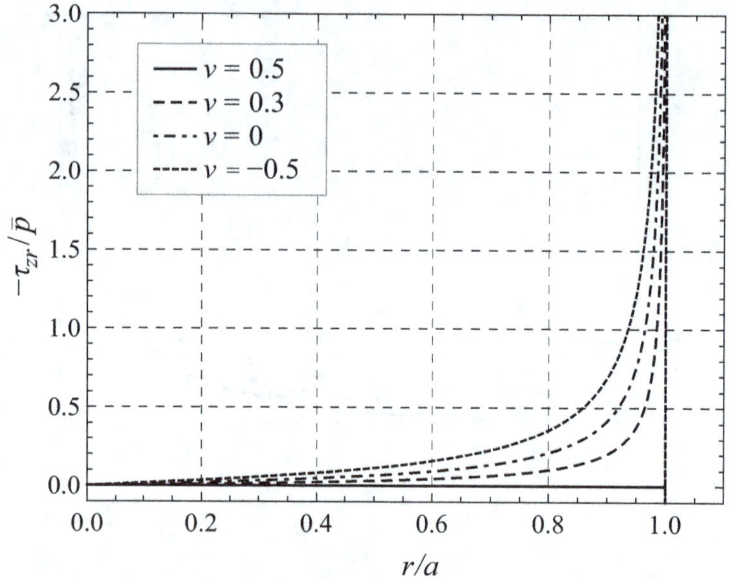

Fig. 2.50 Tangential stresses for the indentation by a flat cylindrical indenter for different Poisson's ratios ν, normalized to the average pressure $\overline{p}$

frictionless contact (compare this to Fig. 2.5). For regular positive Poisson's ratios, there is no distinguishable difference between the graphs. Only once the Poisson's ratios drop to negative values the pressure in the center notably increases. There is characteristic behavior of the stresses solely at the edge of the contact, which is already visible in solution (2.143). Here, stress oscillations occur, even leading to tensile stresses. On the one hand, these fluctuations can be viewed as indicators that the assumption of complete stick is self-contradictory for normal contact, indicating that slip at the contact edge is unavoidable (Zhupanska 2009). On the other hand, the oscillations are localized so close to the contact edge that this zone need not be ascribed significant importance.[1]

The tangential stresses in Fig. 2.50 are zero at the center of the contact and feature a singularity at the edge that is comparable to the one caused by the normal stresses. For the incompressible case, no tangential stresses occur in the contact area since the material resists radial displacements. Expanding the examination to the contact between two elastic materials, the tangential stresses in the contact area are zero when Dundur's second constant, defined in (2.131), vanishes. This condition is, indeed, satisfied for the contact between a rigid and an incompressible medium.

The curve of the normal surface displacement normalized to the respective indentation depth is shown in Fig. 2.51. The curves for $\nu = 1/2$ and $\nu = 0.3$ are nearly indistinguishable, so that their shape can be approximated by the displacements for a frictionless normal contact described by (2.22). A more detailed analysis of the displacements can be referenced in the publication of Fabrikant (1986).

[1] In the plane case, the normal stresses initially changes signs at $x = 0.9997a$.

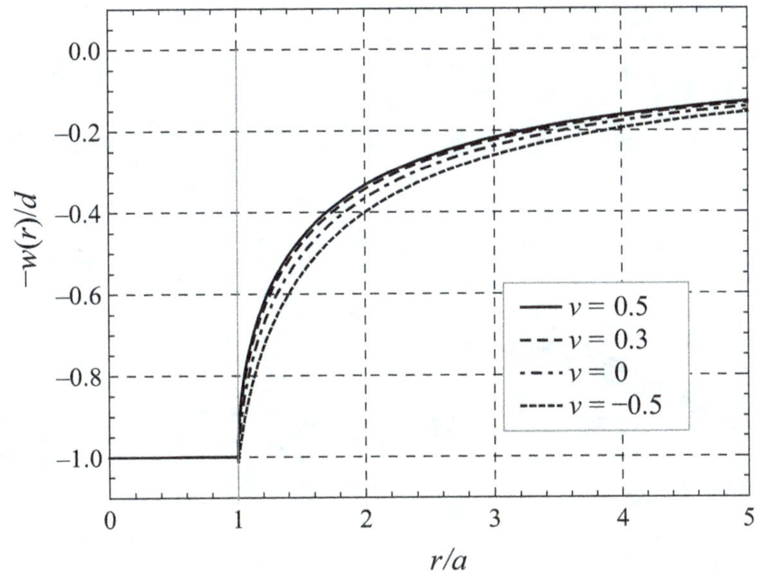

Fig. 2.51 Normal surface displacement of the half-space for indentation by a flat cylindrical indenter for different Poisson's ratios ν, normalized to the indentation depth d

2.6.2 The Profile in the Form of a Power-Law

The normal contact of a profile in the shape of a power-law was examined for the case of complete stick by both Mossakovskii (1963) and Spence (1968). As shown for the frictionless contact, a power-law profile is mapped to a power-law profile:

$$f(r) = c_n r^n \quad \mapsto \quad g(x) = \kappa(n,\nu) f(|x|) \quad \text{for } n \in \mathbb{R}^+. \tag{2.144}$$

The scaling factor $\kappa(n,\nu)$ can be determined by simply substituting (2.144) in (2.133). A complicated integration is unnecessary; a trivial normalization of the integral variables is sufficient. This leads to:

$$\kappa(n,\nu) = \frac{\sqrt{\pi}\,\Gamma\left(1+\frac{n}{2}\right)}{\Gamma\left(\frac{1+n}{2}\right) n I^*(n)} \quad \text{with}$$
$$I^*(n) := \int_0^1 t^{n-1} \cos\left(\vartheta \ln\left(\frac{1-t}{1+t}\right)\right) \mathrm{d}t, \tag{2.145}$$

with the definition of ϑ from (2.133). A comparison of the scaling factors to the ones for the frictionless contact is easily performed by setting $\beta_D = 0$, i.e., by assuming similar elastic materials and yielding $\vartheta = 0$, and consequently, $nI^*(n) = 1$. From (2.145), we obtain the scaling factors for the frictionless contact defined in (2.61). In the following, we will operate under the assumption of *one* rigid body. In this case, the other body must be incompressible to prevent tangential stresses from occurring in the contact surface. The limiting curve for

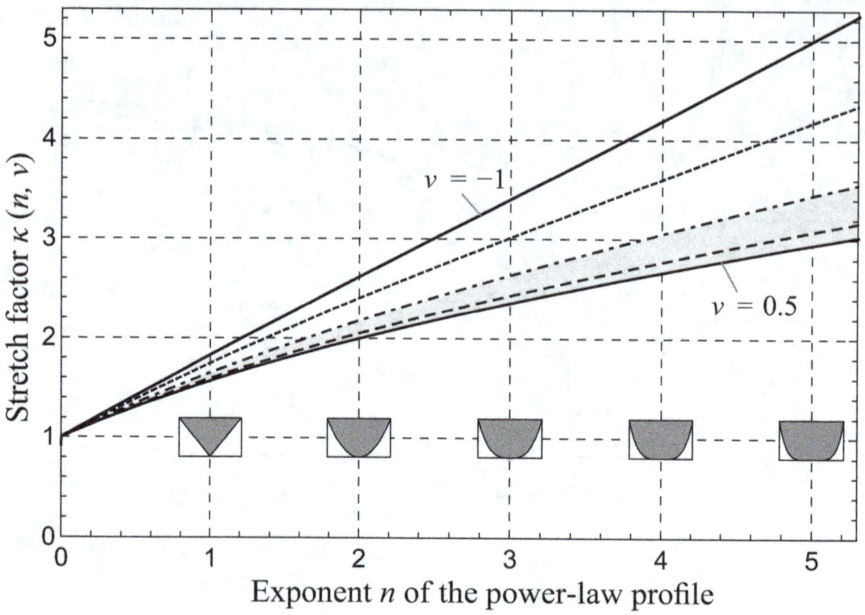

Fig. 2.52 Scaling factors κ as a function of the exponent n of the power-law profile for different Poisson's ratios ν; the area shaded in gray represents common materials ($0 \leq \nu \leq 0.5$)

$\nu = 1/2$ in Fig. 2.52 must exactly coincide with the curve from Fig. 2.21, which displays the scaling factor as a function of the exponents of the profile function for the *frictionless contact*.

In contrast, complete stick in the normal contact between a rigid power-law indenter and a compressible, elastic half-space results in tangential stresses in the contact surface. The full description of such a contact requires greater scaling factors, also given in Fig. 2.52. The smaller the Poisson's ratio, the greater the scaling factor. The area shaded in gray represents the range of values of the scaling factors for common materials.

Since the integral $I^*(n)$ in formula (2.145) of the scaling factor is generally only solvable numerically, the corresponding value for selected Poisson's ratios and exponents is provided in Table 2.2. These are of particular importance for axially symmetric profiles that are either defined by a polynomial or a Taylor expansion.

The equivalent plane profile was already determined in (2.144). Applying formulas (2.135) and (2.136) yields the indentation depth and the normal force as a function of the contact radius:

$$d(a) = \kappa(n, \nu)c_n a^n,$$
$$F_N(a) = \frac{4G\ln(3-4\nu)}{1-2\nu}\frac{n}{n+1}\kappa(n, \nu)c_n a^{n+1}. \tag{2.146}$$

The calculation and graphical representation of surface stresses and displacements will be omitted at this point. However, we will analyze the particular cases of the conical ($n = 1$) and parabolic ($n = 2$) contact.

Table 2.2 Stretch factor κ as a function of the exponent of the power-law and Poisson's ratio

		Poisson's ratio ν				
		−1	−0.5	0	0.3	0.5
Exponent n of the power-law profile	0.5	1.429	1.389	1.348	1.322	1.311
	1	1.831	1.746	1.651	1.594	1.571
	2	2.617	2.402	2.177	2.049	2
	3	3.398	3.014	2.638	2.433	2.356
	4	4.189	3.602	3.056	2.771	2.667
	5	5.000	4.175	3.444	3.077	2.945
	6	5.840	4.739	3.811	3.360	3.2
	7	6.714	5.298	4.159	3.623	3.436
	8	7.626	5.855	4.494	3.872	3.657
	9	8.582	6.412	4.816	4.107	3.866
	10	9.588	6.971	5.123	4.332	4.063

2.6.3 The Cone

We will now consider the normal contact of a rigid cone and a planar elastic half-space under the condition of complete stick. The shape of the profile function is given by:

$$f(r) = r\tan\theta \tag{2.147}$$

(see Fig. 2.7). The equivalent plane profile follows as a special case of (2.144), where the scaling factor (2.145) must be determined. An analytical expression for the integral I^*, and consequently the stretch factor (solely in this case), is published in literature (see Spence 1968):

$$\kappa(1,\nu) = \frac{\pi}{2I^*(1)} = \frac{\pi}{2}\frac{1-2\nu}{\pi\vartheta\sqrt{3-4\nu}} = \frac{\pi(1-2\nu)}{\ln(3-4\nu)\sqrt{3-4\nu}}. \tag{2.148}$$

The equivalent profile is then:

$$g(x) = \frac{\pi(1-2\nu)}{\ln(3-4\nu)\sqrt{3-4\nu}}|x|\tan\theta. \tag{2.149}$$

Substituting this last result into formulas (2.135) and (2.136), followed by a basic calculation, leads to the indentation depth and normal force

$$d(a) = \frac{\pi(1-2\nu)}{\ln(3-4\nu)\sqrt{3-4\nu}}a\tan\theta,$$
$$F_N(a) = \frac{2G\pi}{\sqrt{3-4\nu}}a^2\tan\theta. \tag{2.150}$$

We omit explicitly providing the stresses and normal surface displacements since, once again, the integrals in expression (2.137) can only be solved numerically.

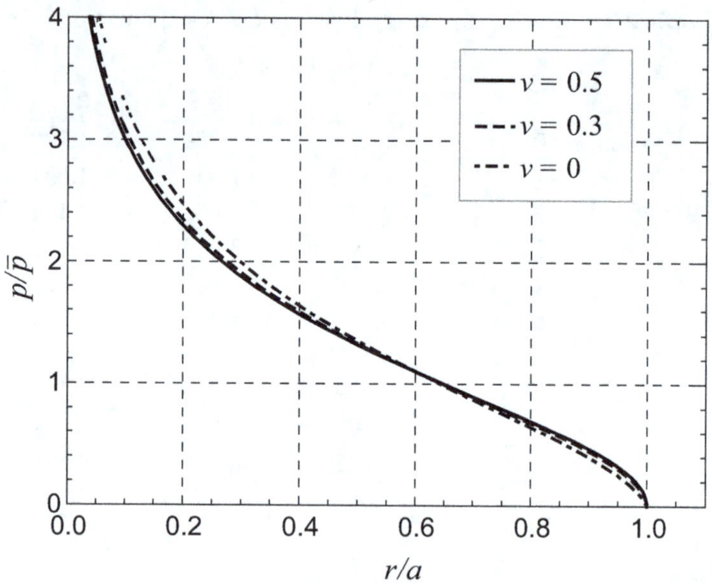

Fig. 2.53 Pressure distribution in the conical contact for different Poisson's ratios ν, normalized to the average pressure $\overline{p}$

The results for the normal and tangential stresses via numerical integration are visualized in Figs. 2.53 and 2.54. Note again that the differences in the pressure distribution compared to the frictionless contact are minimal. The magnitude of the tangential stresses increases towards the center. As expected, they increase with a decreasing Poisson's ratio.

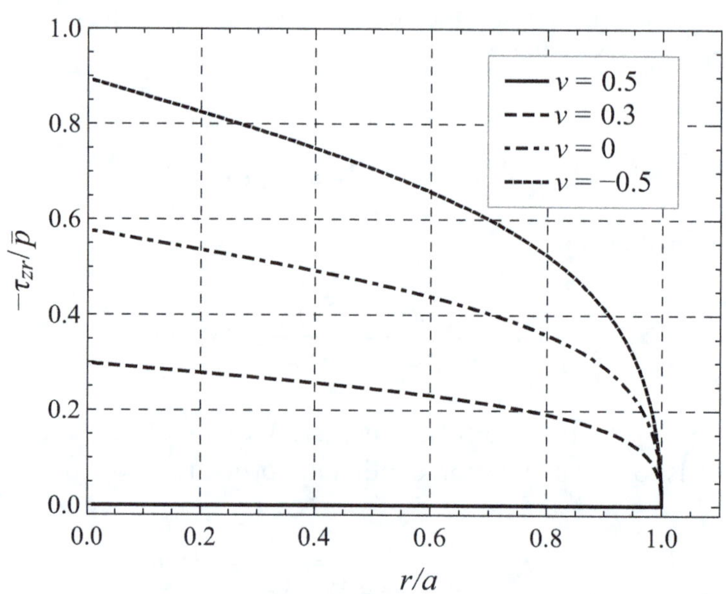

Fig. 2.54 Tangential stresses in the conical contact for different Poisson's ratios ν, normalized to the average pressure $\overline{p}$

2.6.4 The Paraboloid

Finally, we will cover the important parabolic contact with complete stick, which can be considered an approximate solution for various curved surfaces, and was solved by Mossakovskii (1963) as well as by Spence (1968). The axially symmetric profile is given by:

$$f(r) = \frac{r^2}{2R}, \tag{2.151}$$

where R denotes the curvature radius of the paraboloid. The equivalent one-dimensional profile follows from (2.144):

$$g(x) = \kappa(2, \nu)\frac{x^2}{2R}. \tag{2.152}$$

In contrast to the conical contact, the stretch factor can only be calculated numerically. Selected values can be found in Table 2.2. Spence (1968) did provide a good approximation for the scaling factor:

$$\kappa(2, \nu) \approx \frac{2}{1 - 0.6931(2\vartheta)^2 + 0.2254(2\vartheta)^4} \quad \text{with}$$
$$\vartheta(\nu) = \frac{1}{2\pi}\ln(3 - 4\nu). \tag{2.153}$$

The indentation depth can be determined from (2.135), from which we can then derive the condition of the vanishing displacement of the Winkler foundation at the contact edge. Additionally, the normal force can be calculated from the balance of forces in the z-direction in accordance with (2.136). The indentation depth and normal force then follows as:

$$d(a) = \kappa(2, \nu)\frac{a^2}{2R}$$
$$F_N(a) = \frac{4G}{3R}\frac{\ln(3 - 4\nu)}{1 - 2\nu}\kappa(2, \nu)a^3. \tag{2.154}$$

However, analytical solutions of the surface stresses and normal displacements do not appear possible. Although Zhupanska (2009) claimed to have analytically calculated these quantities, her formulas contain series and integral expressions. Taking into account the surface displacement of the Winkler foundation in formulas (2.137), a numerical calculation leads to the solutions for the normal and tangential stresses shown in Figs. 2.55 and 2.56. From the pressure distribution in Fig. 2.55, it is apparent that the pressure maximum in the center of the contact area increases with a decreasing Poisson's ratio. The contact radius decreases simultaneously which, due to the normalization with respect to the contact radius, is not represented in the figure. The tangential stresses are zero in the center and at the edge. As expected, they increase with a decreasing Poisson's ratio. The curve for

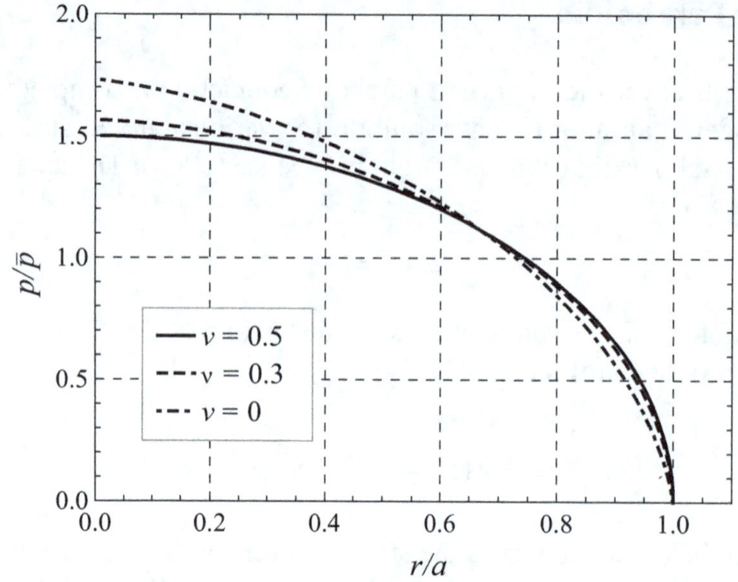

Fig. 2.55 Pressure distribution in the contact with a paraboloid for different Poisson's ratios ν, normalized to the average pressure

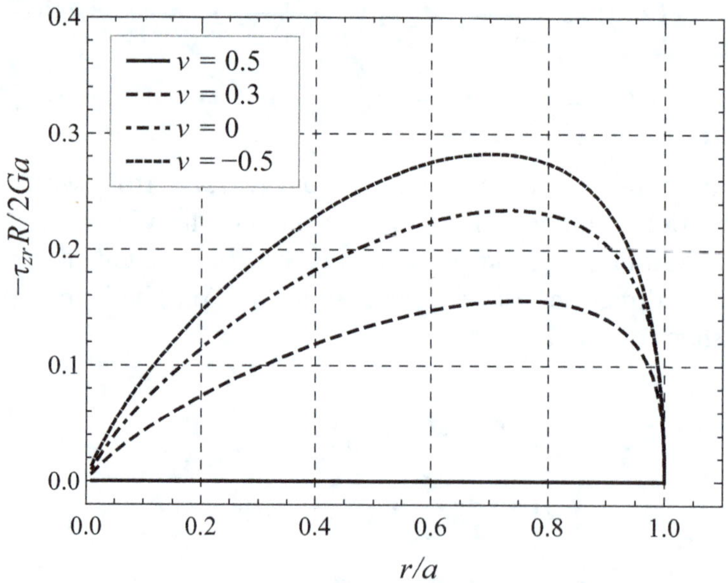

Fig. 2.56 Normalized tangential stresses in the contact with a paraboloid for different Poisson's ratios ν

$\nu = 0.3$ coincides exactly with Zhupanska's (2009), who utilized torus coordinates for the solution.

For the sake of completeness, the normal surface displacement for several Poisson's ratios is given in a graphical representation in Fig. 2.57. The figure offers a clear illustration of the fact that achieving the same contact area requires a greater indentation depth than for the frictionless contact.

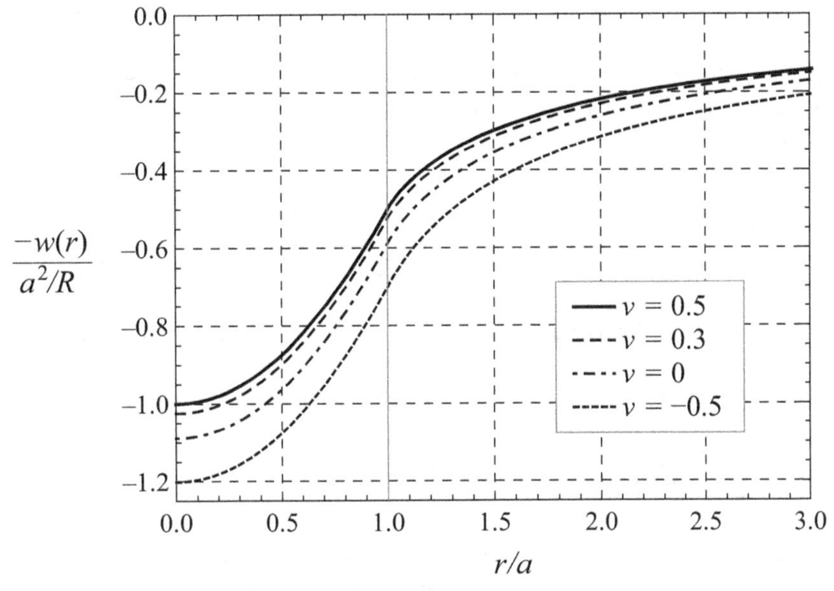

Fig. 2.57 Normalized normal surface displacement of the half-space for indentation by a paraboloid for different Poisson's ratios ν

References

Abramian, B.L., Arutiunian, N.K., Babloian, A.A.: On two-contact problems for an elastic sphere. PMM J. Appl. Math. Mech. **28**(4), 769–777 (1964)

Argatov, I.I., Guinovart-Díaz, R., Sabina, F.J.: On local indentation and impact compliance of isotropic auxetic materials from the continuum mechanics viewpoint. Int. J. Eng. Sci. **54**, 42–57 (2012)

Barber, J.R.: Indentation of the semi-infinite elastic solid by a concave rigid punch. J. Elast. **6**(2), 149–159 (1976)

Barber, J.R.: A four-part boundary value problem in elasticity: indentation by a discontinuously concave punch. Appl. Sci. Res. **40**(2), 159–167 (1983)

Borodich, F.M., Keer, L.M.: Evaluation of elastic modulus of materials by adhesive (no-slip) nano-indentation. Proc. R. Soc. London Ser. A **460**, 507–514 (2004)

Boussinesq, J.: Application des Potentiels a L'etude de L'Equilibre et du Mouvement des Solides Elastiques. Gauthier-Villars, Paris (1885)

Ciavarella, M.: Indentation by nominally flat or conical indenters with rounded corners. Int. J. Solids Struct. **36**(27), 4149–4181 (1999)

Ciavarella, M., Hills, D.A., Monno, G.: The influence of rounded edges on indentation by a flat punch. Proc. Inst. Mech. Eng. C J. Mech. Eng. Sci. **212**(4), 319–327 (1998)

Collins, W.D.: On the solution of some Axi-symmetric boundary value problems by means of integral equations. VIII. Potential problems for a circular annulus. Proc. Edinb. Math. Soc. Ser. 2 **13**(3), 235–246 (1963)

Ejike, U.B.C.O.: Contact problem for an elastic half-space and a rigid conical frustum. Proceedings of the First National Colloquium on Mathematics and Physics. (1969)

Ejike, U.B.C.O.: The stress on an elastic half-space due to sectionally smooth-ended punch. J. Elast. **11**(4), 395–402 (1981)

Fabrikant, V.I.: Four types of exact solution to the problem of an axi-symmetric punch bonded to a transversely isotropic half-space. Int. J. Eng. Sci. **24**(5), 785–801 (1986)

Föppl, L.: Elastische Beanspruchung des Erdbodens unter Fundamenten. Forsch. Gebiet Ingenieurwes. A **12**(1), 31–39 (1941)

Galin, L.A.: Three-dimensional contact problems of the theory of elasticity for punches with a circular planform. Prikladnaya Matem. Mekhanika **10**, 425–448 (1946)

Guduru, P.R.: Detachment of a rigid solid from an elastic wavy surface: theory. J. Mech. Phys. Solids **55**(3), 445–472 (2007)

Hamilton, G.M., Goodman, L.E.: The stress field created by a circular sliding contact. J. Appl. Mech. **33**(2), 371–376 (1966)

Hertz, H.: Über die Berührung fester elastischer Körper. J. Reine Angew. Math. **92**, 156–171 (1882)

Huber, M.T.: Zur Theorie der Berührung fester elastischer Körper. Ann. Phys. **14**, 153–163 (1904)

Hunter, S.C.: Energy absorbed by elastic waves during impact. J. Mech. Phys. Solids **5**(3), 162–171 (1957)

Jordan, E.H., Urban, M.R.: An approximate analytical expression for elastic stresses in flat punch problem. Wear **236**, 134–143 (1999)

Lamb, H.: On Boussinesq's problem. Proc. London Math. Soc. **34**, 276–284 (1902)

Love, A.E.H.: Boussinesq's problem for a rigid cone. Q. J. Math. **10**(1), 161–175 (1939)

Maugis, D., Barquins, M.: Adhesive contact of sectionally smooth-ended punches on elastic half-spaces: theory and experiment. J. Phys. D Appl. Phys. **16**(10), 1843–1874 (1983)

Mossakovskii, V.I.: The fundamental mixed problem of the theory of elasticity for a half-space with a circular line separating the boundary conditions. Prikladnaya Matem. Mekhanika **18**(2), 187–196 (1954)

Mossakovskii, V.I.: Compression of elastic bodies under conditions of adhesion (axi-symmetric case). PMM J. Appl. Math. Mech. **27**(3), 630–643 (1963)

Pharr, G.M., Oliver, W.C., Brotzen, F.R.: On the generality of the relationship among contact stiffness, contact area, and elastic modulus during indentation. J. Mater. Res. **7**(3), 613–617 (1992)

Popov, V.L., Heß, M.: Methode der Dimensionsreduktion in Kontaktmechanik und Reibung. Springer, Heidelberg (2013). ISBN 978-3-642-32672-1

Popov, V.L., Heß, M.: Method of dimensionality reduction in contact mechanics and friction. Springer, Heidelberg (2015). ISBN 978-3-642-53875-9

Popov, V.L., Heß, M., Willert, E., Li, Q.: Indentation of concave power-law profiles with arbitrary exponents (2018). https://arXiv.org/abs/1806.05872. cond-mat.soft

Rao, A.K.: Stress concentrations and singularities at interface corners. Z. Angew. Math. Mech. **31**, 395–406 (1971)

Schubert, G.: Zur Frage der Druckverteilung unter elastisch gelagerten Tragwerken. Ing. Arch. **13**(3), 132–147 (1942)

Segedin, C.M.: The relation between load and penetration for a spherical punch. Mathematika **4**(2), 156–161 (1957)

Shtaerman, I.Y.: On the Hertz theory of local deformations resulting from the pressure of elastic bodies. Dokl. Akad. Nauk. SSSR **25**, 359–361 (1939)

Shtaerman, I.Y.: Contact problem of the theory of elasticity. Gostekhizdat, Moscow (1949)

Sneddon, I.N.: The relation between load and penetration in the axi-symmetric Boussinesq problem for a punch of arbitrary profile. Int. J. Eng. Sci. **3**(1), 47–57 (1965)

Spence, D.A.: Self-similar solutions to adhesive contact problems with incremental loading. Proc. R. Soc. London A Math. Phys. Eng. Sci. **305**, 55–80 (1968)

Spence, D.A.: The Hertz contact problem with finite friction. J. Elast **5**(3–4), 297–319 (1975)

Zhupanska, O.I.: Axi-symmetric contact with friction of a rigid sphere with an elastic half-space. Proc. R. Soc. London Ser. A **465**, 2565–2588 (2009)

3 Normal Contact with Adhesion

3.1 Introduction

Between any two electrically neutral bodies there exist relatively weak interaction forces which rapidly decline with increasing distance between the bodies. These forces are known as *adhesive forces* and, in most cases, cause a mutual attraction. Adhesive forces play an essential role in many technical applications. It is the adhesive forces that are responsible for the behavior of glue, for instance. Adhesive tape, self-adhesive envelopes, etc., are further examples of adhesive forces. They are of particular importance for applications where one of the following conditions is met:

- The surfaces of the bodies are very smooth (e.g., the magnetic disc of a hard drive).
- One of the contact partners is made of a soft material (rubber or biological structures).
- We are dealing with a microscopic system, in which adhesive forces generally have a larger influence than body forces, because the body and surface forces are scaled differently (micro-mechanical devices, atomic force microscope, biological structures, etc.).

At a microscopic scale, the adhesive forces are determined by the type of the interaction potential. It is possible to define a characteristic "range" of adhesive forces based on the specific type of the potential. However, as Griffith already demonstrated in his famous work on the theory of cracks (Griffith 1921), the most important parameter is not solely the magnitude of the interactions or their range, but instead the product of both; i.e., the work required separation of the surfaces. This work per unit surface area is referred to as the work of adhesion or effective *relative surface energy*, $\Delta\gamma$, of the contacting bodies.

V.L. Popov, M. Heß, E. Willert, *Handbook of Contact Mechanics*,
https://doi.org/10.1007/978-3-662-58709-6_3

Griffith's theory is based on the energy balance between the elastic energy released due to an advancement of the crack boundary and the required work of adhesion. It is assumed that there are no interaction forces beyond the contact area, which corresponds to the assumption of a vanishingly low range of the adhesive forces. This assumption is valid for real adhesive interactions if the range of the adhesive forces is much smaller than any characteristic length of the contact. The theory of adhesive contact published in 1971 by Johnson, Kendall, and Roberts is valid under the same conditions as Griffith's theory: for the vanishingly low range of the interactions. In this chapter, when we refer to the theory of adhesive contacts in the "JKR approximation", it is understood to mean the vanishingly low range of the adhesive forces. In JKR theory, the work of adhesion is also the sole parameter of the adhesive interaction. Among other results, the adhesive force between a sphere of radius R and an elastic half-space is given by the equation:

$$F_A = \frac{3}{2}\Delta\gamma\pi R, \tag{3.1}$$

In micro-systems, situations can arise where the range of the adhesive forces is of the same length as the smallest characteristic length of the contact (usually the indentation depth), or even greater than the characteristic contact length. The simplest of such cases is the contact of hard spheres with weak interaction forces, where the elastic deformation is negligible. This case was examined and published by Bradley (1932). Bradley calculated the adhesive force between a rigid plane and a rigid sphere of radius R. He assumed the existence of van der Waals forces acting between the molecules of both bodies, decreasing proportionally to $1/r^7$ with increasing distance between the molecules (which corresponds to the attractive component of the Lennard-Jones potential). For the contact of a rigid plane and a rigid sphere, the adhesive force equals:

$$F_A = 2\Delta\gamma\pi R. \tag{3.2}$$

This equation also only features the separation energy and does not account for the coordinate dependency of the interaction. It can be easily shown that the result from Bradley's approximation is invariant with regards to the exact coordinate dependency of the interaction potential, as long as the half-space approximation is valid.

The logical extension of Bradley's model lies in the consideration of the elastic deformations caused by long-range adhesive interactions. An approximate solution for this problem was only discovered about 40 years after Bradley's publication by Derjaguin, Muller, and Toporov (1975) (DMT theory). While the approximation does take into account the adhesive forces, the elastic deformation of the surfaces is, nonetheless, calculated with the solution by Hertz (1882) for non-adhesive contacts. Under these assumptions, Derjaguin, Muller, and Toporov arrived at the same equation for the adhesive force as Bradley, stating:

> The van der Waals' forces are capable of increasing the area of elastic contact of the ball with the plane, yet it has been shown that the force, as required for overcoming the van der Waals' forces and breaking up the contact, does not increase thereby and may be calculated, if one considers the point contact of a non-deformed ball with a plane.

To judge whether the adhesive interactions are short or long-ranged (thus determining whether the adhesive contact is of the "JKR type" or "DMT type"), the characteristic displacement of the bodies until the separation of the adhesive contact (the characteristic value for the height of the adhesive "neck") $(R(\Delta\gamma)^2/(E^*)^2)^{1/3}$ (see (3.45)) should be compared to the characteristic range of the adhesive interactions, z_0. This leads to the parameter:

$$\mu := \frac{R(\Delta\gamma)^2}{(E^*)^2 z_0^3}, \tag{3.3}$$

which was initially introduced by Tabor (1977) and is known as the "Tabor parameter". For neck heights much greater than the range of the interactions (large Tabor parameter), the range can be considered vanishingly low. This limiting case leads to the JKR theory. The other limiting case is the DMT theory.

Maugis's (1992) theory of a contact with Dugdale's (1960) simple model for the interaction potential was of great methodological interest for the theory of adhesive contacts. Maugis assumed that the adhesive stress between surfaces remains constant up to a certain distance h and then drops abruptly. For this case the specific work of adhesion equals:

$$\Delta\gamma = \sigma_0 h. \tag{3.4}$$

While the coordinate dependency of the Dugdale potential is not realistic, this bears little importance for most adhesive problems since both limiting cases—JKR and DMT—are independent of the exact type of the interaction potential. Under these conditions, even the simplest model of interaction is valid and informative. The great advantage of the Dugdale potential lies in the fact that it allows a mostly analytical solution of the problem. Maugis's theory not only provided a representation of the two limiting cases but also an explanation of the transition between the JKR and DMT theories.

Since the exact form of the interaction potentials is insignificant for the adhesion (as long as the work of adhesion is defined and remains constant), Greenwood and Johnson (1998) developed a theory which represented the stress distribution in the adhesive contact as the superposition of two Hertzian stress distributions of different radii. Compared to Maugis's theory, this represented a vast "trivialization" of the involved contact mechanics. It should be noted that the "double Hertz" solution corresponds to a rather strange interaction potential. But since the exact form of the interaction potential is insignificant, the theory of Greenwood and Johnson represents an interesting alternative to Maugis's theory. It too features the JKR and DMT models as its limiting cases.

The two most well-known theories of adhesion by JKR and DMT both lead to an adhesive force which is independent of the elastic properties of the contacting media. To avoid any misunderstanding, it should be expressly noted that this simple property only applies to parabolic contacts. In no way can this lack of dependency on the elastic moduli be generalized to arbitrary adhesive contacts.

The consideration of the tangential stresses in the adhesive contact requires another general remark. Both JKR and DMT theory are based on the Hertzian theory of *frictionless* contact. It is surely a valid and self-consistent model assumption. From the physical point of view, on the contrary, this assumption is rather questionable. Physically, the JKR limiting case implies an infinitely strong yet infinitely short-ranged interaction. This means that the surfaces of the adhesive JKR contact are pressed together under infinitely low ranged but infinitely strong stress, which undermines the notion of a "frictionless" contact. However, for practical applications, the difference between frictional or frictionless adhesive contact is relatively limited and can be considered negligible in most cases.

The following two sections will present two alternative approaches to the solution of the adhesive normal contact problem. The first approach is the reduction of the adhesive normal contact problem to the non-adhesive one, and the second approach is the direct solution utilizing the MDR.

3.2 Solution of the Adhesive Normal Contact Problem by Reducing to the Non-Adhesive Normal Contact Problem

The basic idea of the theory of adhesive contact by Johnson, Kendall, and Roberts (1971) is the same as the one of Griffith's theory of cracks. In their frequently cited paper they write:

> … the approach followed in this analysis, is similar to that used by Griffith in his criterion for the propagation of a brittle crack.

The idea is based on the consideration of energy balance between the elastic energy and the surface energy during the propagation of the crack or the boundary of the adhesive contact. Since the surface energy of an axisymmetric contact is trivially determined from the contact area, the only remaining non-trivial problem lies in calculating the elastic energy of the adhesive contact. This can always be done if the solution of the respective non-adhesive normal contact problem is known. The JKR method to calculate the energy is the second important point of the classic paper. It is also ingeniously simple and based on the assertion that the adhesive contact can be represented as a superposition of a non-adhesive contact and a rigid body translation. Perhaps the easiest way to imagine this is to consider the contact between a rigid indenter with the profile $\tilde{z} = f(r)$ and an elastic half-space. We obtain the configuration of the adhesive contact by initially indenting the elastic half-space (without regard for adhesion), causing it to form a contact area of radius a, and raising the entire contact area after that, without change in contact radius.

Furthermore, we will assume the solution of the non-adhesive contact problem, particularly the relationships between the indentation depth, the contact radius, and the normal force. Any quantity of the triple $\{F_N, a, d\}$ uniquely defines the others. It will prove advantageous to describe the normal force and the indentation depth as functions of the contact radius:

$$F_N = F_{N,\text{n.a.}}(a), \quad d = d_{\text{n.a.}}(a). \tag{3.5}$$

The indices "n.a." indicate that these are the solutions of the non-adhesive contact problem. These equations also imply that the dependency of the force on the indentation depth is known. We obtain the incremental contact stiffness $k_{\text{n.a.}}$ by differentiating the force with respect to d and the elastic energy $U_{\text{n.a.}}$ by integrating with respect to d. These quantities can also be rewritten as functions of the contact radius:

$$k_{\text{n.a.}} = k_{\text{n.a.}}(a), \quad U_{\text{n.a.}} = U_{\text{n.a.}}(a). \tag{3.6}$$

In the following all functions for (3.5) and (3.6) are considered to be known.

Let us now indent the profile to a contact radius a. The elastic energy of this state is $U_{\text{n.a.}}(a)$, the indentation depth $d_{\text{n.a.}}(a)$, and the force $F_{N,\text{n.a.}}(a)$. In the second step, we lift the profile by Δl *without changing the contact radius*. The stiffness (only dependent on the radius) remains constant for this process and equals $k_{\text{n.a.}}(a)$. The force is then given by:

$$F_N(a) = F_{N,\text{n.a.}}(a) - k_{\text{n.a.}}(a)\Delta l, \tag{3.7}$$

and the potential energy is:

$$U(a) = U_{\text{n.a.}}(a) - F_{N,\text{n.a.}}(a)\Delta l + k_{\text{n.a.}}(a)\frac{\Delta l^2}{2}. \tag{3.8}$$

The new indentation depth at the end of the process equals:

$$d = d_{\text{n.a.}}(a) - \Delta l. \tag{3.9}$$

Obtaining Δl from (3.9) and by substituting it into (3.8), we obtain the potential energy:

$$U(a) = U_{\text{n.a.}}(a) - F_{N,\text{n.a.}}(a)(d_{\text{n.a.}}(a) - d) + k_{\text{n.a.}}(a)\frac{(d_{\text{n.a.}}(a) - d)^2}{2}. \tag{3.10}$$

The total energy (under consideration of the surface energy) now equals:

$$\begin{aligned} U_{\text{tot}}(a) = {} & U_{\text{n.a.}}(a) - F_{N,\text{n.a.}}(a)(d_{\text{n.a.}}(a) - d) \\ & + k_{\text{n.a.}}(a)\frac{(d_{\text{n.a.}}(a) - d)^2}{2} - \pi a^2 \Delta\gamma. \end{aligned} \tag{3.11}$$

The equilibrium value of the contact radius follows from the total energy minimum condition (for constant indentation depth d):

$$\begin{aligned}\frac{\partial U_{\text{tot}}(a)}{\partial a} &= \frac{\partial U_{\text{n.a.}}(a)}{\partial a} - \frac{\partial F_{N,\text{n.a.}}(a)}{\partial a}\Delta l - F_{N,\text{n.a.}}(a)\frac{\partial d_{\text{n.a.}}(a)}{\partial a} \\ &\quad + \frac{\partial k_{\text{n.a.}}(a)}{\partial a}\frac{\Delta l^2}{2} + k_{\text{n.a.}}(a)\Delta l\frac{\partial d_{\text{n.a.}}(a)}{\partial a} - 2\pi a\Delta\gamma \\ &= \left(\frac{\partial U_{\text{n.a.}}(a)}{\partial a} - F_{N,\text{n.a.}}(a)\frac{\partial d_{\text{n.a.}}(a)}{\partial a}\right) \\ &\quad - \left(\frac{\partial F_{N,\text{n.a.}}(a)}{\partial a}\Delta l - k_{\text{n.a.}}(a)\Delta l\frac{\partial d_{\text{n.a.}}(a)}{\partial a}\right) \\ &\quad + \frac{\partial k_{\text{n.a.}}(a)}{\partial a}\frac{\Delta l^2}{2} - 2\pi a\Delta\gamma \\ &= 0. \end{aligned} \tag{3.12}$$

The terms in parentheses disappear and the equation takes the form:

$$\frac{\partial k_{\text{n.a.}}(a)}{\partial a}\frac{\Delta l^2}{2} = 2\pi a\Delta\gamma. \tag{3.13}$$

It follows that:

$$\Delta l = \sqrt{\frac{4\pi a\Delta\gamma}{\frac{\partial k_{\text{n.a.}}(a)}{\partial a}}} \quad \text{(for the general axially symmetric case).} \tag{3.14}$$

Substituting this quantity into (3.9) and (3.7) yields an equation for determining the relationship between the indentation depth, the contact radius, and the normal force:

$$d = d_{\text{n.a.}}(a) - \sqrt{\frac{4\pi a\Delta\gamma}{\frac{\partial k_{\text{n.a.}}(a)}{\partial a}}} \quad \text{(for an arbitrary medium),} \tag{3.15}$$

$$F_N(a) = F_{N,\text{n.a.}}(a) - k_{\text{n.a.}}(a)\sqrt{\frac{4\pi a\Delta\gamma}{\frac{\partial k_{\text{n.a.}}(a)}{\partial a}}} \quad \text{(for an arbitrary medium).} \tag{3.16}$$

It becomes apparent that the three functions which directly (and in its entirety) determine the solution of the adhesive contact problem are the three dependencies of the non-adhesive contact: indentation depth as a function of the contact radius, normal force as a function of the contact radius, and therefore the incremental stiffness as a function of the contact radius. The latter quantity equals the stiffness for the indentation by a circular cylindrical indenter of radius a. *It should be noted that these are general equations and not subject to the homogeneity of the medium (neither in-depth nor in the radial direction). As such, they also apply to layered or functionally graded media.* The sole condition for the validity of (3.15) and (3.16) is the conservation of rotational symmetry during indentation.

For a homogeneous medium, the equations can be simplified even further. Here, the stiffness is given by $k_{\text{n.a.}} = 2E^*a$, and we obtain:

$$\Delta l(a) = \sqrt{\frac{2\pi a \Delta\gamma}{E^*}} \quad \text{(for a homogeneous medium).} \tag{3.17}$$

The determining equations (3.15) and (3.16) take on the form:

$$d = d_{\text{n.a.}}(a) - \sqrt{\frac{2\pi a \Delta\gamma}{E^*}} \quad \text{(for a homogeneous medium),}$$
$$F_N(a) = F_{N,\text{n.a.}}(a) - \sqrt{8\pi E^* \Delta\gamma a^3} \quad \text{(for a homogeneous medium).} \tag{3.18}$$

Naturally, the pressure distribution in an adhesive contact and the displacement field outside the contact area are also composed of the two solutions of the non-adhesive contact problem: the solutions for the non-adhesive indentation by $d_{\text{n.a.}}(a)$ and the subsequent rigid retraction by Δl. Let us denote the stress distribution and the displacement field for the non-adhesive contact problem by $\sigma(r;a)_{\text{n.a.}}$ and $w(r;a)_{\text{n.a.}}$, respectively. With the stress and the displacement field for a rigid translation given by (2.22), the stress distribution and the displacement for the adhesive contact problem are then represented by the following equations:

$$\sigma(r;a) = \sigma(r;a)_{\text{n.a.}} + \frac{E^* \Delta l}{\pi \sqrt{a^2 - r^2}}, \quad r < a,$$
$$w(r;a) = w(r;a)_{\text{n.a.}} - \frac{2\Delta l}{\pi} \arcsin\left(\frac{a}{r}\right), \quad r > a, \tag{3.19}$$

or after inserting (3.17):

$$\sigma(r;a) = \sigma(r;a)_{\text{n.a.}} + \sqrt{\frac{2E^* a \Delta\gamma}{\pi}} \frac{1}{\sqrt{a^2 - r^2}},$$
$$r < a \text{ (for a homogeneous medium),}$$
$$w(r;a) = w(r;a)_{\text{n.a.}} - \sqrt{\frac{8\pi a \Delta\gamma}{\pi E^*}} \arcsin\left(\frac{a}{r}\right),$$
$$r > a \text{ (for a homogeneous medium).} \tag{3.20}$$

Equations (3.18) and (3.20) completely solve the adhesive normal contact problem.

The magnitude of the *force of adhesion* is of particular interest. We will define it as the maximum pull-off force required to separate bodies. In mathematical terms, this means the maximum pull-off force for which there still exists a stable equilibrium solution of the normal contact problem.

Another important quantity to consider is the force in the last possible state of stable equilibrium, after which the contact dissolves entirely. This force depends on the character of the loading conditions. We commonly distinguish between the limiting cases of force-controlled and displacement-controlled trials. The stability condition for force-control is given by:

$$\left.\frac{\mathrm{d}F_N}{\mathrm{d}a}\right|_{a=a_c} = 0, \tag{3.21}$$

and for displacement-control case by:

$$\left.\frac{\mathrm{d}d}{\mathrm{d}a}\right|_{a=a_c} = 0. \tag{3.22}$$

The conditions (3.21) and (3.22) can be consolidated into the condition

$$\left.\frac{\mathrm{d}d_{\text{n.a.}}(a)}{\mathrm{d}a}\right|_{a=a_c} = \xi\sqrt{\frac{\pi\Delta\gamma}{2E^*a_c}}, \qquad \xi = \begin{cases} 3, & \text{force-control,} \\ 1, & \text{displacement-control,} \end{cases} \tag{3.23}$$

from which we can determine the critical contact radius, where the contact detaches (see Sect. 3.3 for a full derivation). The respective critical values of the indentation depth and normal force are obtained by substituting this critical radius into (3.18).

3.3 Direct Solution of the Adhesive Normal Contact Problem in the Framework of the MDR

An alternative to the reduction to the non-adhesive contact problem described in the previous section is provided by the MDR (see Popov and Heß 2015, for example), which presents an immediate solution to the adhesive normal contact problem. This alternative approach can, for example, be of interest for complex dynamic loading conditions requiring a numerical implementation.

The calculation method via the MDR consists of the following steps:

- In the first step the given three-dimensional profile $\tilde{z} = f(r)$ is transformed to an equivalent plane profile $g(x)$ via (2.6):

$$g(x) = |x| \int_0^{|x|} \frac{f'(r)}{\sqrt{x^2 - r^2}}\mathrm{d}r. \tag{3.24}$$

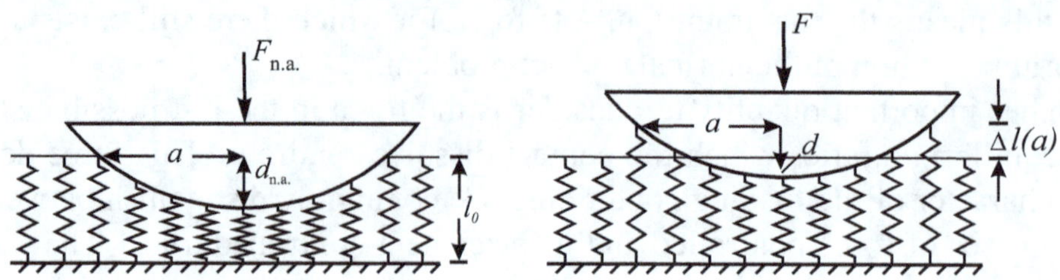

Fig. 3.1 Qualitative representation of the indentation and lifting process of a spherical 1D-indenter with an elastic foundation, which exactly models the properties of the adhesive contact between a rigid spherical indenter and an elastic half-space

- The profile $g(x)$ is now pushed into the one-dimensional elastic foundation defined, according to (2.5):
$$\Delta k_z = E^* \Delta x, \tag{3.25}$$
until a contact radius a is reached. Until this point, the adhesion will not be considered. This process is depicted in Fig. 3.1.
- In the third step, the indenter is lifted up. It is assumed that all springs involved in the contact adhere to the indenter—the contact radius thus remains constant. In this process, the springs at the edge experience the maximum increase in tension. Upon reaching the maximum possible elongation (3.17) of the outer springs
$$\Delta l(a) = \sqrt{\frac{2\pi a \Delta\gamma}{E^*}} \tag{3.26}$$
they detach. This criterion (3.17) was discovered by Heß (2010), and is known as the *rule of Heß*. A derivation of this criterion can be found in the appendix (see (11.31)).

The corresponding equilibrium described by the three quantities $\{F_N, d, a\}$ coincides *exactly* with the one of the three-dimensional adhesive contact.

The only difference to the algorithm for the non-adhesive contact (described in Chap. 2) lies in the modification of the indentation depth formula. The displacement of the outer springs is no longer zero but instead negative, with the absolute value equal to the critical value: $w_{1D}(a) = -\Delta l(a)$. It follows that:

$$d = g(a) - \Delta l(a). \tag{3.27}$$

The normal force is once again given by the equation:

$$F_N = 2E^* \int_0^a [d - g(x)]\mathrm{d}x. \tag{3.28}$$

The only difference to the non-adhesive problem lies in the differing contact radius.

Equation (3.27) determines the equilibrium configuration. The stability of this equilibrium is governed by the sign of the derivative $\mathrm{d}(\Delta l(a) + d - g(a))/\mathrm{d}a$:

$$\begin{aligned} &\frac{\mathrm{d}}{\mathrm{d}a}(\Delta l(a) + d - g(a)) > 0, \quad \text{stable equilibrium,} \\ &\frac{\mathrm{d}}{\mathrm{d}a}(\Delta l(a) + d - g(a)) < 0, \quad \text{unstable equilibrium,} \\ &\frac{\mathrm{d}}{\mathrm{d}a}(\Delta l(a) + d - g(a)) = 0, \quad \text{critical state.} \end{aligned} \tag{3.29}$$

In general, the stability depends on the type of boundary condition employed for the indenter. The two limiting cases are *force-controlled* trials (which correspond to an infinitely soft test system) and *displacement-controlled* trials (which correspond to a rigid test system).

Stability Condition for Displacement-Controlled Trials

In this case, the indentation depth is constant and (3.29) for the critical state is:

$$\left.\frac{\mathrm{d}g(a)}{\mathrm{d}a}\right|_{a=a_c} = \left.\frac{\mathrm{d}\Delta l(a)}{\mathrm{d}a}\right|_{a=a_c} = \sqrt{\frac{\pi \Delta\gamma}{2E^* a_c}}. \tag{3.30}$$

Stability Condition for Force-Controlled Trials

When the force is kept constant we must consider the varying indentation depth. The relationship between the indentation depth and the normal force is given by (3.28). Differentiating (3.28) under the condition $F_N = \text{const}$ yields:

$$\begin{aligned} \mathrm{d}\int_0^a [d - g(x)]\mathrm{d}x &= \mathrm{d}a \cdot \frac{\partial}{\partial a}\int_0^a [d - g(x)]\mathrm{d}x + \mathrm{d}d \cdot \frac{\mathrm{d}}{\mathrm{d}d}\int_0^a [d - g(x)]\mathrm{d}x \\ &= \mathrm{d}a \cdot [d(a) - g(a)] + \mathrm{d}d \cdot a \\ &= -\mathrm{d}a \cdot \Delta l(a) + \mathrm{d}d \cdot a = 0. \end{aligned} \tag{3.31}$$

It follows that:

$$\left.\frac{\mathrm{d}d}{\mathrm{d}a}\right|_{F_N=\text{const}} = \frac{\Delta l(a)}{a}. \tag{3.32}$$

The condition for the critical state (3.29) takes on the form of:

$$\frac{\mathrm{d}\Delta l(a)}{\mathrm{d}a} + \frac{\mathrm{d}d}{\mathrm{d}a} - \frac{\mathrm{d}g(a)}{\mathrm{d}a} = \frac{\mathrm{d}\Delta l(a)}{\mathrm{d}a} + \frac{\Delta l(a)}{a} - \frac{\mathrm{d}g(a)}{\mathrm{d}a} = 0, \tag{3.33}$$

or

$$\left.\frac{\mathrm{d}g(a)}{\mathrm{d}a}\right|_{a=a_c} = \left[\frac{\Delta l(a)}{a} + \frac{\mathrm{d}\Delta l(a)}{\mathrm{d}a}\right]_{a=a_c} = 3\sqrt{\frac{\pi \Delta\gamma}{2E^* a_c}}. \tag{3.34}$$

The stability conditions can now be combined into a single equation:

$$\left.\frac{\mathrm{d}g(a)}{\mathrm{d}a}\right|_{a=a_c} = \xi\sqrt{\frac{\pi\Delta\gamma}{2E^*a_c}}, \qquad \xi = \begin{cases} 3, \text{force-control}, \\ 1, \text{displacement-control}. \end{cases} \tag{3.35}$$

The transformation rules of the MDR for the pressure distribution and the displacements—see (2.13) and (2.14)—remain valid for the adhesive contact.

3.4 Areas of Application

Adhesion can be desirable (such as in bonded connections, in adhesive medical bandages, as well as in many biological systems) or equally undesirable. The systems in the first group, where adhesion is desirable, are often inspired by biological systems and especially concern contacts with concave profiles (see Sect. 3.5.6). The study of the tiny hairs optimized for their adhesive functions on the limbs of (for example) geckos or insects has inspired technical solutions for achieving extreme adhesive effects.

Adhesion is also a major consideration in the design of micro-systems. One example is the contact between a measuring tip and the sample surface in atomic force microscopy. The indenting measuring tip often has a conical (see Sect. 3.5.2) or spherical (see Sects. 3.5.3 and 3.5.4) shape, or imperfect variations of these (see Sects. 3.5.5, 3.5.8, 3.5.9, 3.5.11, or 3.5.12). For assembly of nano- or microscopic systems the smallest flat indenters are utilized, which we will explore in their perfect (see Sect. 3.5.1) or imperfect form (see Sects. 3.5.10 and 3.5.13).

Any given sufficiently differentiable indenter profile can be represented as a Taylor series. Thus, we will provide the solution of the contact problem involving the power-law profile as a basic building block of the solution for such an arbitrary profile (see Sect. 3.5.7).

In addition, real surfaces are unavoidably rough. In Sect. 3.5.14 we will present a simple model of periodic roughness, which quite clearly illustrates the effect roughness has on the adhesive interaction.

3.5 Explicit Solutions for Axially Symmetric Profiles in JKR Approximation

3.5.1 The Cylindrical Flat Punch

The solution for the adhesive normal contact with a cylindrical flat punch was dicovered by Kendall (1971). In the JKR theory, adhesion is interpreted as an indentation by a flat cylindrical punch with a contact radius a, with the superposition of two indenter solutions of identical radius yielding a new indenter solution. Therefore, the solutions of the contact problem concerning the flat cylindrical punch are identical with and without adhesion. The results for the normal force F_N, the normal stresses σ_{zz}, and the displacements of the half-space surface w, are then given by:

$$\begin{aligned} F_N(d) &= 2E^* da, \\ \sigma_{zz}(r; d) &= -\frac{E^* d}{\pi\sqrt{a^2 - r^2}}, \qquad r \le a, \\ w(r; d) &= \frac{2d}{\pi} \arcsin\left(\frac{a}{r}\right), \qquad r > a. \end{aligned} \tag{3.36}$$

The only difference compared to the non-adhesive contact is that the adhesive case also permits negative indentation depths. The critical radius is predetermined by the indenter radius a. The critical indentation depth d_c and critical normal force F_c, where stability of the contact is lost, are calculated from (3.18), independently of whether a force-controlled or displacement-controlled trial is being considered:

$$\begin{aligned} d_c &= -\sqrt{\frac{2\pi a \Delta\gamma}{E^*}}, \\ F_c &= -\sqrt{8\pi a^3 E^* \Delta\gamma}. \end{aligned} \tag{3.37}$$

3.5.2 The Cone

The solution to the contact problem depicted in Fig. 3.2 was first found by Maugis and Barquins (1981). They used the general solution of the non-adhesive problem and the concept of energy release rate from linear fracture mechanics. The contact problem is completely solved by specifying the indentation depth d, the normal force F_N, the stress distribution σ_{zz}, and the displacements w; each as functions of the contact radius a. With the results from the previous Chapter (Sect. 2.5.2) and (3.18) and (3.20) we get:

$$\begin{aligned} d(a) &= \frac{\pi}{2} a \tan\theta - \sqrt{\frac{2\pi a \Delta\gamma}{E^*}}, \\ F_N(a) &= \frac{\pi a^2}{2} E^* \tan\theta - \sqrt{8\pi a^3 E^* \Delta\gamma}, \\ \sigma_{zz}(r; a) &= -\frac{E^* \tan\theta}{2} \operatorname{arcosh}\left(\frac{a}{r}\right) + \sqrt{\frac{2E^*\Delta\gamma}{\pi a}} \frac{a}{\sqrt{a^2 - r^2}}, \qquad r \le a, \\ w(r; a) &= \tan\theta\left(\sqrt{r^2 - a^2} - r\right) + \left(a \tan\theta - \sqrt{\frac{8a\Delta\gamma}{\pi E^*}}\right) \arcsin\left(\frac{a}{r}\right), \\ & \quad r > a. \end{aligned} \tag{3.38}$$

Here, θ denotes the slope angle of the cone. The relationship between contact radius and indentation depth in the case without adhesion is given by:

$$d_{\text{n.a.}}(a) = \frac{\pi}{2} a \tan\theta. \tag{3.39}$$

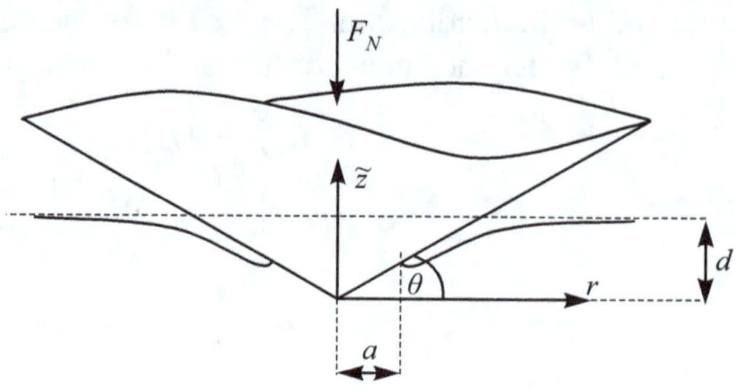

Fig. 3.2 Adhesive normal contact between a rigid conical indenter and an elastic half-space

Therefore, for the critical contact radius, (3.23) results in:

$$a_c = \frac{2\xi^2 \Delta\gamma}{\pi E^* \tan^2\theta}. \tag{3.40}$$

The critical values for the indentation depth and normal force are then:

$$\begin{aligned} d_c &= d_{\text{n.a.}}(a_c) - \sqrt{\frac{2\pi a_c \Delta\gamma}{E^*}} = \frac{\left(\xi^2 - 2\xi\right)\Delta\gamma}{E^* \tan\theta}, \\ F_c &= \frac{2\Delta\gamma^2\left(\xi^4 - 4\xi^3\right)}{\pi E^* \tan^3\theta}. \end{aligned} \tag{3.41}$$

Here it must be specified whether the experiment is carried out under force-controlled or displacement-controlled conditions. For force-controlled experiments it is $\xi = 3$, and in the case of displacement-control, $\xi = 1$. The relationships between the global contact quantities can also be formulated in a normalized form. For this purpose all quantities are normalized to their critical values:

$$\hat{a} := \frac{a}{a_c}, \quad \hat{d} := \frac{d}{|d_c|}, \quad \hat{F} := \frac{F_N}{|F_c|}. \tag{3.42}$$

If we choose the critical values under force-controlled conditions, the normalized relationships are:

$$\begin{aligned} \hat{d}(\hat{a}) &= 3\hat{a} - 2\sqrt{\hat{a}}, \\ \hat{F}(\hat{a}) &= 3\hat{a}^2 - 4\sqrt{\hat{a}^3}. \end{aligned} \tag{3.43}$$

These functions, $\hat{d} = \hat{d}(\hat{a})$ and $\hat{F} = \hat{F}(\hat{a})$, are shown in Fig. 3.3. The resulting, implicitly defined, dependency $\hat{F} = \hat{F}(\hat{d})$ is given in Fig. 3.4.

3.5.3 The Paraboloid

The adhesive contact problem for a parabolic body $\tilde{z} = r^2/(2R)$ (see Fig. 3.5) was solved in the aforementioned classic JKR paper (Johnson et al. 1971). With the

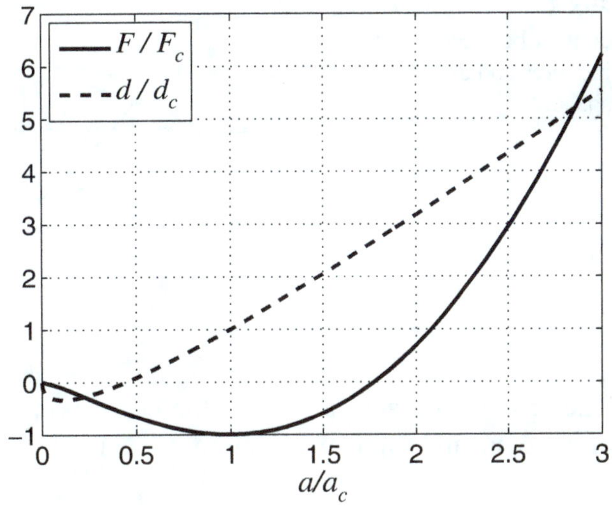

Fig. 3.3 The normalized normal force and indentation depth as functions of the normalized contact radius for adhesive indentation by a cone. All values are normalized to the critical state in the case of force-control

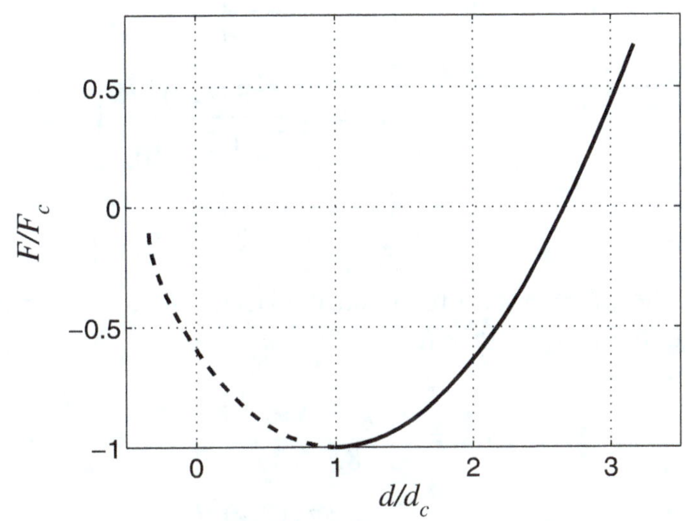

Fig. 3.4 Relationship between the normalized normal force and the normalized indentation depth for the adhesive indentation by a cone. All normalizations refer to the critical state in the case of force-control. The dashed part describes the states that are stable only for displacement-control

results from Chap. 2 (Sect. 2.5.3) and (3.18) and (3.20), we obtain:

$$\begin{aligned}
d(a) &= \frac{a^2}{R} - \sqrt{\frac{2\pi a \Delta\gamma}{E^*}}, \\
F_N(a) &= \frac{4}{3}\frac{E^* a^3}{R} - \sqrt{8\pi a^3 E^* \Delta\gamma}, \\
\sigma_{zz}(r;a) &= -\frac{2E^*}{\pi R}\sqrt{a^2-r^2} + \sqrt{\frac{2E^*\Delta\gamma}{\pi a}}\frac{a}{\sqrt{a^2-r^2}}, \quad r \le a, \\
w(r;a) &= \frac{a^2}{\pi R}\left[\left(2-\frac{r^2}{a^2}\right)\arcsin\left(\frac{a}{r}\right) + \frac{\sqrt{r^2-a^2}}{a}\right] - \sqrt{\frac{8a\Delta\gamma}{\pi E^*}}\arcsin\left(\frac{a}{r}\right), \\
&\quad r > a.
\end{aligned} \tag{3.44}$$

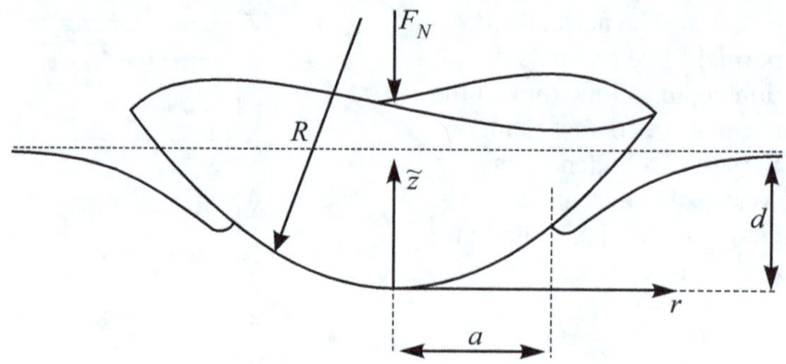

Fig. 3.5 Adhesive normal contact between a parabolic indenter and an elastic half-space

The critical contact radius and the corresponding values for the indentation depth and the normal force are given by (3.23):

$$\begin{aligned} a_c &= \left(\frac{\pi\xi^2 R^2 \Delta\gamma}{8E^*}\right)^{1/3}, \\ d_c &= \left(\frac{\pi^2 \xi\,(\Delta\gamma)^2 R}{(E^*)^2}\right)^{1/3}\left(\frac{\xi}{4}-1\right), \\ F_c &= \pi\xi\Delta\gamma R\left(\frac{\xi}{6}-1\right), \end{aligned} \tag{3.45}$$

where $\xi = 3$ for force-controlled trials and $\xi = 1$ for displacement-controlled trials. In their explicit forms:

$$\left.\begin{aligned} a_c &= \left(\frac{9\pi R^2 \Delta\gamma}{8E^*}\right)^{1/3}, \\ d_c &= -\frac{1}{4}\left(\frac{3\pi^2(\Delta\gamma)^2 R}{(E^*)^2}\right)^{1/3}, \\ F_c &= -\frac{3}{2}\pi\Delta\gamma R \end{aligned}\right. \quad \text{under force-control,} \tag{3.46}$$

and

$$\left.\begin{aligned} a_c &= \left(\frac{\pi R^2 \Delta\gamma}{8E^*}\right)^{1/3}, \\ d_c &= -\frac{3}{4}\left(\frac{\pi^2\,(\Delta\gamma)^2 R}{(E^*)^2}\right)^{1/3}, \\ F_c &= -\frac{5}{6}\pi\Delta\gamma R. \end{aligned}\right. \quad \text{under displacement-control.} \tag{3.47}$$

In this case, the critical force does not depend on E^*; i.e., it does not depend on the elastic properties of the half-space. In Sect. 3.5.7 it is demonstrated that the paraboloid is the only mnemonic indenter for which this is valid. After normalizing

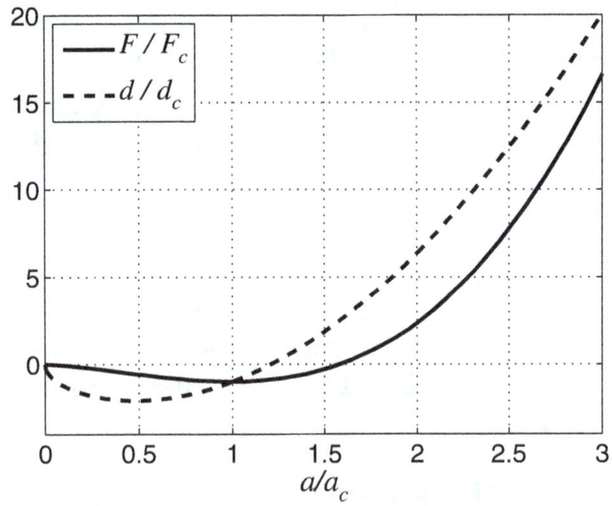

Fig. 3.6 The normalized normal force and indentation depth as functions of the contact radius for the adhesive indentation by a paraboloid. All values are normalized to the critical state under force-controlled boundary conditions

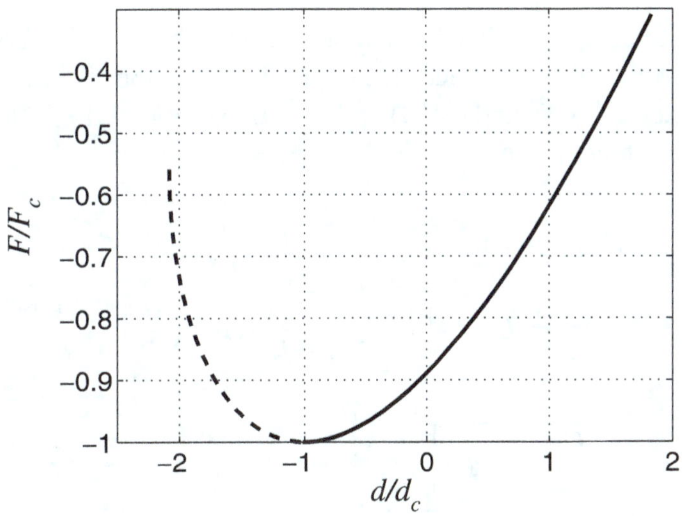

Fig. 3.7 Relationship between the normalized normal force and the normalized indentation depth for the adhesive indentation by a paraboloid. All normalizations are done with respect to the critical state under force-control. The dotted section represents the states that are only stable for displacement-controlled trials

the quantities to the critical values of the force-controlled trial

$$\hat{a} := \frac{a}{a_c}, \quad \hat{d} := \frac{d}{|d_c|}, \quad \hat{F} := \frac{F_N}{|F_c|}, \tag{3.48}$$

the relationships (3.44$_1$), and (3.44$_2$), can be rewritten in a universal, dimensionless form:

$$\begin{aligned} \hat{d} &= 3\hat{a}^2 - 4\sqrt{\hat{a}}, \\ \hat{F} &= \hat{a}^3 - 2\sqrt{\hat{a}^3}. \end{aligned} \tag{3.49}$$

These relationships, $\hat{d} = \hat{d}(\hat{a})$ and $\hat{F} = \hat{F}(\hat{a})$, are illustrated in Fig. 3.6. The implicitly defined function $\hat{F} = \hat{F}(\hat{d})$ is given in Fig. 3.7.

The Adhesive Impact Problem for the Parabolic Indenter

The adhesive normal impact of a parabolic body has been heavily investigated due to its various technical applications. Thornton and Ning (1998) were able to analytically determine the coefficient of restitution for the JKR adhesive normal impact.

Let the body have the mass m and the initial velocity v_0. For the rebound velocity v_e and the corresponding coefficient of restitution e, we obtain the expression:

$$e := \frac{v_e}{v_0} = \sqrt{1-\beta}, \quad \beta \leq 1, \tag{3.50}$$

with

$$\beta := \frac{1}{5mv_0^2}\left[\frac{R^4(\pi\Delta\gamma)^5}{E^{*2}}\right]^{1/3}\left[1+\sqrt[3]{864}\right]. \tag{3.51}$$

For $\beta > 1$ the coefficient of restitution is zero, i.e., low initial velocities will cause the body to stick to the elastic half-space without rebounding.

3.5.4 The Sphere

It has already been discussed in Chap. 2 that this contact problem is very similar to the one described in the previous section. With the non-adhesive solution (see Sect. 2.5.4), and (3.18) and (3.20), we obtain the following solution to the adhesive contact problem:

$$\begin{aligned}
d(a) &= a\,\text{artanh}\left(\frac{a}{R}\right) - \sqrt{\frac{2\pi a\Delta\gamma}{E^*}},\\
F_N(a) &= E^*R^2\left[\left(1+\frac{a^2}{R^2}\right)\text{artanh}\left(\frac{a}{R}\right) - \frac{a}{R}\right] - \sqrt{8\pi a^3 E^*\Delta\gamma},\\
\sigma_{zz}(r;a) &= -\frac{E^*}{\pi}\left[\frac{R}{\sqrt{R^2-r^2}}\,\text{artanh}\left(\frac{\sqrt{a^2-r^2}}{\sqrt{R^2-r^2}}\right) + \int_r^a \text{artanh}\left(\frac{x}{R}\right)\frac{\mathrm{d}x}{\sqrt{x^2-r^2}}\right]\\
&\quad + \sqrt{\frac{2E^*\Delta\gamma}{\pi a}}\frac{a}{\sqrt{a^2-r^2}}, \qquad r \leq a,\\
w(r;a) &= w_{\text{n.a.}}(r;a) - \sqrt{\frac{8a\Delta\gamma}{\pi E^*}}\arcsin\left(\frac{a}{r}\right), \qquad r > a,
\end{aligned} \tag{3.52}$$

with the sphere radius R. Here, $w_{\text{n.a.}}$ denotes the displacements without adhesion:

$$\begin{aligned}
w_{\text{n.a.}}(r;a) = \frac{2}{\pi}\Bigg\{&\text{artanh}\left(\frac{a}{R}\right)\left[a\arcsin\left(\frac{a}{r}\right) + \sqrt{r^2-a^2}\right] - R\arcsin\left(\frac{a}{r}\right)\\
&+ \sqrt{R^2-r^2}\arctan\left(\frac{a\sqrt{R^2-r^2}}{R\sqrt{r^2-a^2}}\right)\Bigg\}.
\end{aligned} \tag{3.53}$$

The critical contact radius a_c, at which the contact loses its stability and detaches, is given by the numerical solution of the transcendental equation

$$\text{artanh}\,\alpha + \frac{\alpha}{1-\alpha^2} - \frac{\beta}{\sqrt{\alpha}} = 0, \tag{3.54}$$

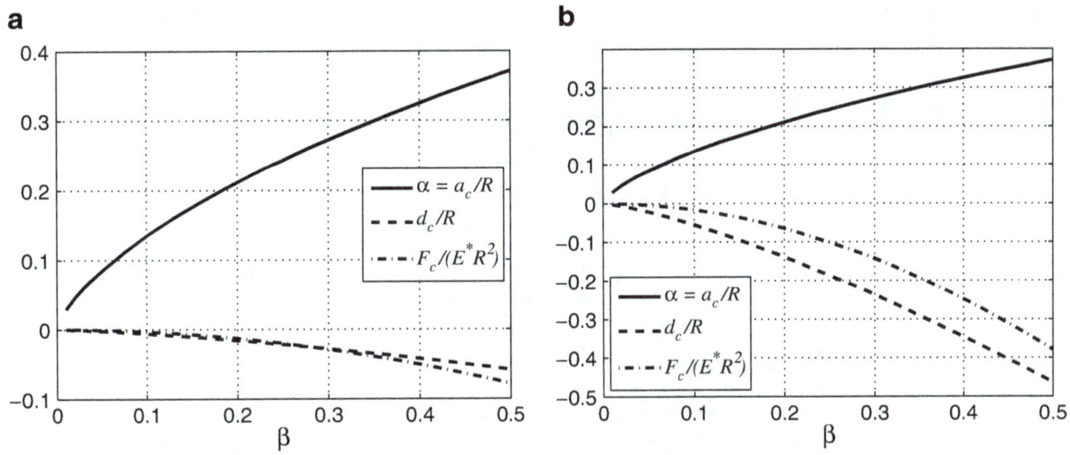

Fig. 3.8 Plots of the normalized critical contact radius a_c/R, the corresponding normalized indentation depth d_c/R, and normal force $F_c/\left(E^* R^2\right)$ as functions of the normalized surface energy β (see (3.55)), according to (3.54) and (3.56); (**a**) force-control, (**b**) displacement-control

with $\alpha := a_c/R$ and

$$\beta := \xi\sqrt{\frac{\pi\Delta\gamma}{2E^*R}}. \tag{3.55}$$

The last parameter describes a normalized surface energy and depends on the type of boundary condition ($\xi = 3$ for force-control, $\xi = 1$ for displacement-control). For the relationships between the critical indentation depth, the critical contact radius, and the critical normal force, one then obtains the equations:

$$\begin{aligned} d_c(\alpha) &= R\alpha\left[\frac{\beta}{\sqrt{a}}\left(1-\frac{2}{\xi}\right)-\frac{\alpha}{1-\alpha^2}\right], \\ F_c(\alpha) &= E^*R^2\left[\frac{\beta}{\sqrt{a}}\left(1+\alpha^2-\frac{4\alpha^2}{\xi}\right)-\frac{2\alpha}{1-\alpha^2}\right]. \end{aligned} \tag{3.56}$$

These results for the critical values of contact radius, indentation depth, and normal force—in normalized form—are shown in Fig. 3.8. The lengths are normalized to the sphere radius and the "adhesion force" to E^*R^2. Since the half-space hypothesis is too severely violated for $a > 0.3R$, it can be seen from the diagram that these results are valid only for $\beta < 0.35$. For example, it means that the effective surface energy in the case of force-control must not exceed $\Delta\gamma = 9\cdot 10^4\,\mathrm{J/m^2}$ for a sphere with $R = 10^{-2}\,\mathrm{m}$ and $E^* = 10^9\,\mathrm{Pa}$. That is an extremely great value though. In the case of displacement-control, this limit is even greater by a factor of 9.

3.5.5 The Ellipsoid

It has been demonstrated in Chap. 2 (Sect. 2.5.5) that there is only a slight difference between the contact problems of an indenting sphere and an ellipsoid of rotation.

The solution to the contact problem is, considering (3.18) and (3.20), given by:

$$\begin{aligned}
d(a) &= a \operatorname{artanh}(ka) - \sqrt{\frac{2\pi a \Delta\gamma}{E^*}}, \\
F_N(a) &= E^* \frac{R}{k}\left[(1 + k^2 a^2)\operatorname{artanh}(ka) - ka\right] - \sqrt{8\pi a^3 E^* \Delta\gamma}, \\
\sigma_{zz}(r;a) &= -\frac{E^* k R}{\pi}\left[\frac{1}{\sqrt{1-k^2 r^2}}\operatorname{artanh}\left(\frac{k\sqrt{a^2-r^2}}{\sqrt{1-k^2 r^2}}\right) + \int_r^a \frac{\operatorname{artanh}(kx)}{\sqrt{x^2-r^2}}\mathrm{d}x\right] \\
&\quad + \sqrt{\frac{2E^*\Delta\gamma}{\pi a}}\frac{a}{\sqrt{a^2-r^2}}, \qquad r \le a, \\
w(r;a) &= kRw_{K,\text{n.a.}}\left(r;a;R = \frac{1}{k}\right) - \sqrt{\frac{8a\Delta\gamma}{\pi E^*}}\arcsin\left(\frac{a}{r}\right), \qquad r > a.
\end{aligned} \tag{3.57}$$

Here, k and R denote the geometric parameters of the indenter profile which can be written as:

$$f(r) = R\left(1 - \sqrt{1 - k^2 r^2}\right). \tag{3.58}$$

$w_{K,\text{n.a.}}(r;a;R)$ is the displacement without adhesion for a spherical indenter of radius R, given in (3.53) of the previous section. The determining equation for the critical contact radius is again given by (3.54), wherein $\alpha := ka_c$ and $\beta := \xi\sqrt{\frac{\pi\Delta\gamma}{2E^* kR^2}}$. The expressions (3.56) for the critical values of indentation depth and normal force remain the same, except that the factor in front of the parenthesis in the normal force must be replaced by $E^* R/k$. It is therefore also possible to directly apply the curves in Fig. 3.8 since normalized curves are shown there.

3.5.6 The Indenter Which Generates a Constant Adhesive Tensile Stress

During their research on biological systems, in which adhesion played a central role (e.g., with a focus on geckos and certain insects), Gao and Yao (2004) came across the problem of the optimal (from a contact mechanical point of view) profile at the hair tips found on, for example, the gecko's feet, which is responsible for the strong adhesive forces in these systems. In this context, optimal means achieving the greatest possible pull-off force with the smallest possible contact area. To determine this optimal profile shape, Gao and Yao set out with a few preliminary considerations. Firstly, the critical state should correspond to a contact of the entire available contact domain. Secondly, the stress at the edges of the adhesive contact usually appears as a singularity since, at least according to the JKR theory, the adhesion itself can be interpreted as an indentation by a cylindrical flat punch. And since, ultimately, the maximum adhesive tension between two surfaces is solely determined by the potential of the van der Waals interaction between said surfaces, i.e., their material properties, the authors concluded that the optimal profile is the one which generates a constant adhesive tensile stress σ_0 in the contact area. The theoretical maximum pull-off force for a given contact area is then attributed to the profile where this stress σ_0 corresponds to the maximum adhesive stress σ_{th}.

The problem of the elastic half-space displacement resulting from the constant circular pressure or tension distribution of radius a was previously considered in Chap. 2 (see Sect. 2.5.6). For a stress distribution

$$\sigma_{zz}(r;a) = \sigma_0, \quad r \le a, \tag{3.59}$$

the resulting displacement of the half-space is:

$$w(r;a,\sigma_0) = -\frac{4\sigma_0 a}{\pi E^*}\mathrm{E}\left(\frac{r}{a}\right), \quad r \le a. \tag{3.60}$$

This means that the optimal profile takes on the form:

$$f(r;a,\sigma_0) = \frac{\sigma_0 a}{E^*}\left[\frac{4}{\pi}\mathrm{E}\left(\frac{r}{a}\right) - 2\right], \quad r \le a. \tag{3.61}$$

Here, $\mathrm{E}(\cdot)$ denotes the complete elliptical integral of the second kind:

$$\mathrm{E}(k) := \int_0^{\pi/2} \sqrt{1 - k^2 \sin^2 \varphi}\, \mathrm{d}\varphi. \tag{3.62}$$

The pull-off force F_c necessary to separate such an indenter profile from complete contact is trivially

$$F_c = \pi \sigma_0 a^2. \tag{3.63}$$

The displacements of the half-space beyond the contact area were also previously calculated and equal

$$w(r;a,p_0) = \frac{4\sigma_0 r}{\pi E^*}\left[\left(1 - \frac{a^2}{r^2}\right)\mathrm{K}\left(\frac{a}{r}\right) - \mathrm{E}\left(\frac{a}{r}\right)\right], \quad r > a, \tag{3.64}$$

with the complete elliptical integral of the first kind:

$$\mathrm{K}(k) := \int_0^{\pi/2} \frac{\mathrm{d}\varphi}{\sqrt{1 - k^2 \sin^2 \varphi}}. \tag{3.65}$$

The maximum pull-off force of this profile can be compared to the flat punch of identical radius a:

$$F_{\max}^{\text{Kendall}} = \sqrt{8\pi a^3 E^* \Delta\gamma} = \sqrt{8\pi a^3 E^* \sigma_0 h}. \tag{3.66}$$

For the optimal profile, the maximum force is:

$$F_{\max}^{\text{opt}} = \pi\sigma_0 a^2. \tag{3.67}$$

Analogous to the theory of Maugis described in Sect. 3.8, the surface energy $\Delta\gamma$ is calculated from the adhesive stresses σ_{th} and the maximum range of the van der Waals interaction:

$$\Delta\gamma = \sigma_0 h. \tag{3.68}$$

The force ratio then equals:

$$\frac{F_{\max}^{\text{Kendall}}}{F_{\max}^{\text{opt}}} = \sqrt{\frac{8}{\pi}\frac{E^*}{\sigma_0}\frac{h}{a}}. \tag{3.69}$$

3.5.7 The Profile in the Form of a Power-Law

We now consider a general indenter with the profile:

$$f(r) = cr^n, \quad n \in \mathbb{R}^+, \tag{3.70}$$

with an arbitrary constant c and a positive real exponent n. In contrast to the results for non-adhesive contact detailed in the previous chapter, for the adhesive contact, we obtain qualitatively different behavior for $n > 0.5$ and $n < 0.5$. We first turn our attention to the case of $n > 0.5$. The contact problem was first investigated by Borodich and Galanov (2004), Spolenak et al. (2005), and Yao and Gao (2006). The solution of the contact problem shown in Fig. 3.9 is, as before, given by the solution of the non-adhesive contact (see Sect. 2.5.8) and (3.18) and (3.20):

$$\begin{aligned}
d(a) &= \kappa(n)ca^n - \sqrt{\frac{2\pi a\Delta\gamma}{E^*}},\\
F_N(a) &= E^*\frac{2n}{n+1}\kappa(n)ca^{n+1} - \sqrt{8\pi a^3 E^*\Delta\gamma},\\
\sigma_{zz}(r;a) &= -\frac{E^*}{\pi}n\kappa(n)c\int_r^a x^{n-1}\frac{\mathrm{d}x}{\sqrt{x^2-r^2}} + \sqrt{\frac{2E^*\Delta\gamma}{\pi a}}\frac{a}{\sqrt{a^2-r^2}}, \quad r \le a,\\
w(r;a) &= \frac{2}{\pi}\kappa(n)c\left[a^n \arcsin\left(\frac{a}{r}\right) - \int_0^a x^n \frac{\mathrm{d}x}{\sqrt{r^2-x^2}}\right] - \sqrt{\frac{8a\Delta\gamma}{\pi E^*}}\arcsin\left(\frac{a}{r}\right),\\
&\quad r > a.
\end{aligned} \tag{3.71}$$

Similarly to Chap. 2, we introduce the scaling factor:

$$\kappa(n) := \sqrt{\pi}\frac{\Gamma(n/2+1)}{\Gamma[(n+1)/2]}, \tag{3.72}$$

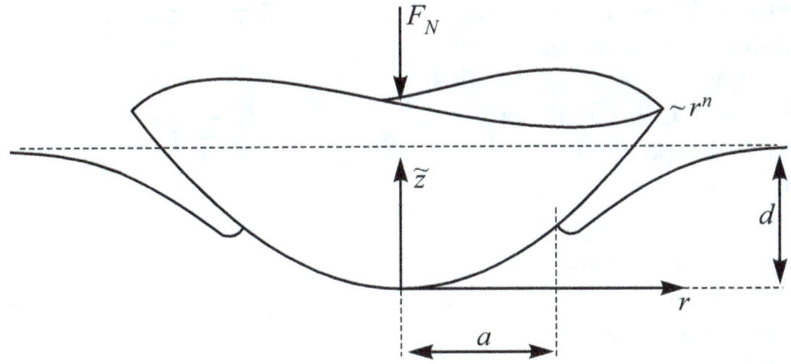

Fig. 3.9 Adhesive normal contact between a rigid indenter with a profile in the form of a power-law and an elastic half-space

with the gamma function $\Gamma(\cdot)$:

$$\Gamma(z) := \int_0^\infty t^{z-1}\exp(-t)\,\mathrm{d}t. \tag{3.73}$$

With regard to the different possibilities of the resolution of the integrals occurring in the stresses and displacements see Sect. 2.5.8, which details the consideration of power profiles. For the critical contact radius, we obtain this with (3.23):

$$a_c = \left(\frac{\xi^2\pi\Delta\gamma}{2E^* n^2 c^2 \kappa^2(n)}\right)^{\frac{1}{2n-1}}. \tag{3.74}$$

The critical indentation depth is:

$$d_c = \left(\frac{\pi\Delta\gamma}{2E^*}\right)^{\frac{n}{2n-1}}\left(\frac{\xi}{nc\kappa(n)}\right)^{\frac{1}{2n-1}}\left[\frac{\xi}{n}-2\right]. \tag{3.75}$$

It is *positive* in the force-controlled case ($\xi = 3$) if $n < 1.5$. Under displacement-control it is $\xi = 1$. The "adhesion force" is given by:

$$F_c = (E^*)^{\frac{n-2}{2n-1}}\left(\frac{\pi\Delta\gamma}{2}\right)^{\frac{n+1}{2n-1}}\left(\frac{\xi}{nc\kappa(n)}\right)^{\frac{3}{2n-1}}\left[\frac{2\xi}{n+1}-4\right]. \tag{3.76}$$

It can be seen that the Hertzian contact for $n = 2$ is the only case in which this force does not depend on E^*. By introducing the quantities

$$\hat{a} := \frac{a}{a_c}, \quad \hat{d} := \frac{d}{|d_c|}, \quad \hat{F} := \frac{F_N}{|F_c|}, \tag{3.77}$$

normalized to the critical values, the relations between the macroscopic quantities can be written as:

$$\begin{aligned}\hat{d} &= \frac{\xi}{|\xi-2n|}\hat{a}^n - \frac{2n}{|\xi-2n|}\sqrt{\hat{a}},\\ \hat{F} &= \frac{\xi}{|\xi-2n-2|}\hat{a}^{n+1} - \frac{2n+2}{|\xi-2n-2|}\sqrt{\hat{a}^3}.\end{aligned} \tag{3.78}$$

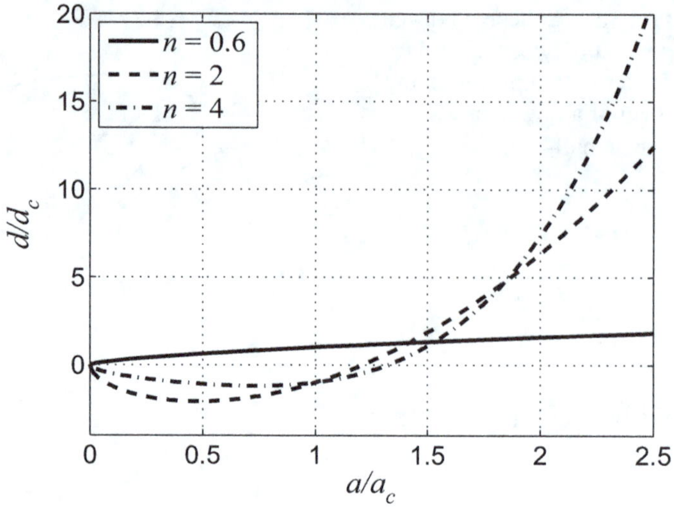

Fig. 3.10 Normalized indentation depth as a function of the normalized contact radius for an indenter with a profile in the shape of a power-law with exponent n. All quantities are normalized by the critical values in the force-controlled experiment. Shown are different values of n

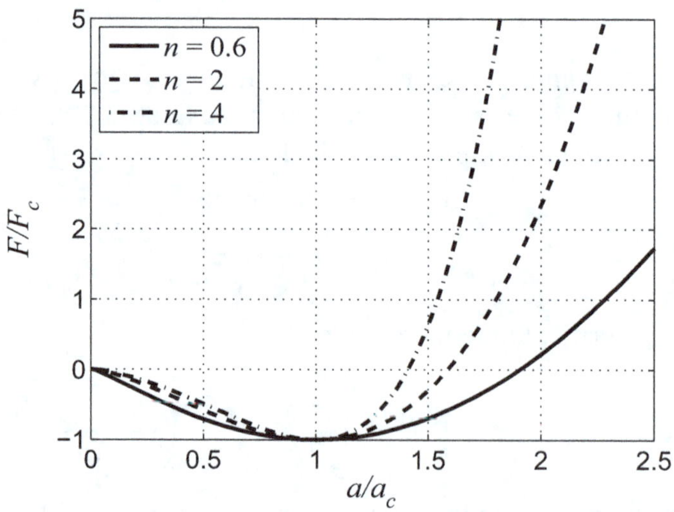

Fig. 3.11 Normalized normal force as a function of the normalized contact radius for an indenter with a profile in the shape of a power-law with exponent n. All quantities are normalized by the critical values in the force-controlled experiment. Shown are different values of n

It is interesting to note that the relations (3.78), except for the type of boundary condition (force- or displacement-control), only depend on the exponent n. The relationships (3.78) are shown for the force-controlled experiment in Figs. 3.10 and 3.11 for three different values of n.

All of the aforementioned results also apply to indenters with $n < 1/2$ (Popov 2017). What changes is just the interpretation of the corresponding quantities. In the case of $n > 1/2$ the critical quantities separate the state of stable adhesive contact with a finite contact radius from the process of unstable shrinkage of the contact area and complete detachment. A stable condition exists with larger forces and an unstable condition with smaller forces. On the other hand, in the case of $n < 1/2$ the critical quantities separate the stable state from the unstable, unlimited propagation of the contact area. The stable state exists for smaller forces and the instability occurs when increasing the force (or indentation). A detailed analysis was given by Popov (2017).

A special case is $n = 0.5$. We want to discuss this case under the condition of a displacement-controlled loading. It is, in this case:

$$g(a) = \kappa(1/2)ca^{1/2}, \quad \text{with} \quad \kappa(1/2) = \frac{1}{4}\frac{\pi^{3/2}\sqrt{2}}{\Gamma(3/4)^2} \approx 1.311,$$
$$\Delta l(a) = \left(\frac{2\pi\Delta\gamma}{E^*}\right)^{1/2} a^{1/2}. \tag{3.79}$$

At the moment of the first contact, $d = 0$, it applies to all a:

$$\text{I. } g(a) > \Delta l(a), \quad \text{if } c > 1.9120\left(\frac{\Delta\gamma}{E^*}\right)^{1/2},$$
$$\text{II. } g(a) < \Delta l(a), \quad \text{if } c < 1.9120\left(\frac{\Delta\gamma}{E^*}\right)^{1/2}. \tag{3.80}$$

In the first case, the radius of the contact will decrease until it disappears. In the second case, it will enlarge until complete contact is established. Furthermore, in the second case, complete contact is formed immediately, as soon as the indenter tip touches the half-space.

The Adhesive Impact Problem for the Indenter with Power-Law Profile

The normal adhesive impact problem, as in the case of the parabolic body, can be solved in general form. The body has the mass m and the initial velocity v_0. For the rebound speed v_e, and thus the coefficient of restitution e, we obtain:

$$e := \frac{v_e}{v_0} = \sqrt{1-\beta}, \quad \beta \leq 1, \tag{3.81}$$

with

$$\beta := \frac{1}{mv_0^2}\left[\left(\frac{\pi\Delta\gamma}{2}\right)^{2n+1}\left(\frac{1}{E^* n^2 c^2 \kappa^2(n)}\right)^2\right]^{\frac{1}{2n-1}} \cdot \frac{8n-4}{2n^2+3n+1}\left[n + (n+1)\,(2n)^{\frac{4}{2n-1}}\right]. \tag{3.82}$$

For $\beta > 1$ it is $e = 0$; that is, the body will stick to the elastic half-space and not rebound for impact velocities below a critical value. For $n = 2$, the known solution from Sect. 3.5.3 is recovered.

3.5.8 The Truncated Cone

The adhesive normal contact problem for a truncated cone (see Fig. 3.12) was first solved by Maugis and Barquins (1983). With the help of the solutions from Chap. 2

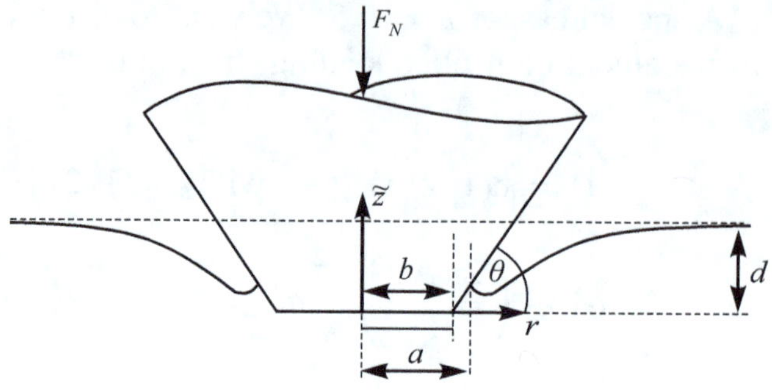

Fig. 3.12 Adhesive normal contact between a rigid truncated cone and an elastic half-space

(see Sect. 2.5.9) and (3.18) and (3.20), the solution of the adhesive normal contact problem can be determined without difficulty:

$$\begin{aligned} d(a) &= a \tan\theta \arccos\left(\frac{b}{a}\right) - \sqrt{\frac{2\pi a \Delta\gamma}{E^*}}, \\ F_N(a) &= E^* \tan\theta a^2 \left[\arccos\left(\frac{b}{a}\right) + \frac{b}{a}\sqrt{1-\frac{b^2}{a^2}}\right] - \sqrt{8\pi a^3 E^* \Delta\gamma}, \\ \sigma_{zz}(r;a) &= \sigma_{zz,\text{n.a.}}(r;a) + \sqrt{\frac{2E^*\Delta\gamma}{\pi a}}\frac{a}{\sqrt{a^2-r^2}}, \quad r \le a, \\ w(r;a) &= w_{\text{n.a.}}(r;a) - \sqrt{\frac{8a\Delta\gamma}{\pi E^*}} \arcsin\left(\frac{a}{r}\right), \quad r > a. \end{aligned} \tag{3.83}$$

Here, b denotes the radius at the blunt end and θ the conical slope angle. The solutions of the stresses and displacements in the case of non-adhesive contact indicated by the index "n.a." can be looked up in Sect. 2.5.9:

$$\begin{aligned} &\sigma_{zz,\text{n.a.}}(r;a) = \\ &-\frac{E^*\tan\theta}{\pi}\begin{cases} \displaystyle\int_b^a \left\{\frac{b}{\sqrt{x^2-b^2}} + \arccos\left(\frac{b}{x}\right)\right\}\frac{dx}{\sqrt{x^2-r^2}}, & r \le b, \\ \displaystyle\int_r^a \left\{\frac{b}{\sqrt{x^2-b^2}} + \arccos\left(\frac{b}{x}\right)\right\}\frac{dx}{\sqrt{x^2-r^2}}, & b < r \le a, \end{cases} \\ &w_{\text{n.a.}}(r;a) = \frac{2\tan\theta}{\pi}\left\{\varphi_0 a \arcsin\left(\frac{a}{r}\right) - \int_b^a x \arccos\left(\frac{b}{x}\right)\frac{dx}{\sqrt{r^2-x^2}}\right\}, \\ &\qquad r > a. \end{aligned} \tag{3.84}$$

The relationship between indentation depth and contact radius in the case without adhesion is described by:

$$d_{\text{n.a.}}(a) = \varphi_0 a \tan\theta, \tag{3.85}$$

with:

$$\varphi_0 := \arccos\left(\frac{b}{a}\right). \tag{3.86}$$

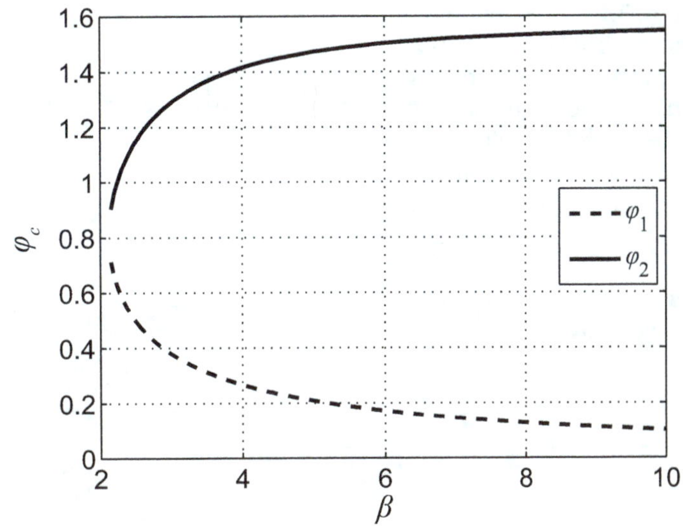

Fig. 3.13 Stability bifurcation for the angle φ_0 as a function of the normalized surface energy β (see (3.88)) for the adhesive indentation by a truncated cone. All points between the two curves indicate unstable configurations

This results in the critical contact radius as a solution to the transcendental equation:

$$\varphi_{0,c} + \cot\varphi_{0,c} - \beta\sqrt{\cos\varphi_{0,c}} = 0, \tag{3.87}$$

with:

$$\beta := \frac{\xi}{\tan\theta}\sqrt{\frac{\pi\Delta\gamma}{2E^*b}}. \tag{3.88}$$

For the critical values of the indentation depth and the normal force, one obtains:

$$\begin{aligned} d_c &= a_c \tan\theta\left[\varphi_{0,c}\left(1-\frac{2}{\xi}\right)-\frac{2}{\xi}\cot\varphi_{0,c}\right], \\ F_c &= E^*\tan\theta a_c^2\left[\varphi_{0,c}\left(1-\frac{4}{\xi}\right)+\cot\varphi_{0,c}\left(\sin^2\varphi_{0,c}-\frac{4}{\xi}\right)\right]. \end{aligned} \tag{3.89}$$

In (3.88) and (3.89), the type of boundary condition must yet be determined. Displacement-controlled trials are characterized by $\xi = 1$, and force-controlled ones by $\xi = 3$. One can easily convince oneself that for $b = 0$, and thus $\varphi_0 \equiv \pi/2$, the solutions of the complete cone from Sect. 3.5.2 are recovered.

However, a closer look at (3.87) reveals a bifurcation of the solution. This is shown in Fig. 3.13. The equation only has solutions for $\beta > 2.125$. For smaller values of β the critical radius a_c is given by the radius b and the critical values of the indentation depth and normal force correspond to those of the flat punch:

$$\begin{aligned} d_c &= -\sqrt{\frac{2\pi b\Delta\gamma}{E^*}}, \\ F_c &= -\sqrt{8\pi b^3E^*\Delta\gamma}. \end{aligned} \tag{3.90}$$

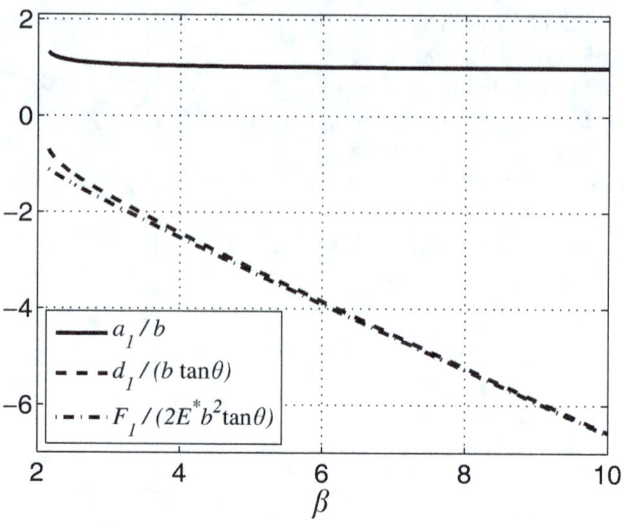

Fig. 3.14 Solution for the normalized critical values for the contact radius a_c/b, indentation depth $d_c/(b\tan\theta)$, and the normal force $F_c/(2E^*b^2\tan\theta)$ as a function of β for the adhesive indentation by a truncated cone for the critical angle φ_1 under force-controlled conditions

For $\beta > 2.125$ there are always two roots of the equation and one can show that all configurations between these two roots (that is, all values $\varphi_{0,c1} < \varphi_0 < \varphi_{0,c2}$) are unstable. For example, for $b = 10^{-2}$ m, $\theta = 0.1$, $E^* = 10^9$ Pa, $\beta > 2.125$ means that the effective surface energy must (in the force-controlled case) be $\Delta\gamma > 3\cdot 10^4\,\mathrm{J/m^2}$. This is a very big value. In this case, the two solutions for the angle φ_0 are shown in Fig. 3.13. Figure 3.14 shows the curves of the contact radius, the indentation depth, and the normal force in normalized form for the first critical solution under force-control as a function of the normalized surface energy β.

3.5.9 The Truncated Paraboloid

With the results of Sect. 2.5.10 and (3.18) and (3.20), we come to the following solution first found by Maugis and Barquins (1983) regarding the adhesive normal contact problem for the truncated paraboloid (see Fig. 3.15):

$$
\begin{aligned}
d(a) &= \frac{a}{R}\sqrt{a^2-b^2} - \sqrt{\frac{2\pi a\Delta\gamma}{E^*}},\\
F_N(a) &= \frac{2E^*}{3R}(2a^2+b^2)\sqrt{a^2-b^2} - \sqrt{8\pi a^3 E^*\Delta\gamma},\\
\sigma_{zz}(r;a) &= \sigma_{zz,\text{n.a.}}(r;a) + \sqrt{\frac{2E^*\Delta\gamma}{\pi a}}\frac{a}{\sqrt{a^2-r^2}}, \qquad r \le a,\\
w(r;a) &= w_{\text{n.a.}}(r;a) - \sqrt{\frac{8a\Delta\gamma}{\pi E^*}}\arcsin\left(\frac{a}{r}\right), \qquad r > a.
\end{aligned}
\tag{3.91}
$$

Here, b denotes the radius at the flat tip. The base paraboloid has the radius of curvature R. The solutions for the stresses and displacements in the non-adhesive

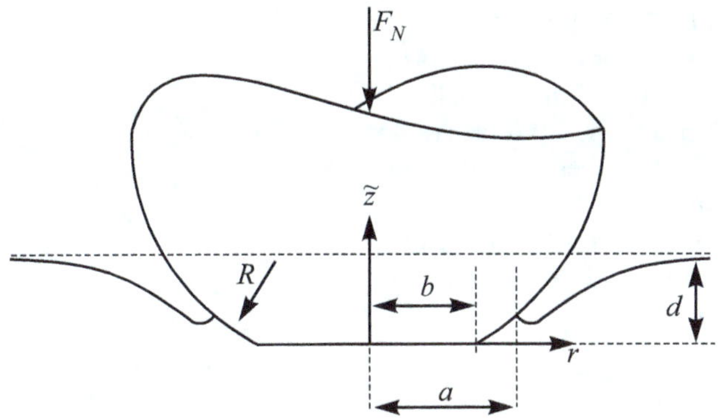

Fig. 3.15 Adhesive normal contact between a rigid truncated paraboloid and an elastic half-space

case, characterized by the index "n.a.", are:

$$\sigma_{zz,\text{n.a.}}(r;a) = -\frac{E^*}{\pi R}\begin{cases}\displaystyle\int_b^a \frac{(2x^2-b^2)\mathrm{d}x}{\sqrt{x^2-b^2}\sqrt{x^2-r^2}}, & r \leq b,\\ \displaystyle\int_r^a \frac{(2x^2-b^2)\mathrm{d}x}{\sqrt{x^2-b^2}\sqrt{x^2-r^2}}, & b < r \leq a,\end{cases}$$

$$w_{\text{n.a.}}(r;a) = \frac{2a}{\pi R}\sqrt{a^2-b^2}\arcsin\left(\frac{a}{r}\right) - \frac{1}{\pi R}\left[\left(r^2-b^2\right)\arcsin\left(\frac{\sqrt{a^2-b^2}}{\sqrt{r^2-b^2}}\right) - \sqrt{a^2-b^2}\sqrt{r^2-a^2}\right],$$
$$r > a. \tag{3.92}$$

Between indentation depth and contact radius in the case of non-adhesive contact, the following relationship applies:

$$d_{\text{n.a.}}(a) = \frac{a}{R}\sqrt{a^2-b^2} = \frac{a^2}{R}\sin\varphi_0, \tag{3.93}$$

with

$$\varphi_0 := \arccos\left(\frac{b}{a}\right). \tag{3.94}$$

Equation (3.23) then provides the transcendental equation that determines the critical contact radius:

$$\frac{1+\sin^2\varphi_{0,c}}{\sin\varphi_{0,c}\cos\varphi_{0,c}} - \beta\sqrt{\cos\varphi_{0,c}} = 0, \tag{3.95}$$

with

$$\beta := \frac{\xi R}{b}\sqrt{\frac{\pi\Delta\gamma}{2E^*b}}. \tag{3.96}$$

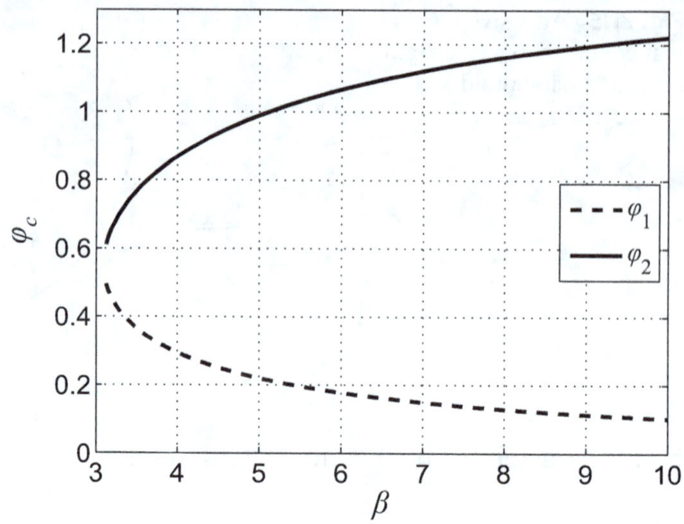

Fig. 3.16 Stability bifurcation for the angle φ_0 as a function of the normalized surface energy β (see (3.96)) for the adhesive indentation by a truncated paraboloid. All points between the two curves indicate unstable configurations

For the critical indentation depth and normal force, we obtain:

$$d_c = \frac{a_c^2}{R}\sin\varphi_{0,c}\left(1 - \frac{2}{\xi} - \frac{2}{\xi\sin^2\varphi_{0,c}}\right),$$
$$F_c = \frac{2E^*a_c^3}{3R}\sin\varphi_{0,c}\left[3 - \sin^2\varphi_{0,c} - \frac{6}{\xi}\left(1 + \frac{1}{\sin^2\varphi_{0,c}}\right)\right], \qquad (3.97)$$

which can be reduced to the results in Sect. 3.5.3 without great difficulty for $b = 0$. In (3.95) and (3.97) , the type of boundary condition has to be defined ($\xi = 1$ for displacement-control, $\xi = 3$ for force-control).

Examining (3.95), one encounters a similar bifurcation as in the previous section. The equation only has solutions for $\beta > 3.095$. For smaller values of β the critical contact radius is given by b and the associated values of indentation depth and normal force are the same as those of the flat punch. For a sufficiently large surface energy and, correspondingly, $\beta > 3.095$, there are always two solutions to (3.95), and one finds that all states between these two solutions are unstable. These solutions as a function of β are shown in Fig. 3.16. For $b = 10^{-3}$ m, $R = 10^{-2}$ m, $E^* = 10^9$ Pa, $\beta > 3.095$ means (as an example) that the effective surface energy in the force-controlled case must be $\Delta\gamma > 7 \cdot 10^3$ J/m^2. Figure 3.17 shows the curves of the contact radius, the indentation depth, and the normal force in normalized form for the first critical solution under force-control as a function of the normalized surface energy β.

3.5.10 The Cylindrical Flat Punch with Parabolic Cap

Consider now the adhesive normal contact between an elastic half-space and a flat punch with a parabolic cap. The punch has the radius b and the cap has the radius of curvature R.

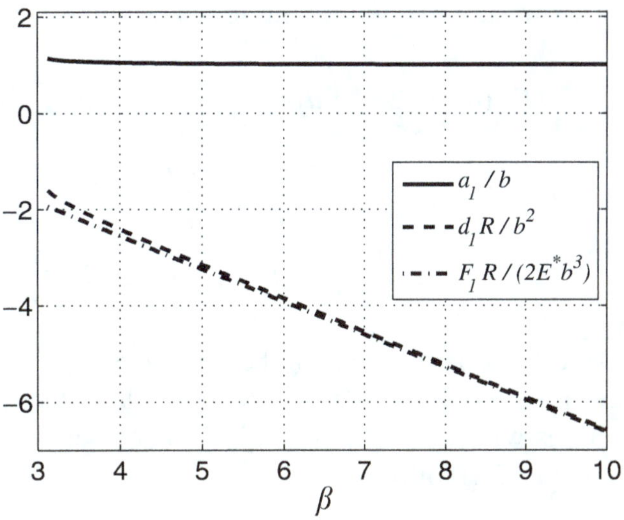

Fig. 3.17 Dependencies of the normalized critical values for the contact radius a_c/b, the indentation depth d_cR/b^2, and the normal force $F_cR/(2E^*b^3)$ as function of β for the adhesive indentation by a truncated paraboloid at force-control for the critical angle φ_1

There may be two outcomes based on these examples. If the contact radius a is smaller than b, the contact assembles simply the one with a parabolic indenter, for which the solution can be looked up in Sect. 3.5.3. With a sufficiently large indentation depth, or normal force, the contact radius is $a = b$. In this case, however, analogous to the problem of the flat punch, there is no difference between the equations of the non-adhesive and the adhesive solution. In Chap. 2 (Sect. 2.5.11), this distinction was not considered further because it had no appreciable consequences. In adhesive contact, however, this results in peculiarities regarding the stability of the contact. We therefore once again repeat the two aforementioned solutions. If $a < b$ is the solution of the contact problem, according to (3.18) and (3.20), it is as follows:

$$\begin{aligned}
d(a) &= \frac{a^2}{R} - \sqrt{\frac{2\pi a \Delta\gamma}{E^*}}, \\
F_N(a) &= \frac{4}{3}\frac{E^* a^3}{R} - \sqrt{8\pi a^3 E^* \Delta\gamma}, \\
\sigma_{zz}(r;a) &= -\frac{2E^*}{\pi R}\sqrt{a^2 - r^2} + \sqrt{\frac{2E^*\Delta\gamma}{\pi a}}\frac{a}{\sqrt{a^2 - r^2}}, \quad r \le a, \\
w(r;a) &= \frac{a^2}{\pi R}\left[\left(2 - \frac{r^2}{a^2}\right)\arcsin\left(\frac{a}{r}\right) + \frac{\sqrt{r^2 - a^2}}{a}\right] \\
&\quad - \sqrt{\frac{8a\Delta\gamma}{\pi E^*}}\arcsin\left(\frac{a}{r}\right), \quad r > a.
\end{aligned} \tag{3.98}$$

In contrast, if $a = b$ the solution is described by:

$$F_N(d) = 2E^*\left(db - \frac{b^3}{3R}\right),$$
$$\sigma_{zz}(r;d) = -\frac{E^*}{\pi R}\frac{b^2 - 2r^2 + dR}{\sqrt{b^2 - r^2}}, \quad r \le b,$$
$$w(r;d) = \frac{1}{\pi R}\left\{(2dR - r^2)\arcsin\left(\frac{b}{r}\right) + b\sqrt{r^2 - b^2}\right\}, \quad r > b. \tag{3.99}$$

Depending on the value of the surface energy, different variants of the critical state are possible. If we denote the critical contact radius under force-controlled conditions with a_c and those with displacement-control with $a_{c,d}$, these cases can be structured as follows:

- $\Delta\gamma < \Delta\gamma_1$: $a_c = a_c^{\mathrm{JKR}}$, $a_{c,d} = a_{c.d}^{\mathrm{JKR}}$
- $\Delta\gamma_1 \le \Delta\gamma < \Delta\gamma_2$: $a_c = b$, $a_{c,d} = a_{c.d}^{\mathrm{JKR}}$
- $\Delta\gamma > \Delta\gamma_2$: $a_c = a_{c,d} = b$

Here, the superscripts "JKR" denote the respective results for the parabolic indenter. $\Delta\gamma_1$ and $\Delta\gamma_2$ indicate the values of the surface energy at which the critical radii of the parabolic solution just coincide with the radius of the punch:

$$a_c^{\mathrm{JKR}}(\Delta\gamma = \Delta\gamma_1) = b,$$
$$a_{c,d}^{\mathrm{JKR}}(\Delta\gamma = \Delta\gamma_2) = b. \tag{3.100}$$

From (3.100) one obtains (with (3.45)):

$$\Delta\gamma_2 = 9\Delta\gamma_1 = \frac{8E^*b^3}{\pi R^2}. \tag{3.101}$$

By introducing the normalized quantities

$$\hat{d} := \frac{d}{|d_c^{\mathrm{JKR}}|}, \quad \hat{a} := \frac{a}{a_c^{\mathrm{JKR}}}, \quad \hat{b} := \frac{b}{a_c^{\mathrm{JKR}}}, \quad \hat{F} := \frac{F_N}{|F_c^{\mathrm{JKR}}|}, \tag{3.102}$$

with the known results for the parabolic indenter (see Sect. 3.5.3),

$$a_c^{\mathrm{JKR}} = \left(\frac{9\pi R^2\Delta\gamma}{8E^*}\right)^{1/3}, \quad d_c^{\mathrm{JKR}} = -\frac{1}{4}\left(\frac{3\pi^2(\Delta\gamma)^2 R}{(E^*)^2}\right)^{1/3},$$
$$F_c^{\mathrm{JKR}} = -\frac{3}{2}\pi\Delta\gamma R, \tag{3.103}$$

the relationships between the global contact quantities can be written as follows:

$$\hat{d} = 3\hat{a}^2 - 4\sqrt{\hat{a}},$$
$$\hat{F} = \begin{cases} \hat{a}^3 - 2\sqrt{\hat{a}^3}, & \hat{a} < \hat{b}, \\ \frac{1}{2}(\hat{d}\hat{b} - \hat{b}^3), & \hat{a} = \hat{b}. \end{cases} \tag{3.104}$$

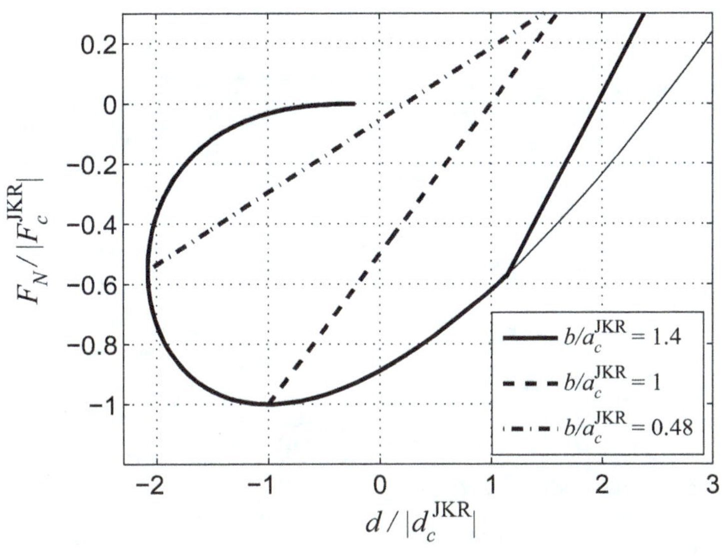

Fig. 3.18 Normalized relationship between normal force and indentation depth for a flat punch with parabolic cap at different normalized punch radii. The thin solid line corresponds to the parabolic JKR-solution

Thus the defined relationship $\hat{F} = \hat{F}(\hat{d})$ is shown in Fig. 3.18 for three different values of $\hat{b}$. Here, $\hat{b} = 1$ corresponds to a surface energy of $\Delta\gamma = \Delta\gamma_1$ and $\hat{b} = 0.48$ of a surface energy of $\Delta\gamma = \Delta\gamma_2$. It is easy to see the corresponding detachment points from the JKR solution for the paraboloid.

3.5.11 The Cone with Parabolic Cap

In Chap. 2 (see Sect. 2.5.12), the solution to the contact problem of a non-adhesive, frictionless, normal contact between an elastic half-space and a rigid cone with a rounded tip was shown. Thus, the solution of the adhesive, frictionless normal contact (see Fig. 3.19) is already known. The solution was first published by Maugis and Barquins (1983). With (3.18) and (3.20), one obtains:

$$
\begin{aligned}
d(a) &= a\tan\theta\left(\frac{1-\sin\varphi_0}{\cos\varphi_0}+\varphi_0\right)-\sqrt{\frac{2\pi a\Delta\gamma}{E^*}},\\
F_N(a) &= E^*a^2\tan\theta\left(\varphi_0+\frac{4}{3}\frac{1-\sin\varphi_0}{\cos\varphi_0}+\frac{1}{3}\sin\varphi_0\cos\varphi_0\right)-\sqrt{8\pi a^3E^*\Delta\gamma},\\
\sigma_{zz}(r;a) &= \sigma_{zz,\mathrm{n.a.}}(r;a)+\sqrt{\frac{2E^*\Delta\gamma}{\pi a}}\frac{a}{\sqrt{a^2-r^2}}, \qquad r\le a,\\
w(r;a) &= w_{\mathrm{n.a.}}(r;a)-\sqrt{\frac{8a\Delta\gamma}{\pi E^*}}\arcsin\left(\frac{a}{r}\right), \qquad r>a,
\end{aligned}
\tag{3.105}
$$

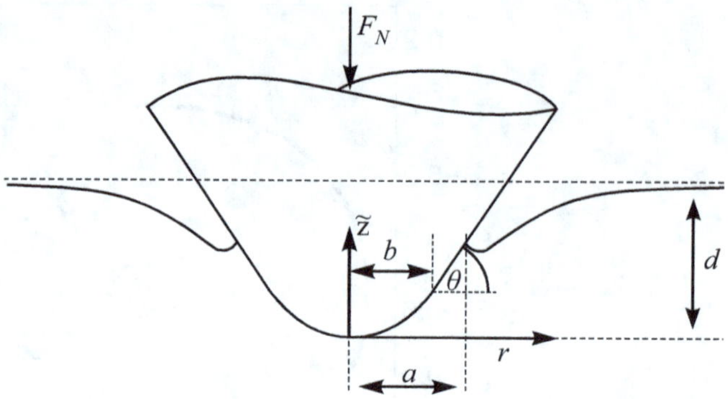

Fig. 3.19 Adhesive normal contact between a rigid cone with parabolic cap and an elastic half-space

with

$$\varphi_0 := \arccos\left(\frac{b}{a}\right), \tag{3.106}$$

where b denotes the radius at which the parabolic tip, which is continuously differentiable, passes into the conical body. θ describes the slope angle of the conical body. The stresses and displacements in the non-adhesive case, denoted by the index "n.a.", are given by:

$$\sigma_{zz,\text{n.a.}}(r;a) = -\frac{E^* \tan\theta}{\pi b}\begin{cases} 2\sqrt{a^2-r^2} + b\displaystyle\int_0^{\varphi_0} \frac{(\varphi - 2\tan\varphi)\tan\varphi \,\mathrm{d}\varphi}{\sqrt{1-k^2\cos^2\varphi}}, & r \le b, \\ 2\sqrt{a^2-r^2} \\ \quad + b\displaystyle\int_0^{\operatorname{arcosh}\left(\frac{a}{r}\right)} \left[\arccos\left(\frac{b}{r\cosh\varphi}\right)\right. \\ \qquad \left. - 2\sqrt{k^2\cosh^2\varphi - 1}\right]\mathrm{d}\varphi, & b < r \le a, \end{cases}$$

$$\begin{aligned} w_{\text{n.a.}}(r;a) = {} & \frac{2d_{\text{n.a.}}(a)}{\pi}\arcsin\left(\frac{a}{r}\right) \\ & - \frac{\tan\theta}{\pi b}\left[r^2\arcsin\left(\frac{a}{r}\right)\right. \\ & \left. - a\sqrt{r^2-a^2} + 2b^2\int_0^{\varphi_0}\frac{(\varphi - \tan\varphi)\tan\varphi \mathrm{d}\varphi}{\cos\varphi\sqrt{k^2\cos^2\varphi - 1}}\right], \quad r > a, \end{aligned} \tag{3.107}$$

with

$$d_{\text{n.a.}}(a) = a\tan\theta\left(\frac{1-\sin\varphi_0}{\cos\varphi_0} + \varphi_0\right). \tag{3.108}$$

The relationships (3.105) to (3.108) are all valid only for $a \ge b$. If $a < b$, the contact resembles the one with a paraboloid, for which results can be looked up in

Sect. 3.5.3. In the previous chapter, this distinction was ignored because of its triviality. For the adhesive contact, however, it does not have quite trivial consequences for the stability of the contact. The non-adhesive indentation depth is completely described by:

$$d_{\text{n.a.}}(a) = \begin{cases} \dfrac{a^2 \tan\theta}{b}, & a \le b, \\ a \tan\theta \left(\dfrac{1-\sin\varphi_0}{\cos\varphi_0} + \varphi_0 \right), & a > b. \end{cases} \tag{3.109}$$

According to (3.23), the contact radius at which the contact loses its stability results as a solution of the equation:

$$2\frac{1-\sin\varphi_{0,c}}{\cos\varphi_{0,c}} + \varphi_{0,c} - \beta\sqrt{\cos\varphi_{0,c}} = 0, \tag{3.110}$$

with

$$\beta := \frac{\xi}{\tan\theta}\sqrt{\frac{\pi\Delta\gamma}{2E^*b}}, \tag{3.111}$$

if that equation has solutions. The corresponding indentation depth is given by:

$$d_c = a\tan\theta\left[\frac{1-\sin\varphi_{0,c}}{\cos\varphi_{0,c}}\left(1-\frac{4}{\xi}\right) + \varphi_{0,c}\left(1-\frac{2}{\xi}\right)\right], \tag{3.112}$$

and the corresponding normal force by:

$$F_c = E^*a^2\tan\theta\left[\varphi_{0,c}\left(1-\frac{4}{\xi}\right) + \frac{1-\sin\varphi_{0,c}}{\cos\varphi_0, c}\left(\frac{4}{3}-\frac{8}{\xi}\right) + \frac{1}{3}\sin\varphi_{0,c}\cos\varphi_{0,c}\right]. \tag{3.113}$$

As expected, the solutions of the cone and the paraboloid from Sects. 3.5.2 and 3.5.3 are obtained by setting $b = 0$ or $b = a$ correspondingly (in this case it is $R := b/\tan\theta$). ξ is a parameter which determines the type of boundary condition (force or displacement-control; see (3.23)).

It turns out that, when solving (3.110), there are three different regimes: For $\beta < 1.795$ there is no solution and the critical contact radius is smaller than b. This means that the relationships for the critical state match those of the parabolic indenter that can be looked up in Sect. 3.5.3. If $1.795 \le \beta \le 2$, there are two solutions for the equation which are both shown in Fig. 3.20 and, correspondingly, two critical states. For values $\beta > 2$ only the larger of the two solutions remains. The smaller one becomes negative and thus unphysical. In this case, the curves of the normalized critical values of contact radius, indentation depth, and normal force in the case of force-control are given in Fig. 3.21. For formatting reasons, a sign change was made during the normalization of the "adhesion force" F_c.

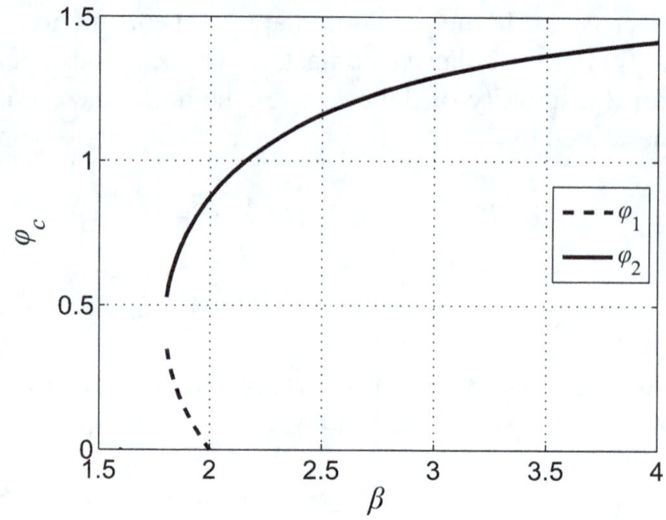

Fig. 3.20 Stability bifurcation of the angle φ_0 as a function of the normalized surface energy β (see (3.111)) for the adhesive indentation by a cone with a rounded tip

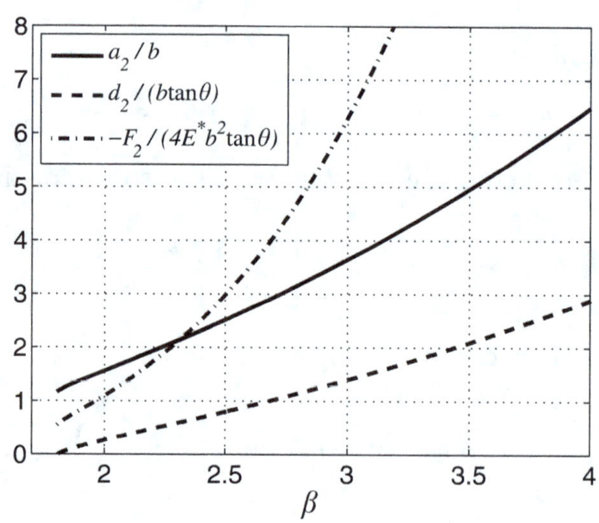

Fig. 3.21 Curves of the normalized critical values for the contact radius a_c/b, the indentation depth $d_c/\left(b\tan\theta\right)$, and the normal force $-F_c/\left(4E^*b^2\tan\theta\right)$ as a function of β for the indentation by a cone with a rounded tip during force-control. Note the negative sign of the normalized force

3.5.12 The Paraboloid with Parabolic Cap

Also, for the paraboloid with a spherical cap whose radius is greater than the radius of curvature of the parabolic base body, Chap. 2 presented the complete solution of the non-adhesive, frictionless, normal contact problem (see Sect. 2.5.13). Thus, from (3.18) and (3.20), the following solution which was first found by Maugis and Barquins (1983) results in the adhesive, frictionless, normal contact:

$$d(a) = \frac{a^2}{R_1} + \frac{a}{R^*}\sqrt{a^2 - b^2} - \sqrt{\frac{2\pi a \Delta\gamma}{E^*}},$$

$$F_N(a) = \frac{2E^*}{3}\left[\frac{2a^3}{R_1} + \frac{1}{R^*}\left(2a^2 + b^2\right)\sqrt{a^2 - b^2}\right] - \sqrt{8\pi a^3 E^* \Delta\gamma},$$

$$\sigma_{zz}(r;a) = \sigma_{zz,\text{n.a.}}(r;a) + \sqrt{\frac{2E^*\Delta\gamma}{\pi a}}\frac{a}{\sqrt{a^2-r^2}}, \quad r \le a,$$
$$w(r;a) = w_{\text{n.a.}}(r;a) - \sqrt{\frac{8a\Delta\gamma}{\pi E^*}}\arcsin\left(\frac{a}{r}\right), \quad r > a. \tag{3.114}$$

Here, b denotes the radius at which the cap goes over into the base body; R^* is an effective radius that can be determined from R_1 and R_2 (see Fig. 3.22):

$$R^* := \frac{R_1 R_2}{R_1 - R_2}. \tag{3.115}$$

The stresses and displacements indicated by the index "n.a." correspond to the solutions of the non-adhesive problem (see Sect. 2.5.13):

$$\sigma_{zz,\text{n.a.}}(r;a) = -\frac{E^*}{\pi}\begin{cases} \dfrac{2\sqrt{a^2-r^2}}{R_1} + \displaystyle\int_b^a \frac{(2x^2-b^2)\,\mathrm{d}x}{R^*\sqrt{x^2-b^2}\sqrt{x^2-r^2}}, & r \le b, \\ \dfrac{2\sqrt{a^2-r^2}}{R_1} + \displaystyle\int_r^a \frac{(2x^2-b^2)\,\mathrm{d}x}{R^*\sqrt{x^2-b^2}\sqrt{x^2-r^2}}, & b < r \le a, \end{cases}$$
$$w_{\text{n.a.}}(r;a) = w_{\text{n.a.},P}(r;a;R=R_1) + w_{\text{n.a.},PS}(r;a;R=R^*), \quad r > a, \tag{3.116}$$

with the corresponding solutions for the paraboloid and the truncated paraboloid:

$$w_{\text{n.a.},P}(r;a;R_1) := \frac{a^2}{\pi R_1}\left[\left(2-\frac{r^2}{a^2}\right)\arcsin\left(\frac{a}{r}\right) + \frac{\sqrt{r^2-a^2}}{a}\right],$$
$$\begin{aligned} w_{\text{n.a.},PS}(r;a;R^*) :=& \frac{2a}{\pi R^*}\sqrt{a^2-b^2}\arcsin\left(\frac{a}{r}\right) \\ &- \frac{1}{\pi R^*}\left[(r^2-b^2)\arcsin\left(\frac{\sqrt{a^2-b^2}}{\sqrt{r^2-b^2}}\right)\right. \\ &\left. - \sqrt{a^2-b^2}\sqrt{r^2-a^2}\right]. \end{aligned} \tag{3.117}$$

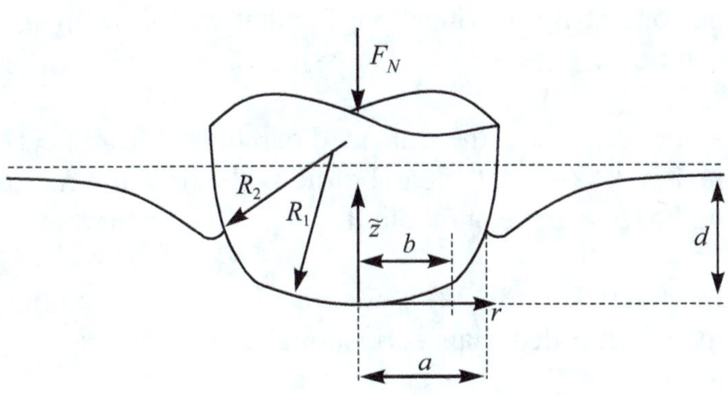

Fig. 3.22 Adhesive normal contact between a rigid paraboloid with parabolic cap and an elastic half-space

All relations in (3.114) to (3.117) are valid only for $a \geq b$. If $a < b$ contact is made only with a parabolic indenter of radius R_1. The indentation depth in the case without adhesion is thus completely given by:

$$d_{\text{n.a.}}(a) = \begin{cases} \dfrac{a^2}{R_1}, & a \leq b, \\ \dfrac{a^2}{R_1} + \dfrac{a}{R^*}\sqrt{a^2 - b^2}, & a > b, \end{cases} = \begin{cases} \dfrac{a^2}{R_1}, & a \leq b, \\ \dfrac{a^2}{R_1}\left(1 + \dfrac{R_1}{R^*}\sin\varphi_0\right), & a > b, \end{cases} \tag{3.118}$$

where we introduced the angle:

$$\varphi_0 := \arccos\left(\frac{b}{a}\right). \tag{3.119}$$

The critical configuration in which the adhesive normal contact loses its stability results from the solution to the transcendental equation:

$$\frac{1}{\cos\varphi_{0,c}}\left(2 + \frac{R_1}{R^*}\frac{1 + \sin^2\varphi_{0,c}}{\sin\varphi_{0,c}}\right) - \beta\sqrt{\cos\varphi_{0,c}} = 0, \tag{3.120}$$

with

$$\beta := \frac{\xi R_1}{b}\sqrt{\frac{\pi\Delta\gamma}{2E^*b}}. \tag{3.121}$$

The corresponding values of the indentation depth and the normal force are determined by:

$$\begin{aligned} d_c &= \frac{a_c^2}{R_1}\left[1 - \frac{4}{\xi} + \frac{R_1}{R^*}\sin\varphi_{0,c}\left(1 - \frac{2}{\xi} - \frac{2}{\xi\sin^2\varphi_{0,c}}\right)\right], \\ F_c &= \frac{4E^*a_c^3}{3R_1}\left\{1 - \frac{6}{\xi} + \frac{R_1}{2R^*}\sin\varphi_{0,c}\left[3 - \sin^2\varphi_{0,c}\right.\right. \\ &\qquad \left.\left. - \frac{6}{\xi}\left(1 + \frac{1}{\sin^2\varphi_{0,c}}\right)\right]\right\}, \end{aligned} \tag{3.122}$$

and one easily convinces oneself that the following limiting cases emerge from these solutions:

- For $R_1 \to \infty$, the truncated paraboloid from Sect. 3.5.9
- For $R^* \to \infty$, the paraboloid with curvature radius R_1 from Sect. 3.5.3
- For $b = 0$ the paraboloid with radius of curvature R_2

In (3.120) and (3.122), ξ (i.e., the boundary condition) has to be defined: displacement-controlled trials corresponds to the value $\xi = 1$, and force-controlled ones to $\xi = 3$.

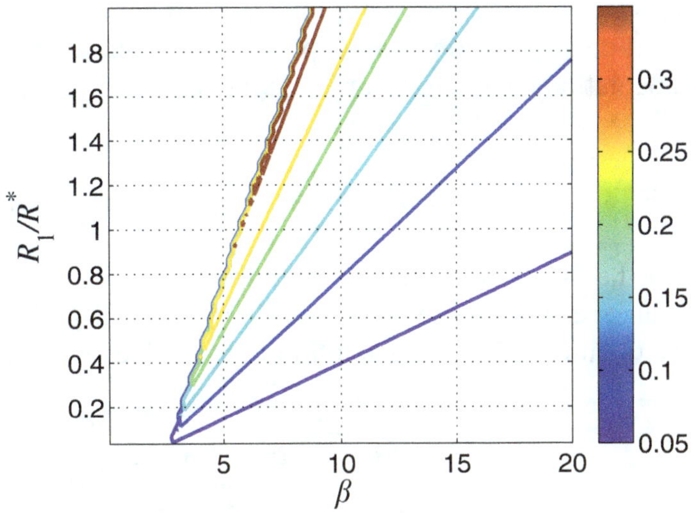

Fig. 3.23 First solution for the critical angle $\varphi_{0,c}$, in which the contact becomes unstable, depending on the parameters R_1/R_1 and β (see (3.121)), for adhesive indentation by a paraboloid with parabolic cap

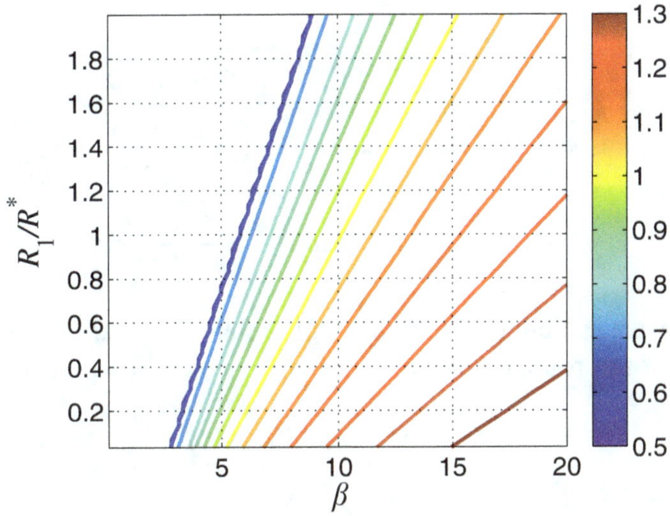

Fig. 3.24 Second solution for the critical angle $\varphi_{0,c}$ in which the contact becomes unstable, depending on the parameters R_1/R_1 and β, for adhesive indentation by a paraboloid with parabolic cap

In (3.120), two dimensionless parameters occur; R_1/R^* and β. The equation has, with one exception, either no solution or two solutions. These two solutions are shown in Figs. 3.23 and 3.24. It can be seen that there are only solutions below the line:

$$\beta > \beta_0\left(\frac{R_1}{R^*}\right) \approx 2.2 + 3.2\frac{R_1}{R^*}. \tag{3.123}$$

For smaller values of the surface energy, and thus smaller values of β, the critical contact radius is smaller than b and the critical state corresponds to that of the complete paraboloid with the radius of curvature R_1, for which the results can be found in Sect. 3.5.3. From Sect. 3.5.9 it can be deduced that:

$$\beta_0\left(\frac{R_1}{R^*} \to \infty\right) \approx 3.1\frac{R_1}{R^*}. \tag{3.124}$$

This agrees relatively well with the estimate (3.123). For $R_1/R^* = 0$, the first solution $\varphi_{0,c1}$ becomes an unphysical artifact, leaving the one critical state for the indentation by a purely parabolic indenter.

3.5.13 The Cylindrical Flat Punch with a Rounded Edge

For the flat cylindrical punch with a round edge, the solution of the non-adhesive Boussinesq problem was derived in Chap. 2 (Sect. 2.5.14). From (3.18) and (3.20), we obtain the following solution for the adhesive problem (see Fig. 3.25):

$$
\begin{aligned}
d(a) &= \frac{a}{R}\left[\sqrt{a^2-b^2} - b\arccos\left(\frac{b}{a}\right)\right] - \sqrt{\frac{2\pi a\Delta\gamma}{E^*}},\\
F_N(a) &= \frac{E^*}{3R}\left[\sqrt{a^2-b^2}\,(4a^2-b^2) - 3ba^2\arccos\left(\frac{b}{a}\right)\right]\\
&\quad - \sqrt{8\pi a^3 E^*\Delta\gamma},\\
\sigma_{zz}(r;a) &= \sigma_{zz,\text{n.a.}}(r;a) + \sqrt{\frac{2E^*\Delta\gamma}{\pi a}}\frac{a}{\sqrt{a^2-r^2}}, \quad r \le a,\\
w(r;a) &= w_{\text{n.a.}}(r;a) - \sqrt{\frac{8a\Delta\gamma}{\pi E^*}}\arcsin\left(\frac{a}{r}\right), \quad r > a.
\end{aligned} \tag{3.125}
$$

Here, b denotes the radius of the flat base of the punch and R the radius of curvature of the rounded corners. The stresses and displacements are indicated by the index

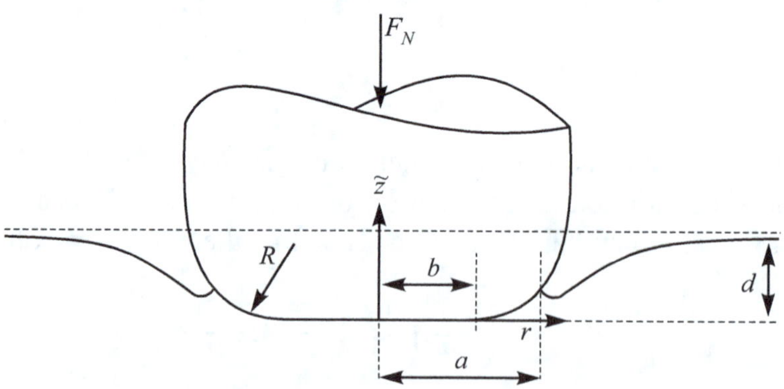

Fig. 3.25 Adhesive normal contact between a rigid cylindrical punch with round corners and an elastic half-space

"n.a." for the problem without adhesion are (see Sect. 2.5.14):

$$\sigma_{zz,\text{n.a.}}(r;a) = -\frac{E^*}{\pi R}\begin{cases}\displaystyle\int_b^a \left(2\sqrt{x^2-b^2} - b\arccos\left(\frac{b}{x}\right)\right)\frac{\mathrm{d}x}{\sqrt{x^2-r^2}}, & r \le b,\\ \displaystyle\int_r^a \left(2\sqrt{x^2-b^2} - b\arccos\left(\frac{b}{x}\right)\right)\frac{\mathrm{d}x}{\sqrt{x^2-r^2}}, & b < r \le a,\end{cases}$$

$$w_{\text{n.a.}}(r;a) = \frac{2d_{\text{n.a.}}(a)}{\pi}\arcsin\left(\frac{a}{r}\right) - \frac{2}{\pi}\left[\int_b^a \frac{x}{R}\left(\sqrt{x^2-b^2} - b\arccos\left(\frac{b}{x}\right)\right)\frac{\mathrm{d}x}{\sqrt{r^2-x^2}}\right],$$
$$r > a, \tag{3.126}$$

with the non-adhesive relationship between indentation depth and contact radius being:

$$d_{\text{n.a.}}(a) = \frac{a}{R}\left\{\sqrt{a^2-b^2} - b\arccos\left(\frac{b}{a}\right)\right\} = \frac{a^2}{R}\left(\sin\varphi_0 - \varphi_0\cos\varphi_0\right), \tag{3.127}$$

with the angle:

$$\varphi_0 := \arccos\left(\frac{b}{a}\right). \tag{3.128}$$

In contrast to Sects. 3.5.11 and 3.5.12, here a can never be smaller than b. The critical contact radius is given as a solution of:

$$2\tan\varphi_{0,c} - \varphi_{0,c} - \beta\sqrt{\cos\varphi_{0,c}} = 0, \tag{3.129}$$

with:

$$\beta := \frac{\xi R}{b}\sqrt{\frac{\pi\Delta\gamma}{2E^*b}}. \tag{3.130}$$

The corresponding values of the indentation depth and the normal force are determined by:

$$d_c = \frac{a_c^2}{R}\left[\sin\varphi_{0,c}\left(1-\frac{4}{\xi}\right) - \varphi_{0,c}\cos\varphi_{0,c}\left(1-\frac{2}{\xi}\right)\right],$$
$$F_c = \frac{4E^*a_c^3}{3R}\left[\sin\varphi_{0,c}\left(1-\frac{6}{\xi}-\frac{1}{4}\cos^2\varphi_{0,c}\right) - \frac{3}{4}\varphi_{0,c}\cos\varphi_{0,c}\left(1-\frac{4}{\xi}\right)\right]. \tag{3.131}$$

For $b = 0$ the solution for the parabolic indenter from Sect. 3.5.3 is reproduced. ξ is a parameter that, depending on the nature of the boundary condition, can take the values one (displacement-control) and three (force-control).

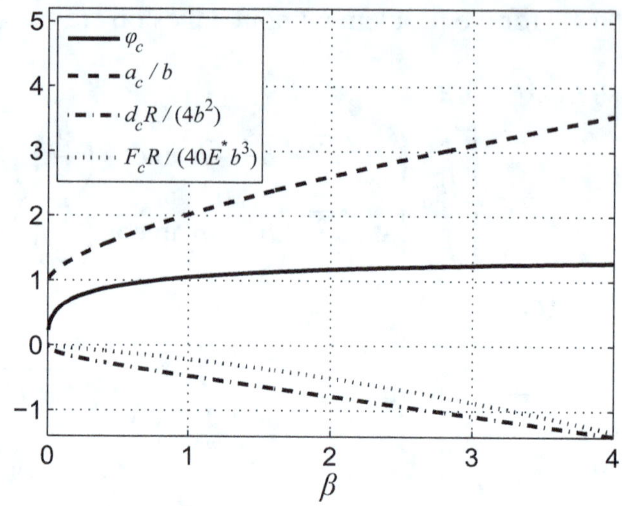

Fig. 3.26 Normalized critical values of the angle φ_0 the contact radius, the indentation depth, and normal force as a function of the parameter β (see (3.130)) for adhesive indentation by a cylindrical flat punch with rounded corners

Interestingly, (3.129) has only one solution for each positive β. This and the associated normalized values of a_c, d_c, and F_c are shown in Fig. 3.26. For $\beta \to 0$ the critical values of the indentation depth and the "adhesion force" disappear, whereas the critical contact radius does not disappear but strives towards its smallest possible value, b. This is due to the fact that the rounded corners, which are continuously differentiable (in contrast to the truncated paraboloid), pass into the flat base parallel to the half-space.

3.5.14 The Paraboloid with Small Periodic Roughness (Complete Contact)

It is established that the adhesive properties of a surface are greatly affected by the surface roughness. In general, increasing roughness sees a reduction in the effective surface energy and, therefore, the adhesive forces. However, there exists much proof in existing literature, such as that by Briggs and Briscoe (1977) demonstrating that the opposite can also be the case. A theoretical yet experimentally validated approach explaining this phenomenon stems from Guduru (2007). With the aid of both the classical approach of minimizing the potential energy by Johnson et al. (1971) as well as the idea of elastic energy release rates borrowed from linear-elastic fracture mechanics, he examined the adhesive normal indentation of a parabolic indenter with periodic roughness. It can be represented by the profile:

$$f(r) = \frac{r^2}{2R} + h\left(1 - \cos\left(\frac{2\pi}{\lambda}r\right)\right), \tag{3.132}$$

with the amplitude h and wavelength λ of the roughness, and the radius R of the paraboloid. For his solution, Guduru assumed a simply connected contact area requiring a monotonically increasing indenter profile. This poses a limitation for the roughness parameters:

$$f'(r) \geq 0 \Rightarrow \frac{1}{\hat{h}} := \frac{\lambda^2}{hR} \geq 4\pi^2 \sup\left[-\frac{\sin(x)}{x}\right] \approx 8.576. \tag{3.133}$$

Here, sup [·] denotes the global maximum of a function. However, a sufficiently great normal force can overcome this to generate a connected contact area, even for profiles that violate this condition. Taking into account the solution of the non-adhesive contact problem from Chap. 2 (Sect. 2.5.17) and (3.18) and (3.20), the solution for the adhesive problem is then as follows:

$$\begin{aligned}
d(a) &= \frac{a^2}{R} + \frac{\pi^2 h a}{\lambda}\mathrm{H}_0\left(\frac{2\pi}{\lambda}a\right) - \sqrt{\frac{2\pi a\Delta\gamma}{E^*}},\\
F_N(a) &= \frac{4E^*a^3}{3R} + E^*\pi a h\left[\frac{2\pi}{\lambda}a\mathrm{H}_0\left(\frac{2\pi}{\lambda}a\right) - \mathrm{H}_1\left(\frac{2\pi}{\lambda}a\right)\right]\\
&\quad - \sqrt{8\pi a^3 E^*\Delta\gamma},\\
\sigma_{zz}(r;a) &= \sigma_{zz,\text{n.a.}}(r,a) + \sqrt{\frac{2E^*\Delta\gamma}{\pi a}}\frac{a}{\sqrt{a^2-r^2}}, \quad r \leq a,\\
w(r;a) &= w_{\text{n.a.}}(r,a) - \sqrt{\frac{8a\Delta\gamma}{\pi E^*}}\arcsin\left(\frac{a}{r}\right), \quad r > a.
\end{aligned} \tag{3.134}$$

Here, $H_n(\cdot)$ denotes the n-th order Struve function. For a short explanation of its properties, the reader can be referred to the solution of the non-adhesive problem detailed in Sect. 2.5.17. The stresses and displacements for the non-adhesive case can also be referenced in Sect. 2.5.17 and are repeated below:

$$\begin{aligned}
&\sigma_{zz,\text{n.a.}}(r;a) =\\
&-\frac{E^*}{\pi}\left\{\frac{2\sqrt{a^2-r^2}}{R} + \frac{\pi^2 h}{\lambda}\int_r^a\left[\mathrm{H}_0\left(\frac{2\pi}{\lambda}x\right) + \frac{2\pi}{\lambda}x\mathrm{H}_{-1}\left(\frac{2\pi}{\lambda}x\right)\right]\frac{\mathrm{d}x}{\sqrt{x^2-r^2}}\right\},\\
&\quad r \leq a,\\
&w_{\text{n.a.}}(r;a) = \frac{2d_{\text{n.a.}}(a)}{\pi}\arcsin\left(\frac{a}{r}\right) - \frac{1}{\pi R}\left[r^2\arcsin\left(\frac{a}{r}\right) - a\sqrt{r^2-a^2}\right]\\
&\quad - \frac{2\pi}{\lambda}h\int_0^a x\mathrm{H}_0\left(\frac{2\pi}{\lambda}x\right)\frac{\mathrm{d}x}{\sqrt{r^2-x^2}}, \quad r > a,
\end{aligned} \tag{3.135}$$

with

$$d_{\text{n.a.}}(a) = \frac{a^2}{R} + \frac{\pi^2 h a}{\lambda}\mathrm{H}_0\left(\frac{2\pi}{\lambda}a\right). \tag{3.136}$$

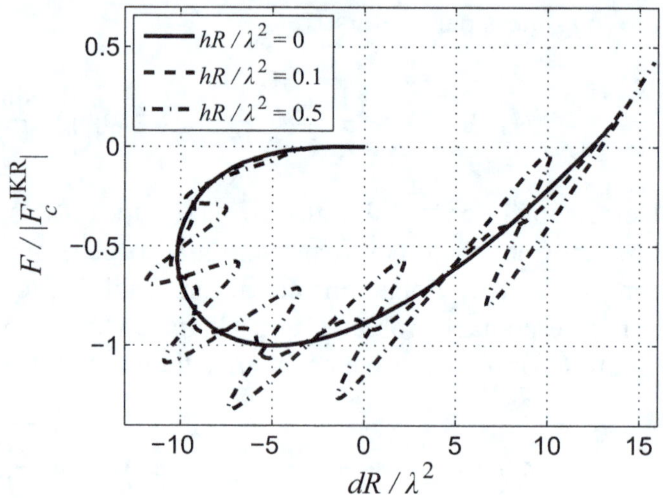

Fig. 3.27 Relationship between the normalized normal force and the normalized indentation depth for the adhesive contact with a paraboloid with periodic roughness for different normalized values of roughness

By introducing the dimensionless quantities

$$\hat{d} := \frac{dR}{\lambda^2}, \quad \hat{a} := \frac{a}{\lambda}, \quad \hat{h} := \frac{hR}{\lambda^2}, \quad \hat{F} := \frac{2F_N}{3\pi R \Delta\gamma}, \quad \overline{\Delta\gamma} := \frac{2\pi\Delta\gamma R^2}{E^*\lambda^3}, \tag{3.137}$$

the relationships (3.134₁), and (3.134₂), can be rewritten as:

$$\hat{d} = \hat{a}^2 + \pi^2 \hat{h}\hat{a}\mathrm{H}_0(2\pi\hat{a}) - \sqrt{\overline{\Delta\gamma}\hat{a}},$$
$$\hat{F} = \frac{16}{9\overline{\Delta\gamma}}\hat{a}^3 + \frac{4\pi}{3}\frac{\hat{h}\hat{a}}{\overline{\Delta\gamma}}\left[2\pi\hat{a}\mathrm{H}_0(2\pi\hat{a}) - \mathrm{H}_1(2\pi\hat{a})\right] - \frac{8}{3}\sqrt{\frac{\hat{a}^3}{\overline{\Delta\gamma}}}. \tag{3.138}$$

The implicitly defined relationship $\hat{F} = \hat{F}(\hat{d})$ is partially graphically represented in Fig. 3.27.

The wavelength is kept constant for the three depicted cases: the JKR case without roughness $h = 0$, one curve with medium roughness, and one with very pronounced roughness. We discover that the force oscillates with increasing magnitude. On the one hand, we see significantly greater extremes for the adhesive forces (during force-control) and indentation depth (during displacement-control) being achieved than without roughness (lowest point of the curve with $\hat{h} = 0.5$ lies at $\hat{F} = -1.3$, roughly 30% lower than the minimum of the JKR curve). On the other hand, constantly switching between the stable and unstable regime causes the indenter to bounce, which is a process that dissipates energy.

3.6 Adhesion According to Bradley

Let us consider a rigid paraboloid $\tilde{z} = r^2/(2R)$ in contact with a rigid plane. Further, we assume that the adhesive stress only depends on the distance between the surface: $\sigma = \sigma(s)$. The separation energy is then:

$$\Delta\gamma = \int_0^\infty \sigma(s)\mathrm{d}s. \tag{3.139}$$

For the adhesive force between the spheres we obtain:

$$F_A = \int_0^\infty 2\pi r \sigma[s(r)]\mathrm{d}r = \int_0^\infty 2\pi r \sigma\left(\frac{r^2}{2R}\right)\mathrm{d}r. \tag{3.140}$$

By substituting $s = r^2/(2R)$, $r\mathrm{d}r = R\mathrm{d}s$ the equation is transformed into the following form:

$$F_A = 2\pi R \int_0^\infty \sigma(s)\mathrm{d}s = 2\pi R\Delta\gamma. \tag{3.141}$$

This result was derived by Bradley (1932), specifically for the van der Waals interaction, while in reality it is independent of the type of interaction potential.

3.7 Adhesion According to Derjaguin, Muller, and Toporov

Derjaguin, Muller, and Toporov (1975) examined the adhesive normal contact between a paraboloid of radius R and an elastic half-space. They assumed that the stresses within the contact area and the deformation outside the contact area matched those found in the non-adhesive Hertzian solution. Of course, this implied that the body forms an ideal sphere in the last moment of the contact. As a result, the pull-off force in this state is identical to the force calculated by Bradley:

$$F_A = 2\pi R\Delta\gamma. \tag{3.142}$$

3.8 Adhesion According to Maugis

As explained in the introduction to this chapter, the adhesive forces between two electrically neutral bodies decrease rapidly with increasing distance. Therefore, integrating with respect to distance (i.e., the work of adhesion) returns a definite value. Adhesive forces that are of much shorter range than any other characteristic length of the problem have a negligible impact and are assumed to be zero. In this case—corresponding to the JKR approximation—the work of adhesion is the sole determinant of the adhesive behavior. The adhesive forces attain distinct significance once the range of the adhesive forces is of the order of the smallest characteristic length of the contact problem. The adhesive behavior is now determined not only by the work of adhesion but also independently by both the intensity *and* range of the adhesive forces. In examining these influences in a qualitative fashion, it is sensible to begin with the easiest model of finitely ranged adhesive forces: by assuming that the adhesive stresses (adhesive force divided by area) are constant up to a certain distance and zero past that point. This approximation was introduced by Dugdale (1960) to analyze crack problems and has since seen heavy usage due to its simplicity. The theory of adhesion based on the aforementioned adhesive force was developed by Maugis (1992). In the following section, the theory of Maugis will be derived via the MDR.

3.8.1 General Solution for the Adhesive Contact of Axisymmetric Bodies in Dugdale Approximation

We will consider the adhesive contact between a rigid rotationally symmetric profile with the shape $f(r)$ and an elastic half-space. The solution is found via MDR and (except for the interpretation of the MDR model) it is identical to the original solution by Maugis. Executing the MDR algorithm, we define an effective profile $g(x)$ (see (2.6)):

$$g(x) = |x| \int_0^{|x|} \frac{f'(r)}{\sqrt{x^2 - r^2}} \mathrm{d}r. \tag{3.143}$$

This will play a central role in the following solution steps. This profile is brought into contact with a one-dimensional Winkler foundation, defined by (2.5). In contrast to the non-adhesive contact, there now exists effective adhesive forces acting between the indenter and the Winkler foundation.

While pressure is the relevant quantity of load in three-dimensional space, the one-dimensional MDR representation only allows the definition of the linear force density (distributed load):

$$q_z(x) = \frac{\Delta F_N(x)}{\Delta x} = E^* w_{1D}(x). \tag{3.144}$$

Between these two quantities, the following transformations are valid:

$$p(r) = -\frac{1}{\pi} \int_r^{\infty} \frac{q_z{}'(x)}{\sqrt{x^2 - r^2}} \mathrm{d}x,$$

$$q_z(x) = 2 \int_x^{\infty} \frac{r p(r)}{\sqrt{r^2 - x^2}} \mathrm{d}r. \tag{3.145}$$

These are described in detail in Chap. 2 (Sect. 2.3) and in the appendix (Chap. 11). The displacements of the surface points in the original three-dimensional problem $w(r)$ and in the one-dimensional MDR representation $w_{1D}(x)$ are linked via the usual MDR transformation:

$$w(r) = \frac{2}{\pi} \int_0^r \frac{w_{1D}(x)}{\sqrt{r^2 - x^2}} \mathrm{d}x. \tag{3.146}$$

Following Maugis's (1992) lead, we will examine the case of a constant adhesive stress with finite range, i.e., we assume that the adhesive pressure in the original three-dimensional system remains constant and equal to the magnitude of σ_0 up to the distance h, and vanishing thereafter:

$$p_{\mathrm{adh}}(r) = \begin{cases} -\sigma_0, & \text{for } f(r) - d + w(r) \le h, \\ 0, & \text{for } f(r) - d + w(r) > h. \end{cases} \tag{3.147}$$

Using the transformations (3.145) we can determine the corresponding "adhesive distributed load" in the one-dimensional MDR model:

$$q_{\mathrm{adh},z}(x) = \begin{cases} -2 \int_x^b \frac{r\sigma_0}{\sqrt{r^2 - x^2}} \mathrm{d}r = -2\sigma_0 \sqrt{b^2 - x^2}, & \text{for } |x| \le b, \\ 0, & \text{for } |x| > b, \end{cases} \tag{3.148}$$

where b denotes the outer radius of the area upon which the adhesive pressure acts (i.e., where the distance between the surfaces does not exceed h). The radius b depends on the deformation of the surfaces and is calculated as part of the solution of the contact problem. The geometry of the contact and the notation is graphically represented in Fig. 3.28.

Within the contact area, the displacement $w_{1D}(x)$ of the 1D model is determined by the shape of the indenter and outside the contact area (but still within the adhesive zone) by the adhesive forces:

$$w_{1D}(x) = \begin{cases} d - g(x), & x \le a, \\ -\frac{2\sigma_0}{E^*} \sqrt{b^2 - x^2}, & a < x \le b, \\ 0, & x > b. \end{cases} \tag{3.149}$$

The radius of the adhesive zone derives from the condition that the gap $w(b) - d + f(b)$ between the indenter and the elastic body must correspond to the range h of the adhesive stresses:

$$w(b) - d + f(b) = h. \tag{3.150}$$

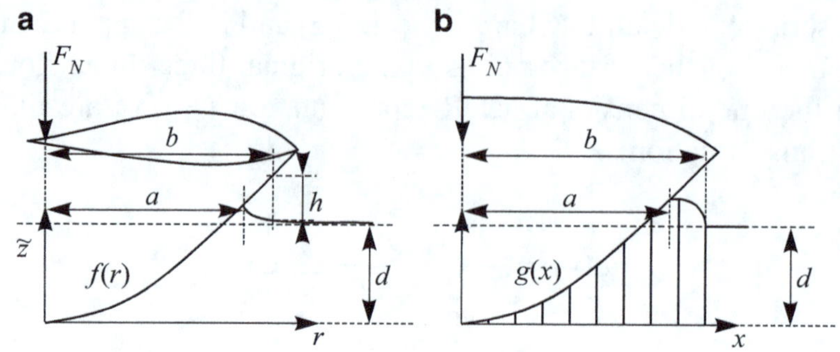

Fig. 3.28 Adhesive contact according to the Dugdale–Maugis model: subfigure (a) represents the real contact with the three-dimensional half-space, and subfigure (b) the contact with the one-dimensional Winkler foundation. In three dimensions the constant attractive stress σ_0 acts wherever the distance between the surfaces is less than h. At greater distances the interactions cease. The radius of the adhesive zone exceeds the contact radius a. In the MDR representation the "adhesive distributed load" $q_{\text{adh}}(x)$ (3.148) takes the place of the adhesive stress

Accounting for the transformation (3.146) and (2.149), we get the displacement $w(b)$ (3.149):

$$\begin{aligned} w(b) &= \frac{2}{\pi}\int_0^b \frac{w_{1D}(x)}{\sqrt{b^2-x^2}}\mathrm{d}x \\ &= d - f(b) - \frac{2}{\pi}\int_a^b \frac{[d-g(x)]}{\sqrt{b^2-x^2}}\mathrm{d}x - \frac{4\sigma_0}{\pi E^*}(b-a). \end{aligned} \tag{3.151}$$

Inserting this into (3.150) then yields the equation for the radius b:

$$\frac{2}{\pi}\int_a^b \frac{[g(x)-d]}{\sqrt{b^2-x^2}}\mathrm{d}x - \frac{4\sigma_0}{\pi E^*}(b-a) = h. \tag{3.152}$$

The force $\Delta F_{\text{cont},z}(x)$ pressing the spring at coordinate x onto the effective profile $g(x)$ consists of the elastic spring force and the adhesive force:

$$\begin{aligned} \Delta F_{\text{cont},z}(x) &= \Delta x\,[q_z(x) - q_{\text{adh},z}(x)] \\ &= \Delta x\,\{E^*[d-g(x)] - q_{\text{adh},z}(x)\}. \end{aligned} \tag{3.153}$$

The contact radius a is obtained from the condition that this contact force disappears:

$$E^*[d-g(a)] + 2\sigma_0\sqrt{b^2-a^2} = 0. \tag{3.154}$$

Furthermore, the total normal force is calculated as the integral over all springs:

$$F_N = 2E^*\int_0^b w_{1D}(x)\mathrm{d}x. \tag{3.155}$$

Substituting (3.149) gives:

$$
\begin{aligned}
F_N &= 2E^* \int_0^b w_{1D}(x)\mathrm{d}x = 2E^* \int_0^a [d - g(x)]\,\mathrm{d}x - 4\sigma_0 \int_a^b \sqrt{b^2 - x^2}\mathrm{d}x \\
&= 2E^* \int_0^a [d - g(x)]\,\mathrm{d}x - \sigma_0 \left[-2a\sqrt{b^2 - a^2} + 2b^2 \arccos\left(\frac{a}{b}\right)\right].
\end{aligned}
\tag{3.156}
$$

Equations (3.152), (3.154), and (3.156) completely solve the normal contact problem. Here, we will summarize them once more:

$$
\begin{aligned}
h &= \frac{2}{\pi} \int_a^b \frac{[g(x) - d]}{\sqrt{b^2 - x^2}}\mathrm{d}x - \frac{4\sigma_0}{\pi E^*}(b - a), \\
0 &= E^*[d - g(a)] + 2\sigma_0 \sqrt{b^2 - a^2}, \\
F_N &= 2E^* \int_0^a [d - g(x)]\mathrm{d}x \\
&\quad - \sigma_0 \left[-2a\sqrt{b^2 - a^2} + 2b^2 \arccos\left(\frac{a}{b}\right)\right].
\end{aligned}
\tag{3.157}
$$

3.8.2 The JKR Limiting Case for Arbitrary Axisymmetric Indenter Shapes

In this case $\sigma_0 \to \infty$ and $h \to 0$, where $\sigma_0 h = \Delta\gamma$ remains finite. The radii a and b are nearly identical: $\varepsilon = b - a \ll a, b$. By writing $b = a + \varepsilon$, then substituting

$$
g(x) \approx g(a) + g'(a)\,(x - a)\,,
\tag{3.158}
$$

and with subsequent integration and expansion to terms of the order $\varepsilon^{1/2}$ and ε, (3.157) can be represented in the following way:

$$
\begin{aligned}
h &\approx \frac{2}{\pi}[g(a) - d]\sqrt{\frac{2\varepsilon}{a}} - \frac{4\sigma_0}{\pi E^*}\varepsilon, \\
d &\approx g(a) - \frac{2\sigma_0 a}{E^*}\sqrt{\frac{2\varepsilon}{a}}, \\
F_N &\approx 2E^* \int_0^a [d - g(x)]\mathrm{d}x.
\end{aligned}
\tag{3.159}
$$

From the first two equations, we obtain the indenter shape independent relationship:

$$\frac{4\sigma_0\varepsilon}{\pi E^*} = h. \tag{3.160}$$

Substituting (3.160) into (3.159) gives the previously established general equations of the JKR approximation for arbitrary rotationally symmetric profiles (compare to (3.27) and (3.28)):

$$d \approx g(a) - \sqrt{\frac{2\pi a \Delta\gamma}{E^*}},$$

$$F_N \approx 2E^* \int_0^a [d - g(x)]\mathrm{d}x. \tag{3.161}$$

3.8.3 The DMT Limiting Case for an Arbitrary Rotationally Symmetric Body

Setting $a = 0$ in (3.157) yields:

$$h = \frac{2}{\pi}\int_0^b \frac{g(x)}{\sqrt{b^2 - x^2}}\mathrm{d}x - d - \frac{4\sigma_0}{\pi E^*}b,$$

$$0 = E^* d + 2\sigma_0 b,$$

$$F_N = -\pi b^2 \sigma_0. \tag{3.162}$$

Applying the familiar MDR inverse transformation (see (2.7))

$$f(b) = \frac{2}{\pi}\int_0^b \frac{g(x)}{\sqrt{b^2 - x^2}}\mathrm{d}x \tag{3.163}$$

to the first two equations provides the relationship:

$$f(b) + \frac{2\sigma_0 b}{E^*}\left(1 - \frac{2}{\pi}\right) = h. \tag{3.164}$$

The classic "DMT limiting case" stands for:

$$f(b) \gg \frac{2\sigma_0 b}{E^*}\left(1 - \frac{2}{\pi}\right). \tag{3.165}$$

It is assumed but not proven that the state $a = 0$ corresponds to a loss of stability. In the following we will provide solutions for specific indenter shapes.

3.8.4 The Paraboloid

In this case, the shape of the indenter is defined by the equation

$$f(r) = \frac{r^2}{2R}, \tag{3.166}$$

with the paraboloid radius R. The MDR transformed profile is:

$$g(x) = \frac{x^2}{R}, \tag{3.167}$$

and (3.157) determining the radii a and b, the normal force F_N, and the indentation depth d, with range h and magnitude of the adhesive stress σ_0, take on the following form:

$$\begin{aligned}
\frac{\pi}{2}h &= \left(\frac{b^2}{2R} - d\right) \arccos\left(\frac{a}{b}\right) + \frac{a}{2R}\sqrt{b^2 - a^2} - \frac{2\sigma_0}{E^*}(b - a), \\
d &= \frac{a^2}{R} - \frac{2\sigma_0}{E^*}\sqrt{b^2 - a^2}, \\
F_N &= 2E^*\left(ad - \frac{a^3}{3R}\right) - \sigma_0\left[-2a\sqrt{b^2 - a^2} + 2b^2 \arccos\left(\frac{a}{b}\right)\right].
\end{aligned} \tag{3.168}$$

We introduce the following dimensionless variables:

$$\tilde{a} := \frac{a}{a_c^{\mathrm{JKR}}}, \quad \tilde{d} := \frac{d}{|d_c^{\mathrm{JKR}}|}, \quad \tilde{F}_N := \frac{F_N}{F_A^{\mathrm{JKR}}}, \quad \kappa := \frac{b}{a} \tag{3.169}$$

where the quantities a_c^{JKR}, d_c^{JKR}, F_A^{JKR} can be referenced in Sect. 3.5.3:

$$\begin{aligned}
a_c^{\mathrm{JKR}} &= \left(\frac{9\pi R^2 \Delta\gamma}{8E^*}\right)^{1/3}, \quad d_c^{\mathrm{JKR}} = -\left(\frac{3\pi^2 R \Delta\gamma^2}{64E^{*2}}\right)^{1/3}, \\
F_A^{\mathrm{JKR}} &= \frac{3}{2}\pi R \Delta\gamma.
\end{aligned} \tag{3.170}$$

Equations (3.168) then take on the following form:

$$\begin{aligned}
\frac{2}{3}\frac{1}{\Lambda} &= \tilde{a}^2\left[\frac{\kappa^2}{2}\arccos\left(\frac{1}{\kappa}\right) - \arccos\left(\frac{1}{\kappa}\right) + \frac{1}{2}\sqrt{\kappa^2 - 1}\right] \\
&\quad + \frac{4}{3}\tilde{a}\Lambda\left[1 - \kappa + \sqrt{\kappa^2 - 1}\arccos\left(\frac{1}{\kappa}\right)\right], \\
\tilde{d} &= 3\tilde{a}^2 - 4\tilde{a}\Lambda\sqrt{\kappa^2 - 1}, \\
\tilde{F}_N &= \tilde{a}^3 - \tilde{a}^2\Lambda\left(\kappa^2 \arccos\left(\frac{1}{\kappa}\right) + \sqrt{\kappa^2 - 1}\right),
\end{aligned} \tag{3.171}$$

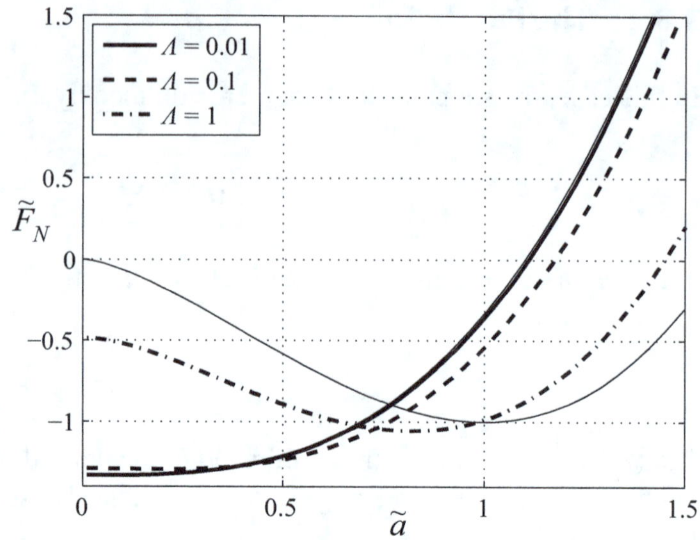

Fig. 3.29 Normalized normal force as a function of the normalized contact radius for the Maugis-adhesive contact of elastic spheres for different values of the Tabor parameter Λ. Thin solid lines denote the JKR and DMT limits

with

$$\Lambda := \left(\frac{3R\sigma_0{}^2}{\pi h E^{*2}}\right)^{1/3}. \tag{3.172}$$

In Figs. 3.29 and 3.30 the dimensionless normal force as a function of the dimensionless contact radius or indentation depth (both relations are implicitly defined by (3.171)) is shown for different values of the Tabor parameter. It is apparent that the curves for small and large values of the Tabor parameter tend to be the respective DMT or JKR limit, denoted in the figures by thin solid lines. Interestingly the convergence towards the JKR limit seems to be more prone in the force-indentation depth-diagram, whereas the convergence towards the DMT limit appears to be "faster" (i.e., achieved for less extreme values of the Tabor parameter) in the force-contact radius-diagram. Also note that, whereas in the compressive branch of the force-indentation-curve there is no significant difference to the JKR limit already for $\Lambda = 1$; the respective tensile branch still exhibits significant deviations from the JKR limit. Hence the adhesion range plays an important role for the stability (and, therefore, the hysteresis) of the contact even for large values of the Tabor parameter (for which most of the force-indentation-curve is practically the same as in the JKR limit), as was pointed out by Wu (2010) and Ciavarella et al. (2017).

For the limiting case of disappearing adhesion, (3.168) resolves to the Hertzian solution:

$$\begin{aligned} d &= \frac{a^2}{R}, \\ F_N &= 2E^*\left(ad - \frac{a^3}{3R}\right) = \frac{4}{3}E^*\frac{a^3}{R}. \end{aligned} \tag{3.173}$$

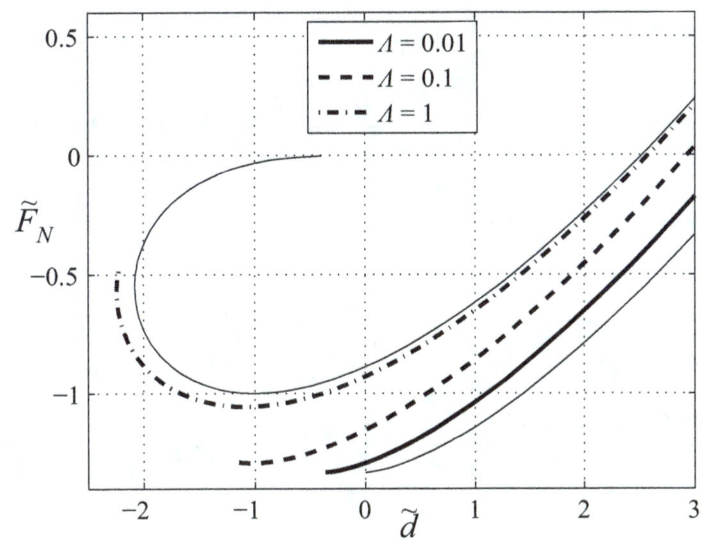

Fig. 3.30 Normalized normal force as a function of the normalized indentation depth for the Maugis-adhesive contact of elastic spheres for different values of the Tabor parameter Λ. Thin solid lines denote the JKR and DMT limits

The JKR Limiting Case

With the aforementioned transition for $\sigma_0 \to \infty$ and $h \to 0$, where $\sigma_0 h = \Delta\gamma$, (3.168) yield:

$$h \approx \frac{2}{\pi}\left(\frac{a^2}{R} - d\right)\sqrt{\frac{2\varepsilon}{a}} - \frac{4\sigma_0\varepsilon}{\pi E^*},$$

$$d \approx \frac{a^2}{R} - \frac{2\sigma_0 a}{E^*}\sqrt{\frac{2\varepsilon}{a}},$$

$$F_N \approx \frac{4}{3}E^*\frac{a^3}{R} - 4\sigma_0 a^2\sqrt{\frac{2\varepsilon}{a}}. \tag{3.174}$$

With (3.160), and taking into account $\Delta\gamma = \sigma_0 h$, we obtain from (3.174) the JKR solution previously detailed in Sect. 3.5.3:

$$d \approx \frac{a^2}{R} - \sqrt{\frac{2\pi a\Delta\gamma}{E^*}},$$

$$F_N \approx \frac{4}{3}E^*\frac{a^3}{R} - \sqrt{8\pi E^* a^3\Delta\gamma}. \tag{3.175}$$

The DMT Limiting Case

For the case $a = 0$, (3.168) yield:

$$h = \frac{b^2}{2R} - d - \frac{4\sigma_0}{\pi E^*}b,$$

$$d = -\frac{2\sigma_0 b}{E^*},$$

$$F_N = -\pi b^2\sigma_0. \tag{3.176}$$

The first two equations provide the relationship:

$$\frac{b^2}{2R} + \frac{2\sigma_0 b}{E^*}\left(1 - \frac{2}{\pi}\right) - h = 0, \tag{3.177}$$

which can be solved for b:

$$b = -\frac{2R\sigma_0}{E^*}\left(1 - \frac{2}{\pi}\right) + \sqrt{\left[\frac{2R\sigma_0}{E^*}\left(1 - \frac{2}{\pi}\right)\right]^2 + 2Rh}. \tag{3.178}$$

If the condition

$$2Rh \gg \frac{4R^2\sigma_0^2}{E^{*2}}\left(1 - \frac{2}{\pi}\right)^2 \tag{3.179}$$

is valid then $b \approx \sqrt{2Rh}$. Substituting into the third of the set of (3.176) then delivers the DMT result:

$$F_{A,\mathrm{DMT}} = -F_N = 2\pi Rh\sigma_0 = 2\pi R\Delta\gamma. \tag{3.180}$$

It is easy to see that the condition (3.179)

$$\frac{2R\Delta\gamma^2}{E^{*2}h^3}\left(1 - \frac{2}{\pi}\right)^2 \ll 1 \tag{3.181}$$

is identical to Tabor's criterion (3.3).

Asymptotic Corrections for the JKR Solution

We introduce the notation

$$\varepsilon_1 = \kappa - 1 \tag{3.182}$$

and expand (3.171) to powers of ε_1:

$$\begin{aligned}
\frac{2}{3\Lambda} &= \frac{4}{3}\tilde{a}\Lambda\varepsilon_1 + \frac{4}{3}\sqrt{2}\tilde{a}^2{\varepsilon_1}^{3/2} + \ldots, \\
\tilde{d} &= 3\tilde{a}^2 - \tilde{a}\Lambda\sqrt{2}\left(4{\varepsilon_1}^{1/2} + {\varepsilon_1}^{3/2} - \frac{1}{8}{\varepsilon_1}^{5/2} + \ldots\right), \\
\tilde{F}_N &= \tilde{a}^3 - \sqrt{2}\tilde{a}^2\Lambda\left(2{\varepsilon_1}^{1/2} + \frac{23}{6}{\varepsilon_1}^{3/2} + \frac{457}{240}{\varepsilon_1}^{5/2} + \ldots\right).
\end{aligned} \tag{3.183}$$

Utilizing second-order perturbation theory, solving the first equation for ε_1 yields:

$$\varepsilon_1^{(1)} \approx \left(\frac{1}{2\tilde{a}\Lambda^2} - \frac{1}{2\tilde{a}^{1/2}\Lambda^4} + \ldots\right) \tag{3.184}$$

Substituting into the second and third equations of (3.183) then gives:

$$\tilde{d} = 3\tilde{a}^2 - 4\tilde{a}^{1/2} + \frac{1}{\Lambda^2}\left(2\tilde{a} - \frac{5}{6}\tilde{a}^{-1/2}\right),$$
$$\tilde{F}_N = \tilde{a}^3 - 2\tilde{a}^{3/2} - \frac{1}{\Lambda^2}\left(-\tilde{a}^2 + \frac{23}{12}\tilde{a}^{1/2}\right). \tag{3.185}$$

Replacing 23 by 24 achieves a precision of approximately 5% for the perturbation term, providing the supremely simple approximation:

$$\tilde{F}_N \approx \left(\tilde{a}^3 - 2\tilde{a}^{3/2}\right)\left(1 + \frac{1}{\tilde{a}\Lambda^2}\right), \quad \Lambda \gg 1. \tag{3.186}$$

In proximity to the critical radius $\tilde{a} = 1$, we can write $\tilde{a} = 1 + \Delta\tilde{a}$ and expand to powers of Δa:

$$\tilde{d} \approx -1 + \frac{7}{6\Lambda^2} + \left(4 + \frac{29}{12\Lambda^2}\right)\Delta\tilde{a},$$
$$\tilde{F}_N \approx -1 - \frac{11}{12\Lambda^2} + \frac{25}{24\Lambda^2}\Delta\tilde{a} + \left(\frac{9}{4} + \frac{119}{96\Lambda^2}\right)\Delta\tilde{a}^2. \tag{3.187}$$

The force reaches a minimum for

$$\frac{\mathrm{d}\tilde{F}_N}{\mathrm{d}(\Delta\tilde{a})} = -\frac{9}{8\Lambda^2} - \left(\frac{9}{2} + \frac{39}{16\Lambda^2}\right)\Delta\tilde{a} = 0 \Leftrightarrow \Delta\tilde{a} = -\frac{50}{216\Lambda^2 + 119}. \tag{3.188}$$

Substituting this result into the second equation in (3.187), and subsequently expanding to $1/\Lambda$ yields:

$$\tilde{F}_A = -\tilde{F}_N \approx 1 + \frac{11}{12\Lambda^2}, \tag{3.189}$$

or (alternatively) in dimensional variables:

$$F_A = \frac{3}{2}\pi R\sigma_0 h + \frac{11}{24}\left(\frac{3\pi^5 E^{*4} R h^5}{\sigma_0}\right)^{1/3}$$
$$= \frac{3}{2}\pi R\Delta\gamma + \frac{11}{24}\left(\frac{3\pi^5 E^{*4} R h^6}{\Delta\gamma}\right)^{1/3}. \tag{3.190}$$

3.8.5 The Profile in the Form of a Power-Law

For an indenting, rotationally symmetric body with the general profile

$$f(r) = cr^n, \quad n \in \mathbb{R}^+, \tag{3.191}$$

and positive numbers n and a constant c, (3.157) take on the form:

$$\begin{aligned} h &= cb^n \left[1 - \frac{2\kappa(n)}{\pi(n+1)}\left(\frac{a}{b}\right)^{n+1} {}_2\mathrm{F}_1\left(\frac{1}{2}, \frac{n+1}{2}; \frac{n+3}{2}; \frac{a^2}{b^2}\right)\right] \\ &\quad - \frac{2d}{\pi}\arccos\left(\frac{a}{b}\right) - \frac{4\sigma_0}{\pi E^*}(b-a), \\ 0 &= E^*\left[d - c\kappa(n)a^n\right] + 2\sigma_0\sqrt{b^2-a^2}, \\ F_N &= 2E^*\left[da - \frac{c\kappa(n)}{n+1}a^{n+1}\right] \\ &\quad - \sigma_0\left[-2a\sqrt{b^2-a^2} + 2b^2\arccos\left(\frac{a}{b}\right)\right]. \end{aligned} \tag{3.192}$$

Consistent with the notation of this subsection, we denote the radius of the contact area by a and the zone radius of the adhesive interaction by $b > a$, the indentation depth by d, the normal force with F_N, the range by h, and the value of the adhesive stress by σ_0. We introduce:

$$\kappa(n) := \sqrt{\pi}\frac{\Gamma(n/2+1)}{\Gamma\left[(n+1)/2\right]} \tag{3.193}$$

and utilize the hypergeometric function:

$${}_2\mathrm{F}_1(a,b;c;z) := \sum_{n=0}^{\infty}\frac{\Gamma(a+n)\Gamma(b+n)\Gamma(c)}{\Gamma(a)\Gamma(b)\Gamma(c+n)}\frac{z^n}{n!}. \tag{3.194}$$

Naturally, setting $n = 2$ returns (3.168) of the preceding section.

The JKR limiting case of the above (3.192) was already discussed in detail in Sect. 3.5.7. As a final note, the adhesive force for this indenter shape in the DMT limiting case ($h = f(b)$, $a = 0$) is given by:

$$F_A^{\mathrm{DMT}} = \frac{\pi\Delta\gamma}{c^{2/n}}h^{\frac{2-n}{n}}, \tag{3.195}$$

with an effective surface energy of $\Delta\gamma = \sigma_0 h$. It is apparent that the parabolic body with $n = 2$ represents the sole case where this force is not explicitly dependent on the range of the adhesion.

3.9 Adhesion According to Greenwood and Johnson

We now consider the adhesive contact between a parabolic indenter $\tilde{z} = r^2/(2R)$ and an elastic half-space. The surface displacement must correspond exactly to the indenter shape within the contact area, regardless of the type of adhesive interaction. Therefore, it is a quadratic function of radius r. Greenwood and Johnson

(1998) rightly noted that this kinematic condition is met when the Hertzian pressure distribution is accompanied by an additional stress distribution. This additional distribution is, in turn, the difference of two Hertzian stress distributions with two varying indentation depths, and thus with the actual contact radius a and another arbitrary fictional contact radius $c > a$, so that:

$$\sigma_{zz}^{ad}(r) = \frac{2E^*}{\pi R}\begin{cases}\sqrt{c^2-r^2}-\sqrt{a^2-r^2}, & r < a,\\ \sqrt{c^2-r^2}, & a < r < c,\\ 0, & r > c.\end{cases} \tag{3.196}$$

A similar stress distribution was used far earlier already in the solution of the non-adhesive tangential contact of spheres to generate a constant displacement of a spherical domain (there will be more on this in the next chapter). Within the contact area, i.e., for $r \le a$, this stress distribution (3.196) causes the constant displacement

$$w(r) = \frac{1}{R}(a^2 - c^2). \tag{3.197}$$

The additional stress (3.196) and the corresponding displacement (3.197) can be modified with an arbitrary factor k for control of the indentation depth without changing the shape of the indenter. The entire stress distribution consisting of the original Hertzian distribution of radius a and also the distribution (3.196) multiplied by the factor k is given by:

$$\sigma_{zz}(r) = -(1+k)\frac{2E^*}{\pi R}\sqrt{a^2-r^2} + k\frac{2E^*}{\pi R}\sqrt{c^2-r^2}, \quad r \le a. \tag{3.198}$$

Hertzian theory provides the distance Δw between the two surfaces within the ring $a < r < c$:

$$\Delta w(r) = (1+k)\frac{a^2}{\pi R}\left[\left(\frac{r^2}{a^2}-2\right)\arccos\left(\frac{a}{r}\right) + \frac{\sqrt{r^2-a^2}}{a}\right], \quad a < r < c. \tag{3.199}$$

The stresses within the ring $a < r < c$

$$\sigma_{zz}(r) = k\frac{2E^*}{\pi R}\sqrt{c^2-r^2}, \quad a < r < c \tag{3.200}$$

then describe the adhesive interactions between the surfaces. Combined with (3.199), they define the separation energy

$$\Delta\gamma = \int \sigma \,\mathrm{d}(\Delta w) = \int_c^a \sigma_{zz}(r)\frac{\mathrm{d}(\Delta w)}{\mathrm{d}r}\mathrm{d}r = k(k+1)\frac{2E^*}{3\pi R^2}(c-a)^2(c+2a). \tag{3.201}$$

Minimizing the total energy provides equilibrium configuration.

While the purely contact mechanical aspect of Greenwood's and Johnson's theory is far simpler than that of Maugis's theory, their model is based on certain assumptions with effects that are not immediately obvious. As an example, the implicitly introduced interaction not only depends on the distance between the surfaces but also on the entire configuration of the contact.

References

Borodich, F.M., Galanov, B.A.: Molecular adhesive contact for indenters of non-ideal shapes. In: ICTAM04, Abstracts book and CD-Rom Proceedings. IPPT PAN, Warsaw (2004)

Bradley, A.I.: The cohesive force between solid surfaces and the surface energy of solids. Philos. Mag. **13**, 853–862 (1932)

Briggs, G.A.D., Briscoe, B.J.: The effect of surface topography on the adhesion of elastic solids. J. Phys. D Appl. Phys. **10**(18), 2453–2466 (1977)

Ciavarella, M., Greenwood, J.A., Barber, J.R.: Effect of Tabor parameter on hysteresis losses during adhesive contact. J. Mech. Phys. Solids **98**, 236–244 (2017)

Derjaguin, B.V., Muller, V.M., Toporov, Y.P.: Effect of contact deformations on the adhesion of particles. J. Colloid Interface Sci. **53**(2), 314–326 (1975)

Dugdale, D.S.: Yielding of steel sheets containing slits. J. Mech. Phys. Solids **8**(2), 100–104 (1960)

Gao, H., Yao, H.: Shape insensitive optimal adhesion of nanoscale fibrillar structures. Proc. Natl. Acad. Sci. U.S.A. **101**(21), 7851–7856 (2004)

Greenwood, J.A., Johnson, K.L.: An alternative to the Maugis model of adhesion between elastic spheres. J. Phys. D Appl. Phys. **31**(22), 3279–3290 (1998)

Griffith, A.A.: The Phenomena of Rupture and Flow in Solids. Philos. Trans. R. Soc. London Ser. A **221**, 163–198 (1921)

Guduru, P.R.: Detachment of a rigid solid from an elastic wavy surface: theory. J. Mech. Phys. Solids **55**(3), 445–472 (2007)

Hertz, H.: Über die Berührung fester elastischer Körper. J. Reine Angew. Math. **92**, 156–171 (1882)

Heß, M.: Über die exakte Abbildung ausgewählter dreidimensionaler Kontakte auf Systeme mit niedrigerer räumlicher Dimension (2010). Dissertation, Technische Universität Berlin

Johnson, K.L., Kendall, K., Roberts, A.D.: Surface Energy and the Contact of Elastic Solids. Proc. R. Soc. London Ser. A **324**, 301–313 (1971)

Kendall, K.: The adhesion and surface energy of elastic solids. J. Phys. D Appl. Phys. **4**(8), 1186–1195 (1971)

Maugis, D.: Adhesion of spheres: the JKR-DMT-transition using a Dugdale model. J. Colloid Interface Sci. **150**(1), 243–269 (1992)

Maugis, D., Barquins, M.: Adhesive contact of a conical punch on an elastic half-space. J. Phys. Lett. **42**(5), 95–97 (1981)

Maugis, D., Barquins, M.: Adhesive contact of sectionally smooth-ended punches on elastic half-spaces: theory and experiment. J. Phys. D Appl. Phys. **16**(10), 1843–1874 (1983)

Popov, V.L.: Surface profiles with zero and finite adhesion force and adhesion instabilities (2017). https://arXiv.org/abs/1707.07867. [cond-mat.soft]

Popov, V.L., Heß, M.: Method of dimensionality reduction in contact mechanics and friction. Springer, Heidelberg (2015). ISBN 978-3-642-53875-9

Spolenak, R., Gorb, S., Gao, H., Arzt, E.: Effects of contact shape on the scaling of biological attachments. Proc. R. Soc. London Ser. A **461**, 305–319 (2005)

Tabor, D.: Surface forces and surface interactions. J. Colloid Interface Sci. **58**(1), 2–13 (1977)

Thornton, C., Ning, Z.: A theoretical model for the stick/bounce behavior of adhesive, elastic-plastic spheres. Powder Technol. **99**(2), 154–162 (1998)

Wu, J.J.: The jump-to-contact distance in atomic force microscopy measurement. J. Adhes. **86**(11), 1071–1085 (2010)

Yao, H., Gao, H.: Optimal shapes for adhesive binding between two elastic bodies. J. Colloid Interface Sci. **298**(2), 564–572 (2006)

4 Tangential Contact

4.1 Introduction

In this chapter, we will consider contacts which are loaded both in the normal direction z (as previously, z points into the half-space and $\tilde{z}$ to out of the half-space) and in the tangential direction x. Although the load now lacks rotational symmetry due to the bias for the x-direction, each of the non-vanishing components of stress tensor or displacement in the contact plane still have an approximately axisymmetric distribution. In this sense, this case can also be considered a rotationally symmetric contact problem.

Initially, we examine the deformation of an elastic half-space under the effect of a concentrated force at a point on the surface, which we define as the origin (Fig. 4.1).

Let the force $\underline{F}$ have a sole component in the x-direction. The components of the displacement vector $\underline{u} = (u, v, w)_{\{x,y,z\}}$ at the surface ($z = 0$) are given by the following (Cerruti 1882; Landau and Lifshitz 1991):

$$\begin{aligned} u &= F_x \frac{1}{4\pi G}\left\{2(1-\nu) + \frac{2\nu x^2}{r^2}\right\}\frac{1}{r}, \\ v &= F_x \frac{1}{4\pi G}\cdot\frac{2\nu}{r^3}xy, \\ w &= F_x \frac{1}{4\pi G}\cdot\frac{(1-2\nu)}{r^2}x, \end{aligned} \tag{4.1}$$

where G stands for the shear modulus. For a stress distribution with a sole stress component $\sigma_{xz}(x, y)$ acting in the surface, the displacements are determined using

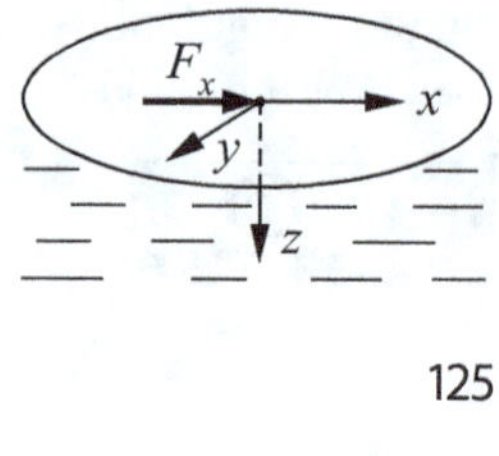

Fig. 4.1 Single tangential force acting on the surface of a half-space

V.L. Popov, M. Heß, E. Willert, *Handbook of Contact Mechanics*,
https://doi.org/10.1007/978-3-662-58709-6_4

the principle of superposition:

$$
\begin{aligned}
u(x, y) &= \frac{1}{4\pi G} \int \left\{ 2(1-\nu) + \frac{2\nu\,(x-x')^2}{r^2} \right\} \frac{1}{r} \sigma_{xz}(x', y') \mathrm{d}x' \mathrm{d}y', \\
v(x, y) &= \frac{2\nu}{4\pi G} \int \frac{1}{r^3} \left(x-x'\right)\left(y-y'\right) \sigma_{xz}(x', y') \mathrm{d}x' \mathrm{d}y', \\
w(x, y) &= \frac{(1-2\nu)}{4\pi G} \int \frac{1}{r^2} \left(x-x'\right) \sigma_{xz}(x', y') \mathrm{d}x' \mathrm{d}y', \\
r &= \sqrt{(x-x')^2 + (y-y')^2}.
\end{aligned} \tag{4.2}
$$

These equations lay the basis of all analytical and also numerical contact mechanical solutions and enable certain general deductions. For two contacting bodies under tangential load, the normal displacements of their surfaces are only then equal and opposite if the coefficients preceding the integral in the third equation of the set (4.2) are equal:

$$
\frac{1-2\nu_1}{G_1} = \frac{1-2\nu_2}{G_2}. \tag{4.3}
$$

The vertical displacements of both bodies are, in this case, "congruent", causing no additional interaction in the normal direction. We refer to these cases as decoupled normal and tangential contact problems. Bodies that meet condition (4.3) are called "elastically similar". This condition is met in two cases (among many others) of practical importance: (a) contact of bodies with the same elastic properties (e.g., wheel-rail contact), or (b) contact of a rigid body ($G_1 \to \infty$) with an incompressible one ($\nu_2 = 0.5$) (e.g., road-tire contact). If (4.3) is not fulfilled, the normal stresses then lead to relative tangential displacements of the contact partners and vice-versa. The mathematical treatment and the complete analytical solutions of the contact problems in particular are rendered far more difficult in this case. However, the solution differs only slightly from the case of elastically similar materials in many cases.

If the single stress component $\sigma_{xz}(x, y)$ is solely dependent on the polar radius r according to the law

$$
\sigma_{xz}(x, y) = \tau(x, y) = \tau_0 (1 - r^2/a^2)^{-1/2}, \tag{4.4}
$$

substituting into (4.2) with subsequent integration yields the result (Johnson 1985):

$$
u = \frac{\pi(2-\nu)}{4G} \tau_0 a = \text{const.} \tag{4.5}
$$

The tangential force is calculated to:

$$
F_x = \int_0^a \frac{2\pi r \tau_0 \mathrm{d}r}{\left(1 - r^2/a^2\right)^{1/2}} = 2\pi a^2 \tau_0. \tag{4.6}
$$

The physical implementation of the stress distribution (4.4) is more complicated than is the case for the analogous normal stress distribution. One might assume it could be achieved by tangentially displacing (4.5) a rigid punch of the radius a. However, this is only true if the resulting normal displacements also vanish. In the case of a rigid indenter, this condition is only fulfilled if the elastic half-space is incompressible. A constant displacement of the contact area is also achieved when two elastically similar half-spaces, stuck together in a circular area of radius a, are displaced relative to one another. The relative displacement of the bodies is then given by the superposition of the displacements of the form (4.5) for both bodies:

$$u^{(2)} - u^{(1)} = \pi \tau_0 a \left(\frac{2-\nu_1}{4G_1} + \frac{2-\nu_2}{4G_2} \right). \tag{4.7}$$

From this, accounting (4.6) we obtain the relationship:

$$F_x = 2G^* a (u^{(2)} - u^{(1)}), \tag{4.8}$$

with

$$\frac{1}{G^*} = \frac{2-\nu_1}{4G_1} + \frac{2-\nu_2}{4G_2}. \tag{4.9}$$

The coefficient connecting the force and the relative displacement (4.8) is the tangential contact stiffness:

$$k_x = 2G^* a. \tag{4.10}$$

Generally, when two bodies with curved surfaces are brought into normal contact and subsequently displaced in a tangential direction relative to one another, the bodies remain stuck to each other in one part of the contact area while in other areas slipping relative to one another. This is already indicated by the fact that the normal pressure vanishes at the boundary, while the tangential stress (4.4) at the boundary of a no-slip contact is singular. Therefore, the no-slip condition for a finite coefficient of friction can generally not be fulfilled in the vicinity of the contact boundary. Cattaneo (1938) and Mindlin (1949) independently solved the associated contact problem with some simplifying assumptions. These assumptions are:

- The existence of a single tangential stress component $\sigma_{xz}(r)$ in the slip plane, which only depends on the polar radius r.
- A unilateral displacement field with a displacement component only in the x-direction.
- Satisfaction of the following boundary conditions:
  - Equal displacements of both bodies in the stick zone, i.e., under the condition that the tangential stress is lower than μ times the normal stress:

$$u^{(1)}(r) = u^{(2)}(r), \quad \text{if } |\sigma_{xz}(r)| \le \mu p(r) \quad \text{(stick)}. \tag{4.11}$$

- Local compliance with Coulomb's law of friction in the slip zone.

$$\sigma_{xz}(r) = -\mu p(r)\,\mathrm{sgn}(\dot{u}_1 - \dot{u}_2), \quad \text{otherwhise (slip).} \tag{4.12}$$

Axially symmetric contacts only approximately satisfy these conditions. A stress component $\sigma_{xz}(r)$ depending on r causes displacements in both x-direction and y-direction. The displacements and the friction forces are therefore not anti-parallel, thus violating the isotropic nature of Coulomb's law of friction. Johnson (1955) was the first to point out this error in the Cattaneo–Mindlin solution. He demonstrated that the maximum deviation of the directions of the displacement and friction force is on the order of $\nu/(4 - \nu)$, and therefore lies between 0.09 for $\nu = 1/3$ and 0.14 for $\nu = 1/2$. Based on this finding, Johnson concluded that the Cattaneo–Mindlin solution provides a good approximation. Indeed, in macroscopic relationships (e.g., dependency of the tangential force on the tangential displacement) it results in an error in the order of 1%. However, local deviations (potentially important for wear, for example) may be significantly higher.

In this book we will examine tangential contacts in Cattaneo–Mindlin approximation, referring to these as "*Cattaneo–Mindlin problems*".

4.2 Cattaneo–Mindlin Problems

In the Cattaneo–Mindlin approximation, the tangential contact problem between two elastic bodies can be reduced to the contact problem of a rigid punch and an elastic half-space. Formulation of the equivalent problem of a rigid indenter and an elastic half-space requires using the previously introduced effective moduli E^* (see (2.1)) and G^* (see (4.9)), which we will list once more in this chapter as follows:

$$\begin{aligned} \frac{1}{E^*} &= \frac{1-\nu_1}{2G_1} + \frac{1-\nu_2}{2G_2}, \\ \frac{1}{G^*} &= \frac{2-\nu_1}{4G_1} + \frac{2-\nu_2}{4G_2}. \end{aligned} \tag{4.13}$$

These effective moduli uniquely define the contact properties of arbitrarily shaped bodies. For flat cylindrical punches of radius a, they directly define the normal and tangential stiffness of the contact (under the assumption of complete stick), as demonstrated in (2.21) and (4.10):

$$\begin{aligned} k_z &= 2E^* a, \\ k_x &= 2G^* a. \end{aligned} \tag{4.14}$$

We will assume that the simplest, local form of Coulomb's law of friction is valid in the contact area, which is given by (4.11) and (4.12). As mentioned in the introduction, small displacements in the y-direction also exist. These will be neglected here; the interested reader can be referred to the works of Vermeulen and Johnson (1964).

In contrast to the normal contact problem, tangential contacts exhibit *hysteresis* and *memory* properties. These properties mean that there is no universal relationship between the tangential force and the tangential displacement. Generally, at least part of the contact area in tangentially loaded contacts is a zone of local slip. The resulting energy dissipation causes hysteresis loops between the global contact quantities: force and displacement. Moreover, the tangentially loaded contact saves a part of its loading history in the form of tangential stresses; in this sense the contact can be said to possess a memory. However, this means that the state expressed by the stresses and displacements, depends on the entire previous loading history, at least for the tangential stresses and displacements. This dependency distinguishes the problems examined in this chapter from the previously discussed purely normal contacts, where the entire current contact configuration is defined by the current value of a single relevant contact quantity, e.g., the indentation depth. Therefore, the consideration of the tangential contact problem theoretically includes not only the specification of the indenter geometry and material properties, but also the complete loading or displacement history. For the sake of brevity in this chapter, we will restrict our consideration to the simplest and most technically relevant loading history, which consists of a constant normal force F_N and a subsequent application of an increasing tangential force F_x. For the consideration of more general loading histories, the reader can be referred to the pioneering publication by Mindlin and Deresiewicz (1953) and to work by Jäger (1993).

Let us consider the indentation of a rigid profile in an elastic half-space with the effective elastic properties given by the effective moduli (4.13). We will use the notation F_N for the normal force and F_x for the tangential force, a for the contact radius, and d and $u^{(0)}$ for the displacement of the rigid indenter in the normal and tangential direction, respectively. As explained earlier in this chapter, the contact area is generally composed of an inner stick zone of radius $c \leq a$ and an outer slip zone at the boundary of the contact area. The mixed boundary conditions at the surface of the elastic half-space at $z = 0$ are then as follows:

$$\begin{aligned}
w(r) &= d - f(r), && r \leq a,\\
u(r) &= u^{(0)}, && r \leq c,\\
\sigma_{xz}(r) &= \mu\sigma_{zz}(r), && c < r \leq a,\\
\sigma_{zz}(r) &= 0, && r > a,\\
\sigma_{xz}(r) &= 0, && r > a.
\end{aligned} \tag{4.15}$$

Here w and u represent the displacement of the half-space in the z and x-direction. The radii of the contact area and stick zone are unknown *a priori* and must be determined as part of the solution process.

4.3 Solution of the Tangential Contact Problem by Reducing to the Normal Contact Problem

Using a principle discovered independently by Ciavarella (1998) and Jäger (1998), the solution for the contact problem described in (4.15) can be determined if the solution of the corresponding frictionless normal contact problem is known. Therefore, we can make use of the solutions of various Boussinesq problems from the first chapters during our consideration of the Cattaneo–Mindlin problem in relation to those same profile shapes.

Let $\sigma_{zz}(r;\tilde{a})$ be the stress distribution in a normal contact of contact radius $\tilde{a}$. And let $w(r;\tilde{a})$ be the normal displacement resulting from this stress distribution. Ciavarella (1998) and Jäger (1998) were able to prove that the distribution of tangential stresses in the form of

$$\sigma_{xz}(r) = \mu \begin{cases} \sigma_{zz}(r;a) - \sigma_{zz}(r;c), & r \le c, \\ \sigma_{zz}(r;a), & c < r \le a \end{cases} \tag{4.16}$$

with the radius of the stick zone from the equation

$$G^*|u^{(0)}| = \mu E^*[d(a) - d(c)] \tag{4.17}$$

satisfy the boundary conditions (4.15), and therefore represent the solution of the corresponding tangential contact problem. The relationship between the forces and the contact radii is obtained from integration of the tangential stresses:

$$F_x = \mu[F_N(a) - F_N(c)]. \tag{4.18}$$

From the superposition of the tangential stresses (4.16), it immediately follows that the principle of superposition equally applies to the tangential displacements in the direction of the tangential force. The unknown tangential displacements outside the stick zone are given by:

$$u(r) = \frac{\mu E^*}{G^*} \begin{cases} d(a) - f(r) - w(r;c), & c < r \le a, \\ w(r;a) - w(r;c), & r > a. \end{cases} \tag{4.19}$$

Equations (4.16), (4.17), (4.18), and (4.19) provide a complete solution of the tangential contact problem via reduction to the normal contact problem.

4.4 Solution of the Tangential Contact Problem Using the MDR

As an alternative to the solution via reduction to the normal contact problem (detailed earlier in this chapter), the tangential contact problem can be solved "directly" (i.e., without knowledge of the solution of the normal contact problem) using the

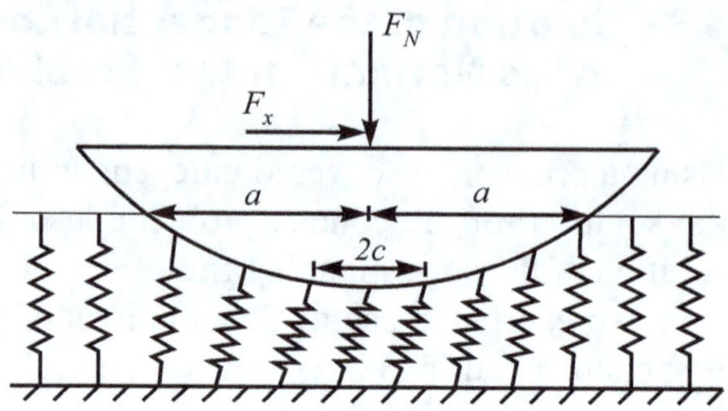

Fig. 4.2 Substitute model of the tangential contact

MDR (see Popov and Heß 2015). While ultimately the solutions do not differ for simple loading histories, this second approach can prove quite valuable for complex loading histories or numerical simulations.

Let us consider the axially symmetric indenter with the profile $\tilde{z} = f(r)$, which is initially pushed into the elastic half-space with the normal force F_N, and subsequently loaded with a tangential force F_x in x-direction. We will assume that the friction in the contact obeys Coulomb's law of friction in its simplest form, described by (4.11) and (4.12).

The application of a tangential force creates a ring-shaped slip zone, which expands inwards for an increasing force until complete slip sets in. We call the inner radius of the slip zone (also the radius of the stick zone) c.

In the MDR, this contact problem is solved as follows in this chapter (we will describe only the solution procedure. The complete derivation can be found in Chap. 11).

As in the case of the normal contact problem, we first determine the modified profile $g(x)$ using the transformation

$$g(x) = |x| \int_0^{|x|} \frac{f'(r)\mathrm{d}r}{\sqrt{x^2 - r^2}}. \tag{4.20}$$

Additionally, we define a Winkler foundation consisting of springs of normal and tangential stiffness

$$\begin{aligned} \Delta k_z &= E^* \Delta x, \\ \Delta k_x &= G^* \Delta x, \end{aligned} \tag{4.21}$$

where Δx is the distance between each spring and E^* and G^* are defined by (4.13). The calculation method involves indenting the Winkler foundation with the profile $g(x)$ under the normal force F_N, and subsequently tangentially displacing the profile by $u^{(0)}$ (see Fig. 4.2). The relationships between the indentation depth, the contact radius, and the normal force from the MDR model correspond exactly to the solution of the original problem, as explained in detail in Chap. 1 (while discussing the normal contact problem).

Each spring sticks to the indenting body and is displaced along with the body as long as the tangential force $\Delta F_x = \Delta k_x u^{(0)}$ of the particular spring is lower than $\mu \Delta F_z$. Upon reaching the maximum static friction force, the spring begins to slip, with the force remaining constant and equal to $\mu \Delta F_z$. This rule can also be expressed in an incremental form for arbitrary loading histories: for small displacements $\Delta u^{(0)}$ of the indenter, the tangential displacements of the springs $u_{1D}(x)$ in the MDR model are given by:

$$\begin{aligned} \Delta u_{1D}(x) &= \Delta u^{(0)}, \quad \text{if } |\Delta k_x u_{1D}(x)| < \mu \Delta F_z, \\ u_{1D}(x) &= \pm \frac{\mu \Delta F_z(x)}{\Delta k_x}, \quad \text{in a state of slip.} \end{aligned} \tag{4.22}$$

The sign in the last equation depends on the direction of the tangential spring displacement, if the spring were sticking. By tracking the incremental difference of the indenter position we can uniquely determine the displacements of all springs in the contact area, thus yielding the values of all tangential forces:

$$\Delta F_x = \Delta k_x u_{1D}(x) = G^* \Delta x \cdot u_{1D}(x), \tag{4.23}$$

and the linear force density (distributed load):

$$q_x(x) = \frac{\Delta F_x}{\Delta x} = G^* u_{1D}(x). \tag{4.24}$$

The distribution of the tangential stress $\tau(r)$ and the displacements $u(r)$ in the original three-dimensional contact are defined by rules which are completely analogous to (2.13) and (2.14) of the normal contact problem:

$$\begin{aligned} \sigma_{xz} &= \tau(r) = -\frac{1}{\pi} \int_r^\infty \frac{q_x{}'(x) \mathrm{d}x}{\sqrt{x^2 - r^2}}, \\ u(r) &= \frac{2}{\pi} \int_0^r \frac{u_{1D}(x) \mathrm{d}x}{\sqrt{r^2 - x^2}}. \end{aligned} \tag{4.25}$$

Equations (4.22) to (4.25) are valid for arbitrary loading histories of the contact (and also for an arbitrary superposition of time-variant normal and tangential forces). In general, these equations must be implemented in a numerical program; this is extremely easy due to the independence of each spring in the Winkler foundation.

For simple loading conditions, the general solution can also be written in an explicit form. Let us illustrate the MDR procedure using the case of an indenter which is first pressed with an initial normal force F_N to generate a contact radius a, which is determined from the equation

$$\begin{aligned} F_N &= 2E^* \int_0^a w_{1D}(x) \mathrm{d}x = 2E^* \int_0^a [d - g(x)] \mathrm{d}x \\ &= 2E^* \int_0^a [g(a) - g(x)] \mathrm{d}x. \end{aligned} \tag{4.26}$$

Subsequently, the indenter is displaced in the tangential direction. The radius of the stick zone c is determined from the condition that the absolute tangential force is equal to the coefficient of friction μ multiplied with the normal force $\Delta k_z w_{1D}(c)$:

$$G^*|u^{(0)}| = \mu E^*[d - g(c)]. \tag{4.27}$$

From (4.27) we can draw an interesting and very general conclusion. The maximum tangential displacement for which the stick zone *just barely vanishes,* i.e., the minimum displacement to see complete slip, is determined by setting $c = 0$ in (4.27) (and therefore also $g(c) = 0$):

$$u_c^{(0)} = \mu \frac{E^*}{G^*} d. \tag{4.28}$$

Thus the displacement that is achieved before complete slip sets in is solely dependent on the indentation depth (and not on the shape of the indenter).

The tangential displacement in the MDR model at a given coordinate x then equals:

$$u_{1D}(x) = \begin{cases} u^{(0)}, & \text{for } x < c, \\ \mu \frac{E^*}{G^*}[d - g(x)], & \text{for } c < x < a, \\ 0, & \text{for } x > a, \end{cases} \tag{4.29}$$

and can also be written in the following simple universal form:

$$u_{1D}(x) = \mu \frac{E^*}{G^*}[w_{1D}(x; a) - w_{1D}(x; c)], \tag{4.30}$$

which corresponds to the principle of superposition by Ciavarella (1998) and Jäger (1998). The distributed load is obtained by multiplying with G^*:

$$q_x(x) = \begin{cases} G^* u^{(0)}, & \text{for } x < c, \\ \mu E^*[d - g(x)], & \text{for } c < x < a, \\ 0, & \text{for } x > a, \end{cases} \tag{4.31}$$

or

$$q_x(x) = \mu \left[q_z(x; a) - q_z(x; c)\right], \tag{4.32}$$

where $q_z(x; a)$ and $q_z(x; c)$ represent the respective distributed load of the normal contact problem with the radius a and c. The tangential force is given by:

$$F_x = 2 \int_0^a q_x(x) \mathrm{d}x = \mu [F_N(a) - F_N(c)], \tag{4.33}$$

where $F_N(a)$ and $F_N(c)$ mean the normal force with respect to the contact radii a or c. The stress distribution is obtained by substituting (4.31) into (4.25):

$$\tau(r) = \frac{\mu E^*}{\pi} \int_c^a \frac{g'(x)\mathrm{d}x}{\sqrt{x^2 - r^2}} = \mu[p(r;a) - p(r;c)]. \tag{4.34}$$

The displacements are calculated by inserting (4.29) into (4.25), resulting in:

$$u(r) = \frac{\mu E^*}{G^*}[w(r;a) - w(r;c)], \tag{4.35}$$

or explicitly:

$$u(r) = \begin{cases} u^{(0)}, & \text{for } r < c, \\ \frac{2}{\pi}\left[u^{(0)} \arcsin\left(\frac{c}{r}\right) + \frac{\mu E^*}{G^*}\int_c^r \frac{d - g(x)}{\sqrt{r^2 - x^2}}\mathrm{d}x\right], & \text{for } c < r < a, \\ \frac{2}{\pi}\left[u^{(0)} \arcsin\left(\frac{c}{r}\right) + \frac{\mu E^*}{G^*}\int_c^a \frac{d - g(x)}{\sqrt{r^2 - x^2}}\mathrm{d}x\right], & \text{for } r > a. \end{cases} \tag{4.36}$$

Equations (4.26)–(4.36) clearly show that this contact problem is completely defined when the shape of the indenter and one macroscopic quantity from each trio $\{d, a, F_N\}$ and $\{u^{(0)}, c, F_x\}$ are known. If the solution of the normal contact problem is known, all macroscopic quantities can be determined from (4.27) and (4.33). For the sake of simplicity and in analogy to Chap. 2, it is assumed that a and c are known quantities. Of course, this is not necessarily true. All other cases require rewriting the equations to solve for the unknown quantities. For instance, the relationship between the tangential force and the radius c of the stick zone is obtained by dividing (4.33) by F_N. Using partial integration, it can be rewritten in the compact form of:

$$\frac{F_x}{\mu F_N} = \frac{\int_c^a xg'(x)\mathrm{d}x}{\int_0^a xg'(x)\mathrm{d}x} = \frac{F_N(a) - F_N(c)}{F_N(a)}. \tag{4.37}$$

In summary, there are two approaches to solving the tangential contact problem for the simplest standard loading case (first normal and, subsequently, tangential):

I. The tangential contact problem is *reduced to the normal contact problem* using (4.33), (4.34), and (4.35) with the radius c of the stick zone being determined either by (4.27) (if the displacement is known) or (4.37) (if the force is known). For the sake of convenience, we will list all relevant equations once more:

$$\begin{aligned}
F_x &= \mu[F_N(a) - F_N(c)],\\
\sigma_{xz}(r) &= \tau(r) = \mu[p(a) - p(c)],\\
u(r) &= \frac{\mu E^*}{G^*}[w(r;a) - w(r;c)], \quad \text{with } c \text{ from}\\
G^* u^{(0)} &= \mu E^* \left[d(a) - d(c)\right], \quad \text{or}\\
\frac{F_x}{\mu F_N} &= \frac{F_N(a) - F_N(c)}{F_N(a)}.
\end{aligned} \tag{4.38}$$

II. The tangential contact problem can also be solved directly, without knowledge of the corresponding solution of the normal contact problem, using (4.26), (4.27), (4.34), and (4.36), which we will also summarize once more:

$$F_N = 2E^* \int_0^a [d - g(x)]\mathrm{d}x \quad \text{with } d = g(a),$$

$$F_x = 2\int_0^a q_x(x)\mathrm{d}x = 2\left(G^* u_x^{(0)} c + \mu E^* \int_c^a [d - g(x)]\,\mathrm{d}x\right),$$

$$\tau(r) = \frac{\mu E^*}{\pi} \int_c^a \frac{g'(x)\mathrm{d}x}{\sqrt{x^2 - r^2}},$$

$$u(r) = \begin{cases} u^{(0)}, & \text{for } r < c,\\[4pt] \dfrac{2}{\pi}\left[u^{(0)} \arcsin\left(\dfrac{c}{r}\right) + \dfrac{\mu E^*}{G^*}\displaystyle\int_c^r \frac{d - g(x)}{\sqrt{r^2 - x^2}}\mathrm{d}x\right], & \text{for } c < r < a,\\[4pt] \dfrac{2}{\pi}\left[u^{(0)} \arcsin\left(\dfrac{c}{r}\right) + \dfrac{\mu E^*}{G^*}\displaystyle\int_c^a \frac{d - g(x)}{\sqrt{r^2 - x^2}}\mathrm{d}x\right], & \text{for } r > a. \end{cases}$$

Stick radius is determined by

$$G^* u^{(0)} = \mu E^*[d - g(c)], \quad \text{or}$$

$$\frac{F_x}{\mu F_N} = \frac{\int_c^a x g'(x)\mathrm{d}x}{\int_0^a x g'(x)\mathrm{d}x}. \tag{4.39}$$

If the stresses and displacements are known, the dissipated friction energy can also be calculated with

$$W_R = \int \Delta u \sigma_{xz} \mathrm{d}A. \tag{4.40}$$

Here, Δu represents the relative displacement between the indenter and the half-space. It vanishes in the stick zone. With the stresses outside of the contact also being zero, only the zone of local slip contributes to this integral. The integral can thus be reformulated to:

$$\begin{aligned} W_R(c,a) &= 2\pi\mu \int_c^a (u^{(0)} - u)\sigma_{zz} r \mathrm{d}r \\ &= \frac{2\pi\mu^2 E^*}{G^*} \int_c^a [f(r) + w(r;c) - d(c)]\sigma_{zz}(r;a) r \mathrm{d}r. \end{aligned} \tag{4.41}$$

However, it is easier to calculate the dissipated energy directly using the MDR model of the contact. Then we get:

$$\begin{aligned} W_R(c,a) &= -2G^* \int_c^a u_{1D} \Delta u_{1D} \mathrm{d}x \\ &= -2G^* \left(\frac{\mu E^*}{G^*}\right)^2 \int_c^a [d(a) - g(x)][g(x) - d(c)] \mathrm{d}x. \end{aligned} \tag{4.42}$$

Here, $d(a) = g(a)$ and $d(c) = g(c)$ are the indentation depths corresponding to the radii a and c.

Once again it should be noted that the detailed MDR algorithm is not restricted to the specifically examined loading case (application of a normal force with subsequent tangential loading). In the context of tangential problems, it is valid for all loading cases, including arbitrarily varying normal and tangential forces. Thus, we can utilize it for simulations of arbitrary loading histories, e.g., in stick-slip drives.

4.5 Areas of Application

The technical applications of mechanical contact problems with friction are virtually uncountable. Even by neglecting rolling contacts (not featured in this book due to their asymmetry) occurring in bearings or other types of transport elements, tangential contact problems are found in a wide range of applications, e.g., friction-based connections, friction-induced damping (such as in leaf springs), surface treatment via a sliding indenter (burnishing), or mechanical stick-slip linear drives which can be miniaturized to an extreme degree. In the latter two applications, the indenters usually appear in the classical shapes of the flat punch, cone, and sphere. Once again we will consider these three bodies in their "pure form" (see Sects. 4.6.1, 4.6.2, and 4.6.3) as well as modifications thereof, owing to imperfections due to manufacturing or wear (see Sects. 4.6.5 to 4.6.10). As always, we will consider the profile in shape of a power-law (see Sect. 4.6.4), which is a basic building block of the solution in the form of a Taylor series of any sufficiently differentiable profile.

4.6 Explicit Solutions for Axially Symmetric Tangential Contact Problems

For the sake of simplicity, we will assume in the following that $u^{(0)} \geq 0$. This condition does not represent a restriction to the general validity of the presented results.

4.6.1 The Cylindrical Flat Punch

The solution of the normal contact problem for the indentation by a flat cylindrical punch of radius a according to Chap. 2 (see Sect. 2.5.1) is given by:

$$\begin{aligned} F_N(d) &= 2E^* da, \\ \sigma_{zz}(r;d) &= -\frac{E^* d}{\pi\sqrt{a^2-r^2}}, \quad r \leq a, \\ w(r;d) &= \frac{2d}{\pi} \arcsin\left(\frac{a}{r}\right), \quad r > a. \end{aligned} \tag{4.43}$$

Here, F_N represents the normal force, d the indentation depth, σ_{zz} the normal stress, and w the normal displacement of the half-space. Under tangential load, the contact is either sticking or slipping completely; there is no limited zone of local slip. The contact starts to slip once the tangential displacement of the punch reaches the critical value (4.28):

$$u_c^{(0)} = \frac{\mu E^*}{G^*} d. \tag{4.44}$$

The shear stress distribution in the contact equals

$$\sigma_{xz}(r;u^{(0)},d) = -\frac{G^*}{\pi\sqrt{a^2-r^2}} \cdot \begin{cases} u^{(0)}, & \text{for } u^{(0)} \leq u_c^{(0)}, \\ u_c^{(0)}, & \text{for } u^{(0)} > u_c^{(0)} \end{cases} \tag{4.45}$$

and the tangential displacements outside the contact area $r > a$ are:

$$u(r;u^{(0)},d) = \frac{2}{\pi} \arcsin\left(\frac{a}{r}\right) \cdot \begin{cases} u^{(0)}, & \text{for } u^{(0)} < u_c^{(0)}, \\ u_c^{(0)}, & \text{for } u^{(0)} > u_c^{(0)}. \end{cases} \tag{4.46}$$

The total tangential force is:

$$F_x(u^{(0)}, d) = 2G^*a \cdot \begin{cases} u^{(0)}, & \text{for } u^{(0)} < u_c^{(0)}, \\ u_c^{(0)}, & \text{for } u^{(0)} > u_c^{(0)}. \end{cases} \tag{4.47}$$

4.6.2 The Cone

The consideration of the indentation by a cone with a slope angle θ in Chap. 2 (Sect. 2.5.2) in the context of frictionless normal contact yielded the following relationships for the indentation depth d, the contact radius a, the normal force F_N, the normal stresses σ_{zz}, and the normal displacements w:

$$\begin{aligned} d(a) &= \frac{\pi}{2} a \tan\theta, \\ F_N(a) &= \frac{\pi a^2}{2} E^* \tan\theta, \\ \sigma_{zz}(r;a) &= -\frac{E^* \tan\theta}{2} \operatorname{arcosh}\left(\frac{a}{r}\right), \quad r \le a, \\ w(r;a) &= \tan\theta \left(\sqrt{r^2 - a^2} - r + a \arcsin\left(\frac{a}{r}\right)\right), \quad r > a. \end{aligned} \tag{4.48}$$

The mean pressure in the contact area is:

$$p_0 = \frac{E^* \tan\theta}{2}. \tag{4.49}$$

The solution of the tangential contact problem depicted in Fig. 4.3 (which is the relationships between the tangential displacement $u^{(0)}$), the radius of the stick zone c, and the tangential force F_x (first published by Truman et al. 1995) is then expressed

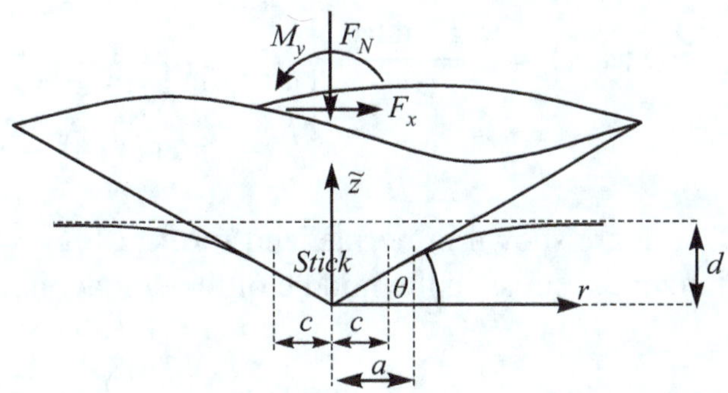

Fig. 4.3 Tangential contact of a rigid conical indenter and elastic half-space

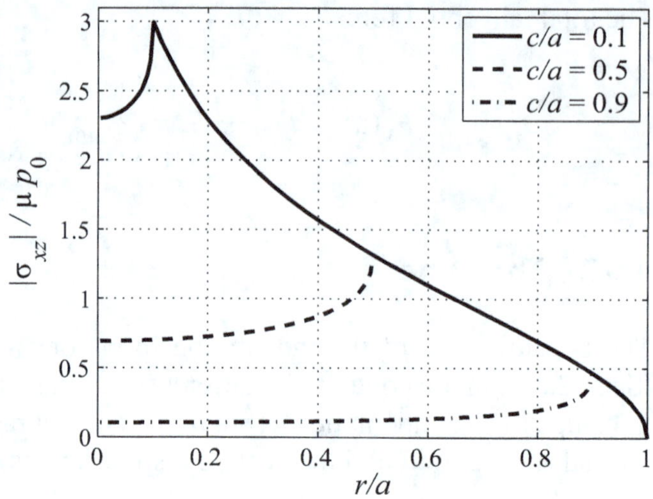

Fig. 4.4 Normalized tangential stresses in contact for different values of the normalized radius of the stick zone c/a when indented by a cone

by (4.38) as follows:

$$
\begin{aligned}
u^{(0)}(a,c) &= \frac{\pi\mu E^*}{2G^*}(a-c)\tan\theta, \\
u_c^{(0)}(a) &= \frac{\pi\mu E^*}{2G^*}a\tan\theta, \\
F_x(a,c) &= \frac{\mu\pi E^*\tan\theta}{2}(a^2-c^2) = \mu F_N(a)\left(1-\frac{c^2}{a^2}\right).
\end{aligned}
\tag{4.50}
$$

For the missing tangential stresses and displacements one gets:

$$
\begin{aligned}
\sigma_{xz}(r;a,c) &= -\frac{\mu E^*\tan\theta}{2}\begin{cases}\operatorname{arcosh}\left(\dfrac{a}{r}\right)-\operatorname{arcosh}\left(\dfrac{c}{r}\right), & r\le c,\\ \operatorname{arcosh}\left(\dfrac{a}{r}\right), & c<r\le a,\end{cases} \\
u(r;a,c) &= \frac{\mu E^*\tan\theta}{G^*}\begin{cases}-c\arcsin\left(\dfrac{c}{r}\right)+\dfrac{\pi a}{2}-\sqrt{r^2-c^2}, & c<r\le a,\\ a\arcsin\left(\dfrac{a}{r}\right)+\sqrt{r^2-a^2} & \\ \quad -c\arcsin\left(\frac{c}{r}\right)-\sqrt{r^2-c^2}, & r>a.\end{cases}
\end{aligned}
\tag{4.51}
$$

These are shown in normalized form in Figs. 4.4 and 4.5. The finite value of the tangential stress in the middle of the contact is:

$$
\lim_{r\to 0}\left(\frac{|\sigma_{xz}|}{\mu p_0}\right) = \lim_{r\to 0}\left[\operatorname{arcosh}\left(\frac{a}{r}\right)-\operatorname{arcosh}\left(\frac{c}{r}\right)\right] = \ln\left(\frac{a}{c}\right). \tag{4.52}
$$

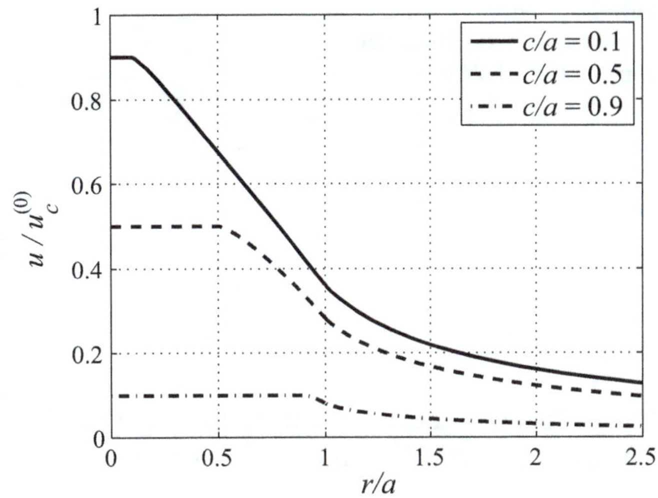

Fig. 4.5 Normalized tangential displacements of the half-space for different values of the normalized radius of the stick zone c/a when indented by a cone

The loss of mechanical energy is calculated according to (4.42):

$$W_R(a,c) = -\frac{(\pi\mu E^* \tan\theta)^2}{12G^*}(a-c)^3. \tag{4.53}$$

In normalized variables, this can be represented as:

$$\eta := \frac{|W_R|}{\mu F_N u_c^{(0)}} = \frac{1}{3}\left(1-\frac{c}{a}\right)^3. \tag{4.54}$$

4.6.3 The Paraboloid

As usual, the paraboloid was the first shape for which a broad class of contact problems was solved. In the case of the tangential contact, the classical solution goes back to Cattaneo (1938) and Mindlin (1949). The pure normal contact problem solution is represented by the following relationships linking indentation depth d, contact radius a, normal force F_N, normal stress σ_{zz}, and normal displacement w (see Sect. 2.5.3)

$$\begin{aligned}
d(a) &= \frac{a^2}{R}, \\
F_N(a) &= \frac{4}{3}\frac{E^* a^3}{R}, \\
\sigma_{zz}(r;a) &= -\frac{2E^*}{\pi R}\sqrt{a^2-r^2}, \quad r \le a, \\
w(r;a) &= \frac{a^2}{\pi R}\left[\left(2-\frac{r^2}{a^2}\right)\arcsin\left(\frac{a}{r}\right) + \frac{\sqrt{r^2-a^2}}{a}\right], \quad r > a.
\end{aligned} \tag{4.55}$$

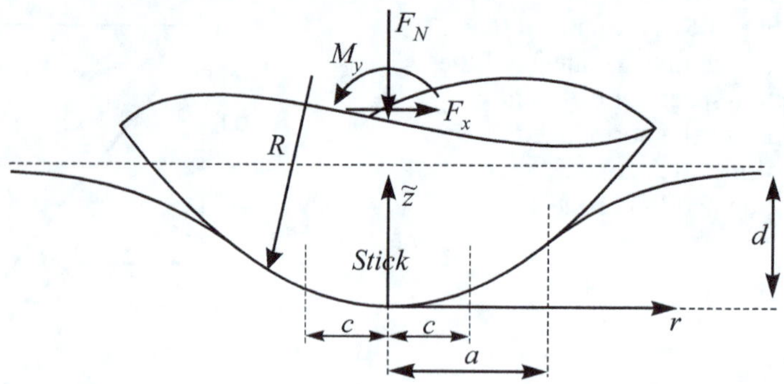

Fig. 4.6 Tangential contact of a rigid parabolic indenter and an elastic half-space

Here, R denotes the curvature radius of the paraboloid in the vicinity of the contact. The average pressure in the contact is:

$$p_0 = \frac{4E^*a}{3\pi R}. \tag{4.56}$$

Taking into account (4.38), the solution of the tangential problem (see Fig. 4.6) is given by the relationships

$$\begin{aligned} u^{(0)}(a,c) &= \frac{\mu E^*}{G^* R}(a^2 - c^2), \\ u_c^{(0)}(a) &= \frac{\mu E^*}{G^* R}a^2, \\ F_x(a,c) &= \frac{4\mu E^*}{3R}(a^3 - c^3) = \mu F_N(a)\left(1 - \frac{c^3}{a^3}\right), \end{aligned} \tag{4.57}$$

with the tangential displacement of the rigid paraboloid $u^{(0)}$, the radius c of the stick zone, and the tangential force F_x. The tangential stresses and displacements of the elastic half-space then amount to:

$$\sigma_{xz}(r;a,c) = -\frac{2\mu E^*}{\pi R}\begin{cases} \sqrt{a^2 - r^2} - \sqrt{c^2 - r^2}, & r \leq c, \\ \sqrt{a^2 - r^2}, & c < r \leq a, \end{cases}$$

$$u(r;a,c) =$$

$$\frac{\mu E^*}{G^* R}\begin{cases} a^2 - \frac{r^2}{2} - \frac{c^2}{\pi}\left[\left(2 - \frac{r^2}{c^2}\right)\arcsin\left(\frac{c}{r}\right) + \frac{\sqrt{r^2 - c^2}}{c}\right], & c < r \leq a, \\ \frac{a^2}{\pi}\left[\left(2 - \frac{r^2}{a^2}\right)\arcsin\left(\frac{a}{r}\right) + \frac{\sqrt{r^2 - a^2}}{a}\right] & \\ \quad -\frac{c^2}{\pi}\left[\left(2 - \frac{r^2}{c^2}\right)\arcsin\left(\frac{c}{r}\right) + \frac{\sqrt{r^2 - c^2}}{c}\right], & r > a. \end{cases} \tag{4.58}$$

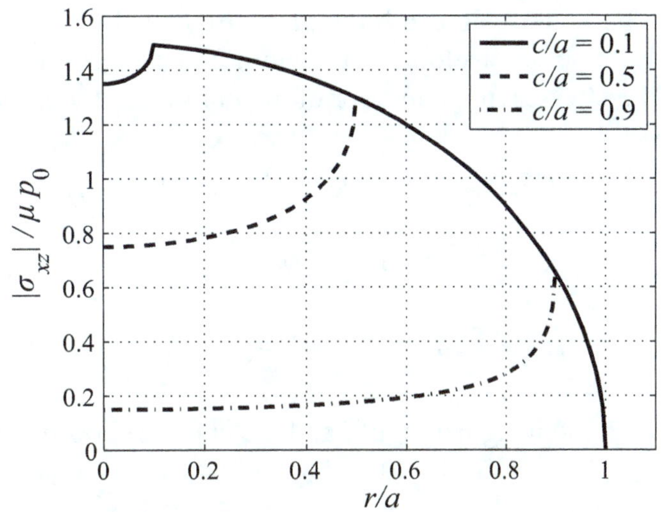

Fig. 4.7 Normalized tangential stresses in the contact for different values of the normalized stick zone radius c/a for indentation by a paraboloid

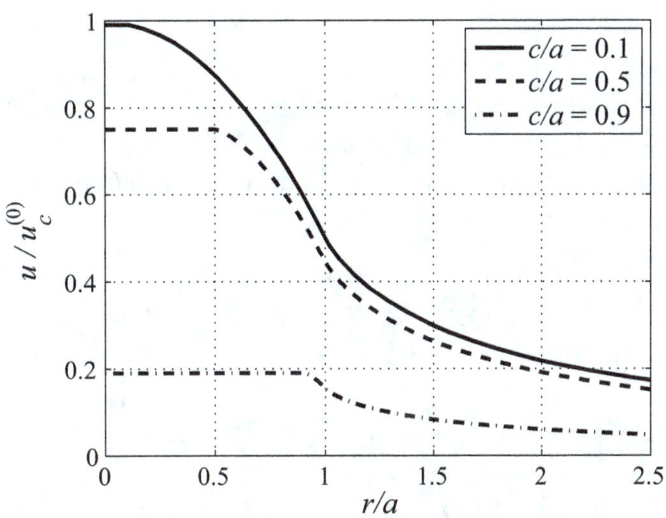

Fig. 4.8 Normalized tangential displacement of the half-space for different values of the normalized stick zone radius c/a for indentation by a paraboloid

These are visualized in normalized form in Figs. 4.7 and 4.8. Using (4.42), the loss of mechanical energy can be written as:

$$W_R(c,a) = -\frac{2(\mu E^*)^2}{G^* R^2} \int_c^a (a^2 - x^2)(x^2 - c^2)\mathrm{d}x. \tag{4.59}$$

When expressed in normalized quantities, it can be rewritten as:

$$\eta := \frac{|W_R|}{\mu F_N u_c^{(0)}} = \frac{1}{5}\left(1 - \frac{c}{a}\right)^3 \left(1 + 3\frac{c}{a} + \frac{c^2}{a^2}\right). \tag{4.60}$$

The Stresses in the Interior of the Half-Space

In the case of global slip, Hamilton and Goodman (1966) were able to determine the stresses in the interior of the half-space by introducing the complex functions

$$
\begin{aligned}
F &:= \frac{1}{2}(\tilde{z} - ia)R_2 + \frac{1}{2}r^2 \ln(R_2 + z_2),\\
G &:= -\frac{1}{3}R_2^3 + \frac{1}{2}\tilde{z}z_2R_2 - \frac{1}{3}ia^3 + \frac{1}{2}\tilde{z}r^2 \ln(R_2 + z_2),\\
H &:= \frac{4}{3}ia^3\tilde{z} - \frac{1}{6}\tilde{z}R_2^3 + \frac{1}{2}iaR_2^3 - \frac{1}{4}\tilde{z}r^2R_2 - \frac{1}{4}r^4 \ln(R_2 + z_2),
\end{aligned} \tag{4.61}
$$

with the imaginary unit i and the complex coordinates:

$$
\begin{aligned}
z_2 &:= \tilde{z} + ia,\\
R_2 &:= \sqrt{z_2^2 + r^2}.
\end{aligned} \tag{4.62}
$$

To avoid any misunderstandings, we will stress once more that $\tilde{z}$ represents the normal axis pointing out of the half-space while the z-axis points inwards. The stress configuration in the half-space is then defined by the imaginary parts of the expressions:

$$
\begin{aligned}
\hat{\sigma}_{xx} &:= -\frac{3}{2}\frac{\mu p_0}{a}\frac{x}{r^4}\left[\left(4\frac{x^2}{r^2} - 3\right)\left(H\nu - \frac{1}{2}\tilde{z}\frac{\partial H}{\partial \tilde{z}}\right) + y\frac{\partial H}{\partial y}\right.\\
&\qquad \left. + (1-\nu)\,x\frac{\partial H}{\partial x} + \frac{1}{2}x\tilde{z}\frac{\partial^2 H}{\partial x \partial \tilde{z}} - 2\nu r^2 F\right],\\
\hat{\sigma}_{yy} &:= -\frac{3}{2}\frac{\mu p_0}{a}\frac{x}{r^4}\left[\left(4\frac{y^2}{r^2} - 1\right)\left(H\nu - \frac{1}{2}\tilde{z}\frac{\partial H}{\partial \tilde{z}}\right) - \nu y\frac{\partial H}{\partial y}\right.\\
&\qquad \left. + \frac{1}{2}y\tilde{z}\frac{\partial^2 H}{\partial y \partial \tilde{z}} - 2\nu r^2 F\right],\\
\hat{\sigma}_{zz} &:= -\frac{3}{2}\frac{\mu p_0}{a}\frac{x\tilde{z}}{r^2}\frac{\partial F}{\partial \tilde{z}},\\
\hat{\sigma}_{yz} &:= -\frac{3}{2}\frac{\mu p_0}{a}\frac{xy\tilde{z}}{2r^4}\frac{\partial^2 H}{\partial \tilde{z}^2},\\
\hat{\sigma}_{xz} &:= -\frac{3}{2}\frac{\mu p_0}{a}\frac{1}{r^2}\left[2G + \frac{1}{2}\frac{\partial H}{\partial \tilde{z}} + \tilde{z}\frac{\partial}{\partial x}(xF) - 2\frac{\tilde{z}x^2}{r^2}F\right],\\
\hat{\sigma}_{xy} &:= -\frac{3}{2}\frac{\mu p_0}{a}\frac{y}{r^4}\left[\left(4\frac{x^2}{r^2} - 1\right)\left(H\nu - \frac{1}{2}\tilde{z}\frac{\partial H}{\partial \tilde{z}}\right) + \frac{1}{2}y\frac{\partial H}{\partial y}\right.\\
&\qquad \left. + \frac{1}{2}x(1-2\nu)\frac{\partial H}{\partial x} + \frac{1}{2}x\tilde{z}\frac{\partial^2 H}{\partial x \partial \tilde{z}}\right].
\end{aligned} \tag{4.63}
$$

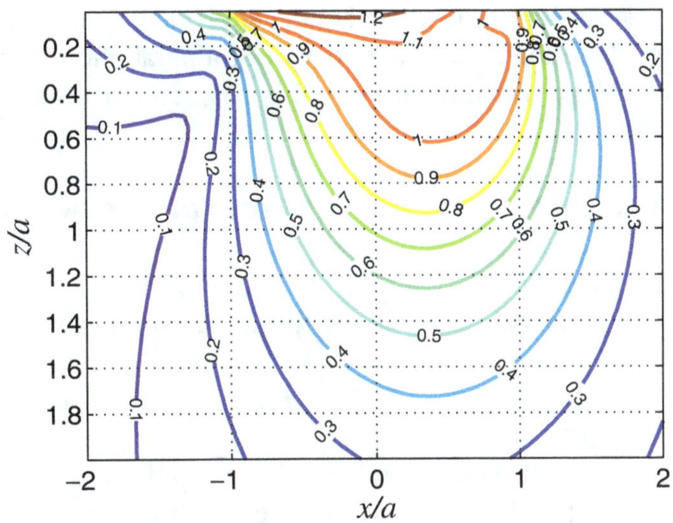

Fig. 4.9 Curve of the equivalent stress according to the von Mises criterion in the x–z-plane under a fully sliding tangential contact with a paraboloid, normalized to the average pressure in the contact

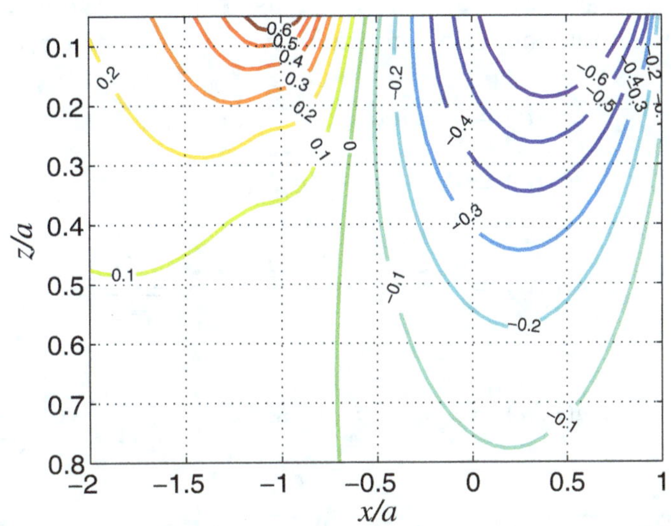

Fig. 4.10 Curve of the greatest principal stress in the x–z-plane under a fully slipping tangential contact with a paraboloid, normalized to the average pressure in the contact

To illustrate these expressions, consider Figs. 4.9 and 4.10. Depicted are the stress curves in the x–z-plane (i.e., the plane of the greatest stresses) of the equivalent stress, according to the von Mises criterion, and the greatest principal stress for a globally sliding contact with $\nu = 0.3$ and $\mu = 0.5$, both normalized to the average pressure in the contact. It is apparent that the leading edge of the contact experiences pressure while the trailing edge is under tension.

It can also be seen that the maximum of the equivalent stress has moved to the surface of the medium. This means that, in this case, the plastic deformation will begin at the surface of the elastic half-space. Since the maximum of the equivalent stress in the case of Hertzian (frictionless) contact lies underneath the surface (detailed in Sect. 2.5.3), there must exist a critical coefficient of friction for which this maximum reaches the surface. This value is of great technical significance for burnishing, a surface treatment which relies on the plastic deformation generated by the global sliding of the indenter. To achieve the desired property change using this process, it is imperative that the equivalent stress maximum is located at the surface level. In the case of the contact of a parabolic indenter with a half-space, Johnson (1985) provided the limit of this coefficient of friction at $\mu_c = 0.3$.

4.6.4 The Profile in the Form of a Power-Law

In Chap. 2 (Sect. 2.5.8) the solution of the normal contact problem for an indenter with a general profile in the form of a power-law:

$$f(r) = br^n, \quad n \in \mathbb{R}^+, \tag{4.64}$$

with a positive real number n, has been derived. In order to not confuse the second constant in this function with the radius of the stick zone c, we call it b (in contrast to what it was called in previous chapters). When considering the normal contact, we found the following expressions ((2.63)–(2.65)):

$$\begin{aligned}
d(a) &= \kappa(n)ba^n, \\
F_N(a) &= E^*\frac{2n}{n+1}\kappa(n)ba^{n+1}, \\
\sigma_{zz}(r;a) &= -\frac{E^*}{\pi}n\kappa(n)b\int_r^a x^{n-1}\frac{\mathrm{d}x}{\sqrt{x^2-r^2}}, \quad r \le a, \\
w(r;a) &= \frac{2}{\pi}\kappa(n)b\left[a^n \arcsin\left(\frac{a}{r}\right) - \int_0^a x^n\frac{\mathrm{d}x}{\sqrt{r^2-x^2}}\right], \quad r > a,
\end{aligned} \tag{4.65}$$

for the indentation depth d, the contact radius a, the normal force F_N and the normal stresses σ_{zz}, and the normal displacements w of the half-space. Here, the stretch factor $\kappa(n)$ is defined as:

$$\kappa(n) := \sqrt{\pi}\frac{\Gamma(n/2+1)}{\Gamma[(n+1)/2]}, \tag{4.66}$$

with the gamma function $\Gamma(\cdot)$:

$$\Gamma(z) := \int_0^\infty t^{z-1}\exp(-t)\mathrm{d}t. \tag{4.67}$$

The solution of the tangential contact problem shown in Fig. 4.11 can be obtained using (4.38); that yields to the following relationships between global contact vari-

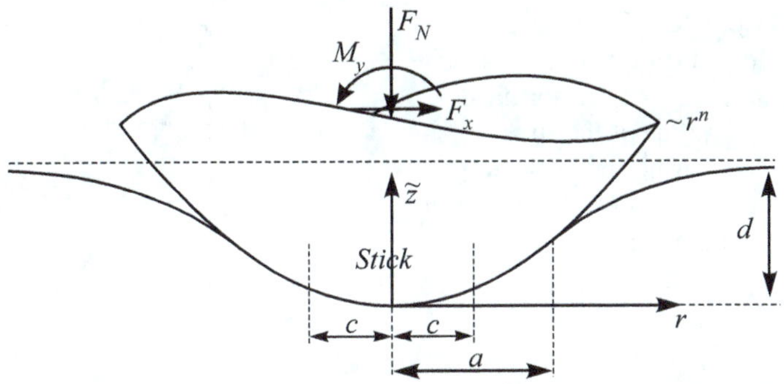

Fig. 4.11 Tangential contact between a rigid indenter with a profile in the form of a power-law and an elastic half-space

ables (displacement $u^{(0)}$ and tangential force F_x):

$$
\begin{aligned}
u^{(0)}(a,c) &= \frac{\mu E^*}{G^*}\kappa(n)b(a^n - c^n),\\
u_c^{(0)}(a) &= \frac{\mu E^*}{G^*}\kappa(n)ba^n,\\
F_x(a,c) &= \mu E^*\frac{2n}{n+1}\kappa(n)b(a^{n+1} - c^{n+1}) = \mu F_N(a)\left(1 - \frac{c^{n+1}}{a^{n+1}}\right).
\end{aligned} \tag{4.68}
$$

The tangential stresses and displacements can also be obtained by (4.38) using (4.65). The explicit representation is extensive, and should therefore be omitted here. For the treatment of integrals occurring in σ_{zz} and w—and thus also in σ_{xz} and u—see Sect. 2.5.8 which deals with power profiles. The loss of mechanical energy is, according to (4.42),

$$
\begin{aligned}
W_R(c,a) &= -\frac{2[\mu E^* b\kappa(n)]^2}{G^*}\int_c^a (a^n - x^n)(x^n - c^n)\mathrm{d}x\\
&= -\frac{2[\mu E^* b\kappa(n)]^2}{G^*}\frac{n}{n+1}\left[\frac{a^{2n+1} - c^{2n+1}}{2n+1} - a^n c^n (a - c)\right].
\end{aligned} \tag{4.69}
$$

This can be shown in normalized variables as

$$
\eta := \frac{|W_R|}{\mu F_N u_c^{(0)}} = \frac{1}{2n+1}\left[1 - \left(\frac{c}{a}\right)^{2n+1}\right] - \left(\frac{c}{a}\right)^n\left(1 - \frac{c}{a}\right). \tag{4.70}
$$

with the maximum

$$
\eta_{\max} = \eta(c = 0) = \frac{1}{2n+1}. \tag{4.71}
$$

Since n must be a positive number, this only assumes values between zero and one, which of course is physically necessary. In Fig. 4.12 the expression $\eta/\eta_{\max}$ is represented as a function of c/a for different exponents n. It can be seen that the curves for larger exponents decrease with increasing radius of the stick zone later and are steeper against zero.

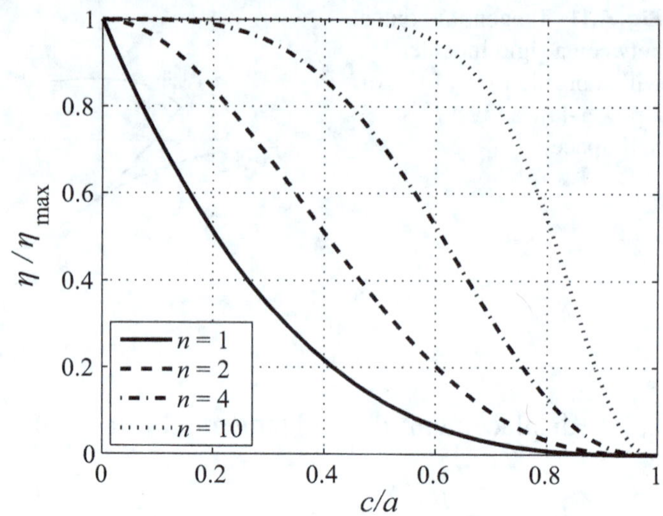

Fig. 4.12 Normalized dissipated friction energy as a function of the normalized radius of the stick zone during indentation by a power-law profile for different exponents n

4.6.5 The Truncated Cone

Now consider profiles that have a flat tip, for example, due to wear.

$$f(r) = \begin{cases} 0, & r \le b, \\ (r - b)\tan\theta, & r > b. \end{cases} \tag{4.72}$$

Here, θ denotes the slope angle of the cone and b the radius at the blunt end. In Chap. 2 (Sect. 2.5.9) the following relationships between the global contact variables (indentation depth d, the contact radius a, and the normal force F_N) were derived for the solution of the normal contact problem:

$$d(a) = a\tan\theta \arccos\left(\frac{b}{a}\right),$$
$$F_N(a) = E^* \tan\theta a^2 \left[\arccos\left(\frac{b}{a}\right) + \frac{b}{a}\sqrt{1 - \frac{b^2}{a^2}}\right]. \tag{4.73}$$

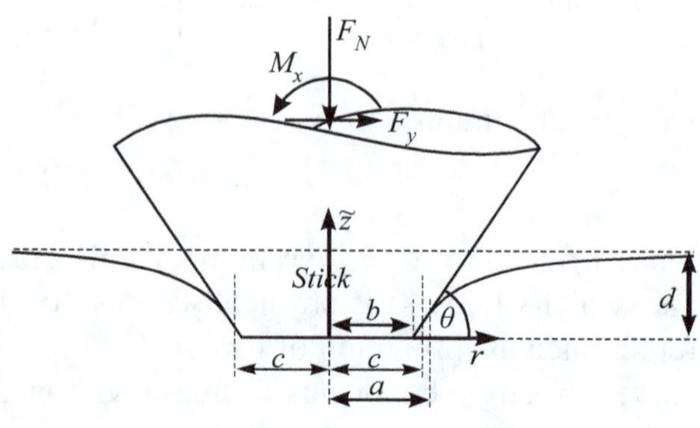

Fig. 4.13 Tangential contact between a rigid truncated cone and an elastic half-space

The mean pressure in contact is thus

$$p_0 = \frac{E^* \tan\theta}{\pi}\left[\arccos\left(\frac{b}{a}\right) + \frac{b}{a}\sqrt{1-\frac{b^2}{a^2}}\right]. \tag{4.74}$$

The equations for the normal stresses σ_{zz} and the displacements w of the half-space outside the contact region are described by the expressions:

$$\sigma_{zz}(r;a) = -\frac{E^* \tan\theta}{\pi}\begin{cases} \mathrm{K}\left(\frac{r}{b}\right) - \mathrm{F}\left(\arcsin\left(\frac{b}{a}\right), \frac{r}{b}\right) \\ \quad + \int_b^a \arccos\left(\frac{b}{x}\right)\frac{\mathrm{d}x}{\sqrt{x^2-r^2}}, & r \le b, \\ \frac{b}{r}\left[\mathrm{K}\left(\frac{b}{r}\right) - \mathrm{F}\left(\arcsin\left(\frac{r}{a}\right), \frac{b}{r}\right)\right] \\ \quad + \int_r^a \arccos\left(\frac{b}{x}\right)\frac{\mathrm{d}x}{\sqrt{x^2-r^2}}, & b < r \le a. \end{cases} \tag{4.75}$$

and

$$w(r;a) = \frac{2\tan\theta}{\pi}\left[\varphi_0 a \arcsin\left(\frac{a}{r}\right) - \int_b^a x \arccos\left(\frac{b}{x}\right)\frac{\mathrm{d}x}{\sqrt{r^2-x^2}}\right], \quad r > a. \tag{4.76}$$

Here, $\mathrm{K}(\cdot)$ und $\mathrm{F}(\cdot,\cdot)$ denote the complete and incomplete elliptic integrals of the first kind:

$$\mathrm{K}(k) := \int_0^{\pi/2} \frac{\mathrm{d}\varphi}{\sqrt{1-k^2\sin^2\varphi}},$$

$$\mathrm{F}(\alpha, k) := \int_0^{\alpha} \frac{\mathrm{d}\varphi}{\sqrt{1-k^2\sin^2\varphi}}. \tag{4.77}$$

By applying (4.38), the tangential contact problem (see Fig. 4.13) can now be solved. For the tangential displacement $u^{(0)}$ and the tangential force F_x, one obtains:

$$u^{(0)}(a,c) = \frac{\mu E^*}{G^*}\tan\theta\left[a\arccos\left(\frac{b}{a}\right) - c\arccos\left(\frac{b}{c}\right)\right],$$

$$F_x(a,c) = \mu E^* \tan\theta\left\{a^2\left[\arccos\left(\frac{b}{a}\right) + \frac{b}{a}\sqrt{1-\frac{b^2}{a^2}}\right] - c^2\left[\arccos\left(\frac{b}{c}\right) + \frac{b}{c}\sqrt{1-\frac{b^2}{c^2}}\right]\right\}. \tag{4.78}$$

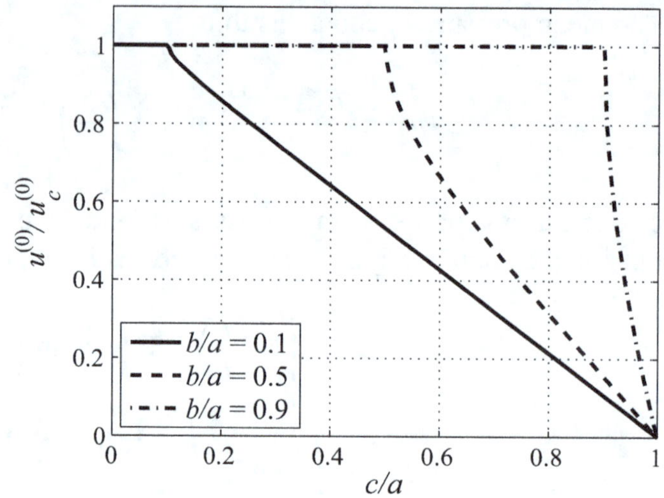

Fig. 4.14 Normalized tangential displacement as a function of the normalized radius of the stick zone for different values b/a during indentation by a truncated cone

The radius of the stick zone c cannot decrease below the value of b; the contact starts to slide completely, if $b = c$. The corresponding maximum tangential displacement is described by:

$$u_c^{(0)}(a) = \frac{\mu E^* a}{G^*} \tan\theta \arccos\left(\frac{b}{a}\right). \tag{4.79}$$

It is easy to see that $b = 0$ reproduces results of the complete cone. In Fig. 4.14 the normalized tangential displacement $u^{(0)}/u_c^{(0)}$ is shown as a function of the normalized radius of the stick zone c/a for different values of b/a. It can be seen that the curves for very small values of b approach the solution of the complete cone

$$\lim_{b\to 0} \frac{u^{(0)}}{u_c^{(0)}} = 1 - \frac{c}{a}. \tag{4.80}$$

Figure 4.15 shows the dependency on c/a for the normalized tangential force $F_R/(\mu F_N)$. Again, the limiting curve corresponding to the whole cone is easily recognizable:

$$\lim_{b\to 0} \frac{F_R}{\mu F_N} = 1 - \frac{c^2}{a^2}. \tag{4.81}$$

The tangential stresses and displacements can be obtained by substituting the results obtained so far in this section into (4.38). Figures 4.16 and 4.17 show the normalized curves of the tangential stress in contact for different values of the normalized radius of the stick zone at $b = 0.09a$ and $b = 0.49a$. For $b = 0.09a$, the result hardly differs from the curves of the complete cone in Fig. 4.4. It can be seen that $c > b$. Thus, before the complete sliding begins, the tangential stresses (in contrast to the pressure distribution) has no singularities. For $b = c$, i.e., at the beginning of complete sliding, is $|\sigma_{xz}| = \mu|\sigma_{zz}|$, which means that the tangential stresses at $r = b$ have the same singular behavior as the normal stresses.

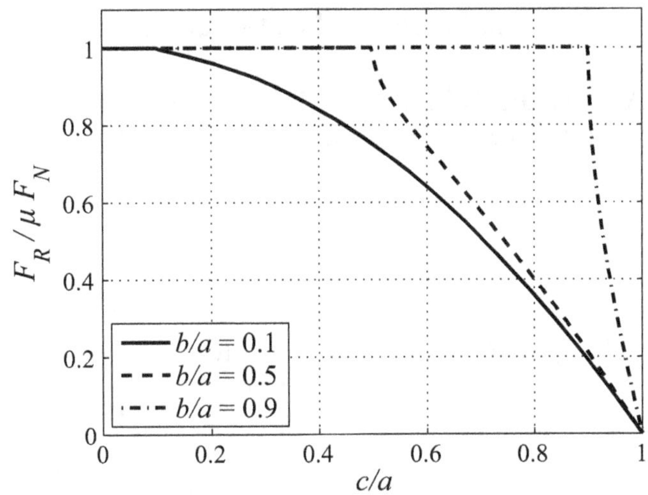

Fig. 4.15 Normalized tangential force as a function of the normalized radius of the stick zone for different values during indentation by a truncated cone

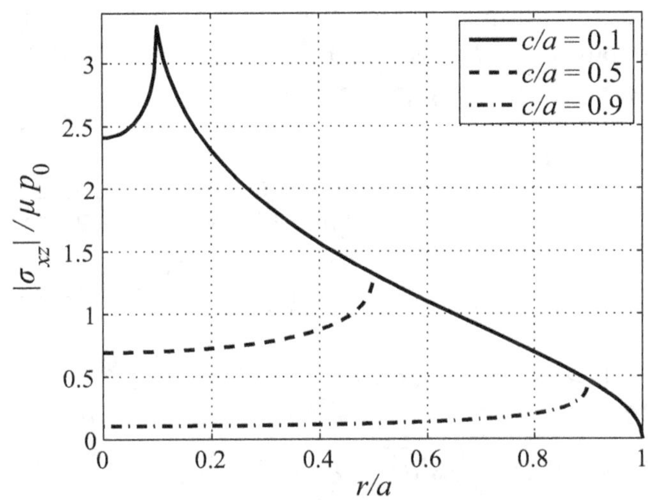

Fig. 4.16 Normalized tangential stresses in contact for different values of the normalized radius of the stick zone c/a during indentation by a truncated cone with $b = 0.09a$

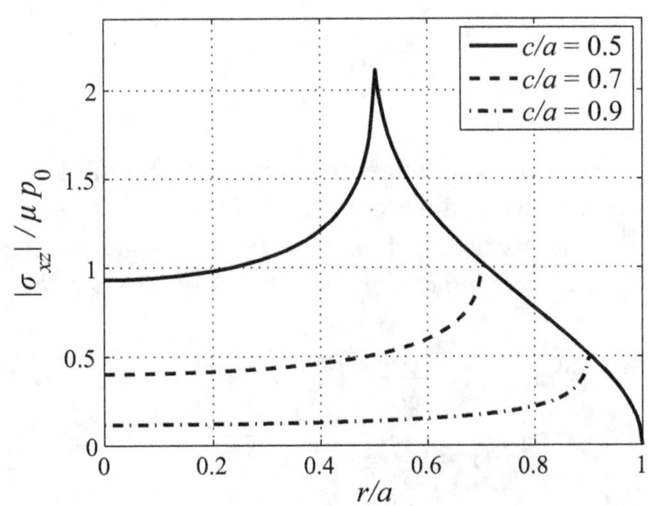

Fig. 4.17 Normalized tangential stresses in contact for different values of the normalized radius of the stick zone c/a during indentation by a truncated cone with $b = 0.49a$

4.6.6 The Truncated Paraboloid

A truncated paraboloid with the radius of curvature R and the radius at the flat tip b can be described by the profile

$$f(r) = \begin{cases} 0, & r \le b, \\ \frac{r^2-b^2}{2R}, & r > b. \end{cases} \tag{4.82}$$

In Chap. 2 in Sect. 2.5.10, the solution

$$\begin{aligned} d(a) &= \frac{a}{R}\sqrt{a^2-b^2}, \\ F_N(a) &= \frac{2E^*}{3R}(2a^2+b^2)\sqrt{a^2-b^2} \end{aligned} \tag{4.83}$$

for the normal contact problem was determined. Here, as always, d denotes the depth of indentation, a the contact radius, and F_N the normal force. The stresses were given in integral form as:

$$\sigma_{zz}(r;a) = -\frac{E^*}{\pi R}\begin{cases} \displaystyle\int_b^a \frac{(2x^2-b^2)\mathrm{d}x}{\sqrt{x^2-b^2}\sqrt{x^2-r^2}}, & r \le b, \\ \displaystyle\int_r^a \frac{(2x^2-b^2)\mathrm{d}x}{\sqrt{x^2-b^2}\sqrt{x^2-r^2}}, & b < r \le a. \end{cases} \tag{4.84}$$

and for the normal off-contact displacements:

$$\begin{aligned} w(r;a) &= \frac{2a}{\pi R}\sqrt{a^2-b^2}\arcsin\left(\frac{a}{r}\right) \\ &\quad - \frac{1}{\pi R}\left[(r^2-b^2)\arcsin\left(\frac{\sqrt{a^2-b^2}}{\sqrt{r^2-b^2}}\right) - \sqrt{a^2-b^2}\sqrt{r^2-a^2}\right], \\ &\quad r > a. \end{aligned} \tag{4.85}$$

was found.

Thus, for the tangential contact problem (see Fig. 4.18), the following relationships between the global contact variables (tangential displacement of the indenter $u^{(0)}$, radius of the stick zone c, and tangential force F_x) can be determined using (4.38) (the contact starts to slide completely at $b = c$):

$$\begin{aligned} u^{(0)}(a,c) &= \frac{\mu E^*}{G^* R}\left(a\sqrt{a^2-b^2} - c\sqrt{c^2-b^2}\right), \\ u_c^{(0)}(a) &= \frac{\mu E^*}{G^* R}a\sqrt{a^2-b^2}, \\ F_x(a,c) &= \frac{2\mu E^*}{3R}\left[(2a^2+b^2)\sqrt{a^2-b^2} - (2c^2+b^2)\sqrt{c^2-b^2}\right]. \end{aligned} \tag{4.86}$$

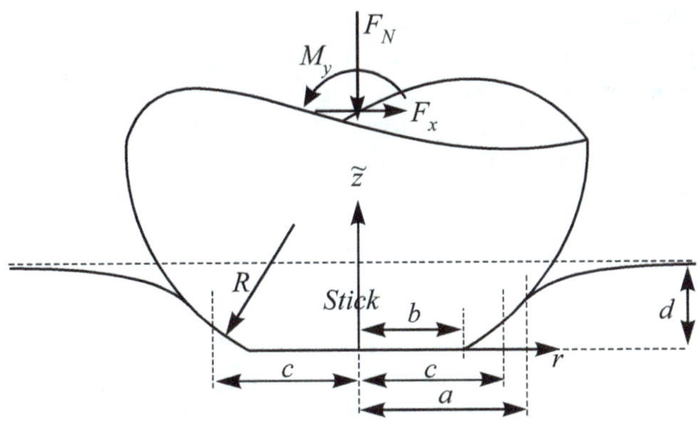

Fig. 4.18 Tangential contact between a rigid truncated paraboloid and an elastic half-space

These are shown in a normalized manner in Figs. 4.19 and 4.20. As in the case of the truncated cone, for small values of b, it is easy to reproduce the known limiting cases of the complete paraboloid; namely:

$$\lim_{b\to 0}\left[\frac{u^{(0)}}{u_c^{(0)}}\right] = 1 - \frac{c^2}{a^2} \tag{4.87}$$

and

$$\lim_{b\to 0}\left[\frac{F_x}{\mu F_N}\right] = 1 - \frac{c^3}{a^3}. \tag{4.88}$$

The tangential stresses and displacements should, again, not be written out for reasons of space but can be obtained by inserting them into the general equations (4.38). Some dependencies of the tangential stresses are shown in Figs. 4.21 and 4.22. These have the same singular behavior as in the case of the truncated cone; that is, the tangential stresses are only singular if $r = b = c$, i.e., in case of full sliding.

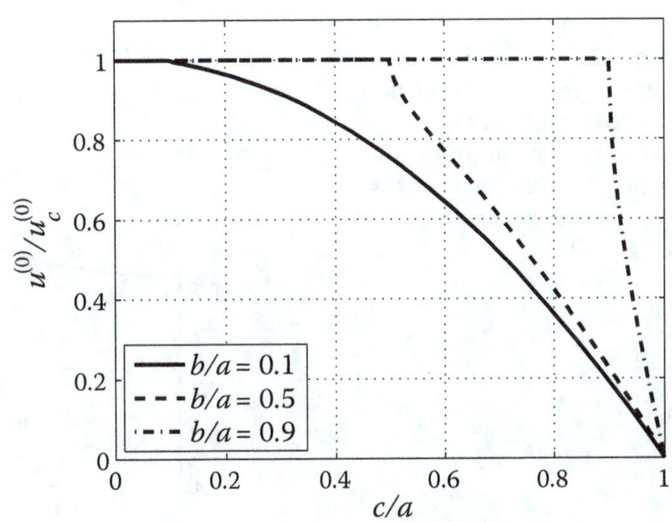

Fig. 4.19 Normalized tangential displacement as a function of the normalized radius of the stick zone for different values b/a when indented by a truncated paraboloid

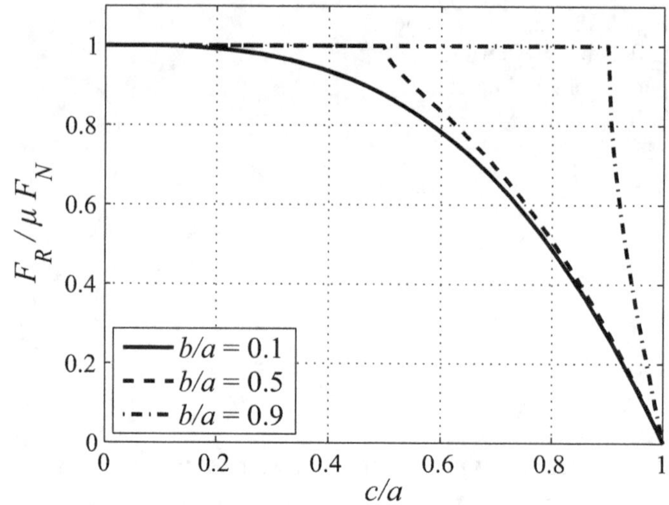

Fig. 4.20 Normalized tangential displacement as a function of the normalized radius of the stick zone for different values b/a when indented by a truncated paraboloid

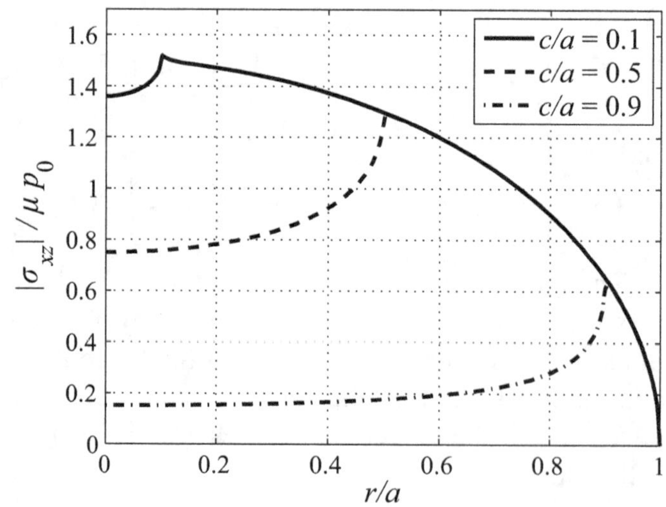

Fig. 4.21 Normalized tangential stresses in the contact for different values of the normalized radius of the stick zone c/a when indented by a truncated paraboloid with $b = 0.09a$

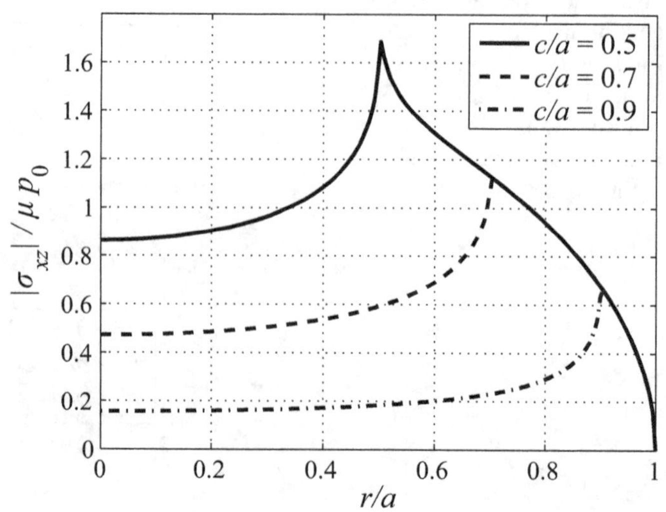

Fig. 4.22 Normalized tangential stresses in the contact for different values of the normalized radius of the stick zone c/a when indented by a truncated paraboloid with $b = 0.49a$

4.6.7 The Cylindrical Flat Punch with Parabolic Cap

For the consideration of the indenter with parabolic cap (see Fig. 4.23) we assume, as always, that the indentation depth d is large enough to actually bring the main body into contact. Otherwise it would be the pure contact with a paraboloid, for which the results can be looked up in Sect. 4.6.3. The flat cylindrical punch with a parabolic cap can be described by the profile:

$$f(r) = \begin{cases} \dfrac{r^2}{2R}, & r \le a, \\ \infty, & r > a. \end{cases} \tag{4.89}$$

Here, a is the radius of the punch and R is the radius of curvature of the cap. In cases where d is indeed sufficiently large, the solution of the friction-free normal contact problem (with the normal force F_N, the stress distribution σ_{zz} within, and the normal displacements of the half-space w outside the contact) was derived in Chap. 2 in Sect. 2.5.11, to which we want to refer again.

$$\begin{aligned} F_N(d) &= 2E^*\left(da - \frac{a^3}{3R}\right), \quad dR \ge a^2, \\ \sigma_{zz}(r;d) &= -\frac{E^*}{\pi R}\frac{a^2 - 2r^2 + dR}{\sqrt{a^2 - r^2}}, \quad r \le a,\ dR \ge a^2, \\ w(r;d) &= \frac{1}{\pi R}\left[\left(2dR - r^2\right)\arcsin\left(\frac{a}{r}\right) + a\sqrt{r^2 - a^2}\right], \\ & \quad r > a,\ dR \ge a^2. \end{aligned} \tag{4.90}$$

The mean pressure in contact is:

$$p_0 = \frac{F_N}{\pi a^2} = \frac{2E^*}{3\pi R a}(3dR - a^2). \tag{4.91}$$

Since the normal stress at the edge of the contact at $r = a$ is singular, the contact can stick completely even after applying a tangential force, as in the case of the flat

Fig. 4.23 Tangential contact between a rigid cylindrical punch with parabolic cap and an elastic half-space

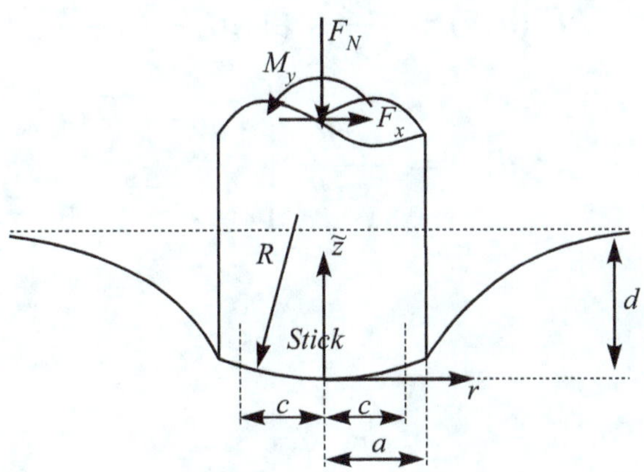

punch. The exact MDR shape of the problem makes it easy to understand how long this condition can last. The springs on the edge of the contact will begin to slide, if

$$u^{(0)} > \frac{\mu E^*}{G^*}\left(d - \frac{a^2}{R}\right) := u_1^{(0)}. \tag{4.92}$$

Thereafter, the area of partial sliding begins to spread from the edge until finally the whole contact slides, if

$$u^{(0)} > \frac{\mu E^*}{G^*} d := u_c^{(0)}. \tag{4.93}$$

In contrast to all previous sections of this chapter, there are three different regimes:

- $u^{(0)} \leq u_1^{(0)}$: complete sticking
- $u_1^{(0)} < u^{(0)} < u_c^{(0)}$: partial sliding
- $u^{(0)} \geq u_c^{(0)}$: complete sliding

The corresponding solutions of the tangential contact problem, i.e., the relationships between the radius of the stick zone c, the tangential displacement of the indenter $u^{(0)}$, the tangential force F_x, and the tangential stresses σ_{xz} and displacements u of the half-space are given by the general relations (3.38) as follows.

Case 1: Complete Sticking

$$\begin{aligned}
F_x(u^{(0)}) &= 2G^* a u^{(0)}, \\
\sigma_{xz}(r; u^{(0)}) &= -\frac{G^* u^{(0)}}{\pi\sqrt{a^2 - r^2}}, \quad r \leq a \\
u(r; u^{(0)}) &= \frac{2u^{(0)}}{\pi}\arcsin\left(\frac{a}{r}\right), \quad r > a.
\end{aligned} \tag{4.94}$$

Case 2: Partial Sliding with Radius of the Stick Zone c

$$\begin{aligned}
u^{(0)}(d,c) &= \frac{\mu E^*}{G^*}\left(d - \frac{c^2}{R}\right), \\
F_x(d,c) &= \frac{2\mu E^*}{3R}\left(3adR - a^3 - 2c^3\right), \\
\sigma_{xz}(r; d, c) &= -\frac{\mu E^*}{\pi R}\begin{cases} 2\sqrt{a^2 - r^2} - 2\sqrt{c^2 - r^2} + \dfrac{dR - a^2}{\sqrt{a^2 - r^2}}, & r \leq c, \\ 2\sqrt{a^2 - r^2} + \dfrac{dR - a^2}{\sqrt{a^2 - r^2}}, & c < r \leq a, \end{cases}
\end{aligned}$$

$$u(r;d,c) = \frac{\mu E^*}{G^* R}\begin{cases} dR - \dfrac{r^2}{2} - \dfrac{1}{\pi}\Big[\left(2c^2 - r^2\right)\arcsin\left(\dfrac{c}{r}\right) \\ \qquad + c\sqrt{r^2 - c^2}\Big], & c < r \le a, \\ \dfrac{1}{\pi}\Big[\left(2dR - r^2\right)\arcsin\left(\dfrac{a}{r}\right) \\ \quad + a\sqrt{r^2 - a^2} - \left(2c^2 - r^2\right)\arcsin\left(\dfrac{c}{r}\right) \\ \quad - c\sqrt{r^2 - c^2}\Big], & r > a. \end{cases} \tag{4.95}$$

The normalized tangential stresses and displacements are shown for the case $dR = 1.2a^2$ in Figs. 4.24 and 4.25. When using the Ciavarella–Jäger principle in the form

$$\sigma_{xz}(r) = \mu\left[\sigma_{zz}(r;a) - \sigma_{zz}(r;c)\right], \tag{4.96}$$

care should be taken. $\sigma_{zz}(r;c)$ denotes the stress distribution which arises when the *same* indenter (i.e., with the punch radius a), is pressed into the half-space up to a contact radius c. But this means only the parabolic cap is in contact, so the stress distribution is the stress distribution of the simple paraboloid going back to Hertz (1882):

$$\sigma_{zz}(r;c) = -\frac{2E^*}{\pi R}\sqrt{c^2 - r^2} \neq \sigma_{zz}(r;a)\big|_{a=c}. \tag{4.97}$$

This blurring of the notation is hard to avoid. The same applies to $w(r;c)$, which is required for the calculation of the tangential displacement distribution according to (4.38).

Case 3: Complete Sliding

In the case of complete sliding, the solution results by substituting $c = 0$ into (4.95). For $R \to \infty$, the case of the flat cylindrical punch results. Then the possibility of partially sliding is eliminated because of $u_1^{(0)} = u_c^{(0)}$.

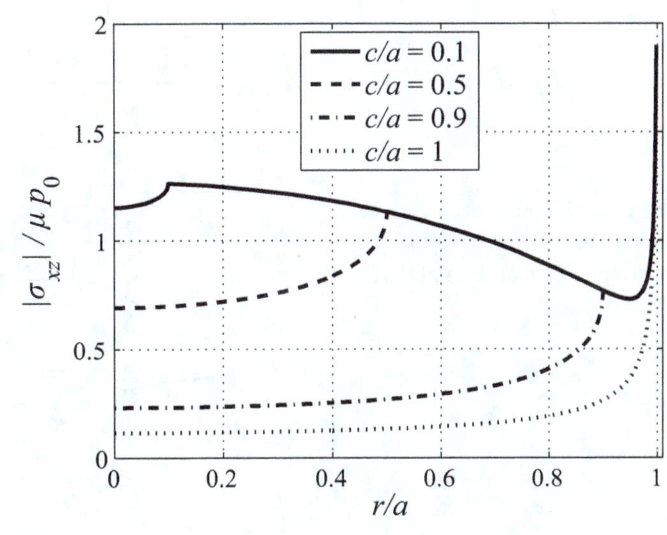

Fig. 4.24 Normalized tangential stresses during indentation by a flat stamp with paraboloidal cap for $dR = 1.2a^2$ and different values of the normalized radius of the stick zone. The dotted line corresponds to the case $u^{(0)} = u_1^{(0)}$

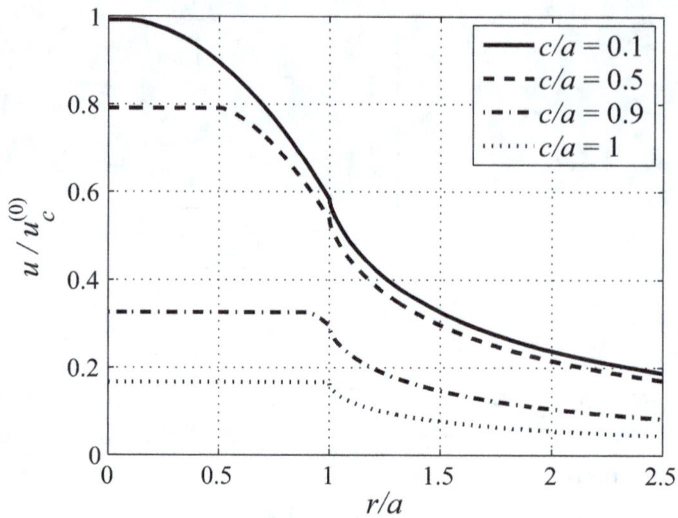

Fig. 4.25 Normalized tangential displacements when indented by a flat punch with paraboloidal cap for $dR = 1.2a^2$ and different values of the normalized radius of the stick zone. The dotted line corresponds to the case $u^{(0)} = u_1^{(0)}$

4.6.8 The Cone with Parabolic Cap

The tangential contact between a cone with a rounded tip and an elastic half-space, shown schematically in Fig. 4.26, was first investigated and solved by Ciavarella (1999). The profile of the indenter with the conical slope angle θ and the value of b of the radial coordinate at which the conical main body differentiably passes into the parabolic cap is described by the rule

$$f(r) = \begin{cases} \dfrac{r^2 \tan\theta}{2b}, & r \le b, \\ r\tan\theta - \dfrac{b}{2}\tan\theta, & r > b. \end{cases} \tag{4.98}$$

In Chap. 2 (see Sect. 2.5.12) the following solution for the normal contact problem was derived:

$$\begin{aligned} d(a) &= a\tan\theta\left(\frac{1-\sin\varphi_0}{\cos\varphi_0} + \varphi_0\right), \\ F_N(a) &= E^* a^2 \tan\theta\left(\varphi_0 + \frac{4}{3}\frac{1-\sin\varphi_0}{\cos\varphi_0} + \frac{1}{3}\sin\varphi_0\cos\varphi_0\right), \end{aligned} \tag{4.99}$$

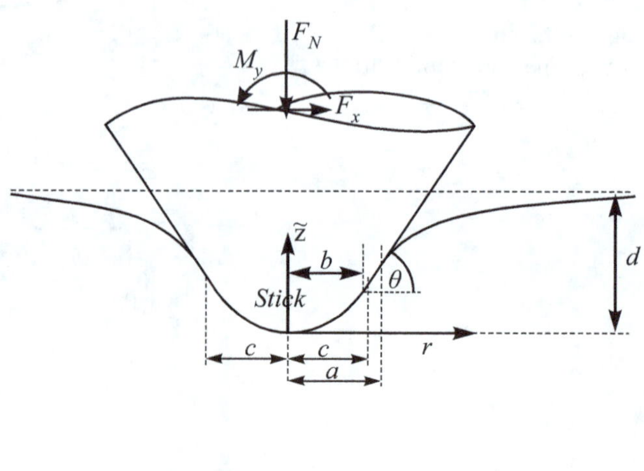

Fig. 4.26 Tangential contact between a rigid cone with a rounded tip and an elastic half-space

with

$$\varphi_0 := \arccos\left(\frac{b}{a}\right), \tag{4.100}$$

The contact is radius a, the indentation depth d, and normal force F_N. The normal stresses in the contact are described by:

$$\sigma_{zz}(r;a) = -\frac{E^* \tan\theta}{\pi b}\begin{cases} 2\sqrt{a^2-r^2} \\ \quad + b\displaystyle\int_0^{\varphi_0} \frac{(\varphi - 2\tan\varphi)\tan\varphi \, \mathrm{d}\varphi}{\sqrt{1-k^2\cos^2\varphi}}, & r \le b, \\ 2\sqrt{a^2-r^2} \\ \quad + b\displaystyle\int_0^{\operatorname{arcosh}\left(\frac{a}{r}\right)} \left[\arccos\left(\frac{b}{r\cosh\varphi}\right)\right. \\ \qquad \left. - 2\sqrt{k^2\cosh^2\varphi - 1}\right]\mathrm{d}\varphi, & b < r \le a \end{cases} \tag{4.101}$$

and the normal displacements out of contact through the distribution:

$$\begin{aligned} w(r;a) = \\ &\frac{2d(a)}{\pi}\arcsin\left(\frac{a}{r}\right) \\ &- \frac{\tan\theta}{\pi b}\left[r^2\arcsin\left(\frac{a}{r}\right) - a\sqrt{r^2-a^2} + 2b^2\int\limits_0^{\varphi_0}\frac{(\varphi-\tan\varphi)\tan\varphi \mathrm{d}\varphi}{\cos\varphi\sqrt{k^2\cos^2\varphi - 1}}\right], \\ &r > a. \end{aligned} \tag{4.102}$$

It is assumed that the contact radius does not fall below the value $a = b$. If $a < b$, it is the contact with a paraboloid with the radius of curvature,

$$R := \frac{b}{\tan\theta}, \tag{4.103}$$

for which the results can be looked up in Sect. 4.6.3. If the normal stress at the edge of the contact disappears, the contact cannot fully stick at a tangential load. There are three different cases for the sliding regime:

- partial sliding with $c > b$
- partial sliding with $c \le b$
- complete sliding

Equations (4.38) can be used to obtain the following solutions to the tangential contact problem (radius of the stick zone c, tangential indenter displacement $u^{(0)}$, tangential force F_R, tangential stresses σ_{xz}, and half-space displacements u):

Case 1: Partial Sliding with $c > b$

$$u^{(0)}(a,c) = \frac{\mu E^*}{G^*}\tan\theta\left[a\left(\frac{1-\sin\varphi_0}{\cos\varphi_0}+\varphi_0\right)-c\left(\frac{1-\sin\psi_0}{\cos\psi_0}+\psi_0\right)\right],$$

$$F_x(a,c) = \mu E^*\tan\theta\left[a^2\left(\varphi_0+\frac{4}{3}\frac{1-\sin\varphi_0}{\cos\varphi_0}+\frac{1}{3}\sin\varphi_0\cos\varphi_0\right)\right.$$
$$\left.-c^2\left(\psi_0+\frac{4}{3}\frac{1-\sin\psi_0}{\cos\psi_0}+\frac{1}{3}\sin\psi_0\cos\psi_0\right)\right],$$

$$\sigma_{xz}(r;a,c) = \mu\begin{cases}\sigma_{zz}(r;a)-\sigma_{zz}(r;c), & r\le c,\\ \sigma_{zz}(r;a), & c<r\le a,\end{cases}$$

$$u(r;a,c) = \frac{\mu E^*}{G^*}\begin{cases}d(a)-f(r)-w(r;c), & c<r\le a,\\ w(r;a)-w(r;c), & r>a,\end{cases} \tag{4.104}$$

with

$$\psi_0 := \arccos\left(\frac{b}{c}\right). \tag{4.105}$$

In this case (in contrast to the previous section), $\sigma_{zz}(r;c)$ actually means $\sigma_{zz}(r;a)|_{a=c}$. This also applies analogously for all other expressions. Therefore, all functions that are still open in (4.104) can be looked up in (4.98) to (4.102). For $b = 0$ and, therefore, $\varphi_0 = \psi_0 = \pi/2$; the solutions for the complete cone from Sect. 4.6.2 are recovered.

Case 2: Partial Sliding with $c \le b$

The principle of Ciavarella and Jäger demands at this point that the solution of the tangential contact problem is given by the difference of solutions of the following two normal contact problems: firstly, the normal indentation by the indenter defined in (4.98) up to a contact radius a, and secondly, the normal indentation of the same indenter up to a contact radius $c \le b$. The latter corresponds to a normal contact with a paraboloid, since only the parabolic tip of the indenter is in contact. So the tangential solution we are looking for is:

$$u^{(0)}(a,c) = \frac{\mu E^*}{G^*}\tan\theta\left[a\left(\frac{1-\sin\varphi_0}{\cos\varphi_0}+\varphi_0\right)-\frac{c^2}{b}\right],$$

$$F_x(a,c) = \mu E^*\tan\theta\left[a^2\left(\varphi_0+\frac{4}{3}\frac{1-\sin\varphi_0}{\cos\varphi_0}+\frac{1}{3}\sin\varphi_0\cos\varphi_0\right)-\frac{4c^3}{3b}\right],$$

$$\sigma_{xz}(r;a,c) = \mu\begin{cases}\sigma_{zz}(r;a)+\dfrac{2E^*}{\pi b}\tan\theta\sqrt{c^2-r^2}, & r\le c,\\ \sigma_{zz}(r;a), & c<r\le a,\end{cases}$$

$$u(r;a,c) = \frac{\mu E^*}{G^*}\begin{cases}d(a)-f(r)-w_p(r;c), & c<r\le a,\\ w(r;a)-w_p(r;c), & r>a,\end{cases} \tag{4.106}$$

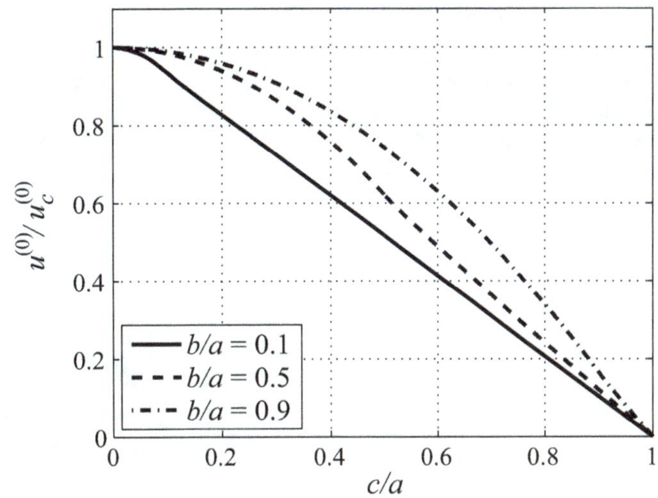

Fig. 4.27 Normalized tangential displacement as a function of the normalized radius of the stick zone for different values b/a when indenting by a cone with a rounded tip

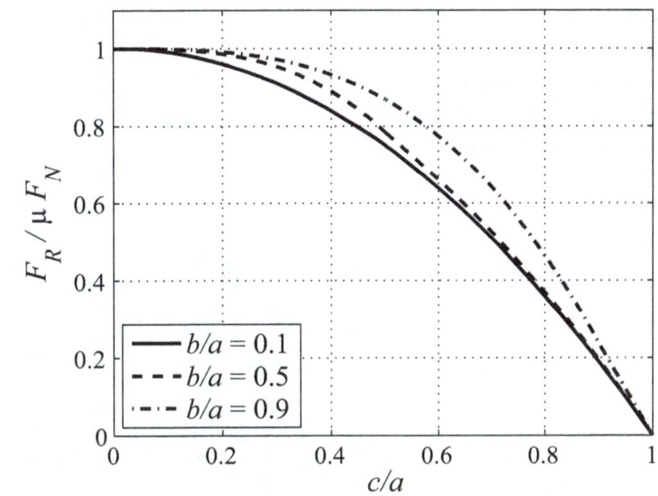

Fig. 4.28 Normalized tangential displacement as a function of the normalized radius of the stick zone for different values b/a when indenting by a cone with a rounded tip

with the displacement:

$$w_p(r;c) = \frac{c^2}{\pi b} \tan\theta \left[\left(2 - \frac{r^2}{c^2}\right) \arcsin\left(\frac{c}{r}\right) + \frac{\sqrt{r^2 - c^2}}{c} \right]. \tag{4.107}$$

Case 3: Complete Sliding

The solution here is a result of inserting $c = 0$ into solutions (4.106).

Figure 4.27 and 4.28 show the normalized values of the global contact sizes $u^{(0)}$ and F_R as a function of the normalized radius of the stick zone. For small values of b, one recognizes very well the limiting case of the ideal cone.

The curves of the normalized tangential stresses for individual normalized values of the two radii b and c are shown in Figs. 4.29 and 4.30. It is immediately apparent that for small values of b the curves of the ideal cone can be found. However, one also sees that the dependency for $c > b$ only weakly depends on b. For example, the curve of $c = 0.9a$ and $b = 0.5a$ is almost exactly that of the ideal cone at $c = 0.9a$, despite the rather large value of b.

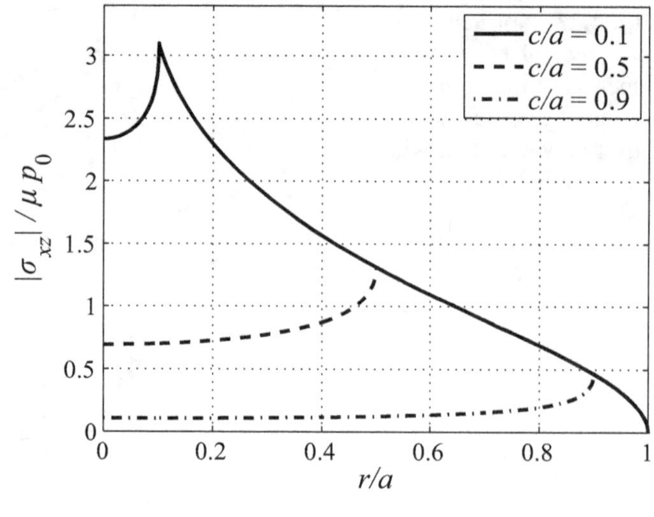

Fig. 4.29 Normalized tangential stresses in the indentation by a cone with a rounded tip for $b = 0.1a$ and different values of the normalized radius of the stick zone

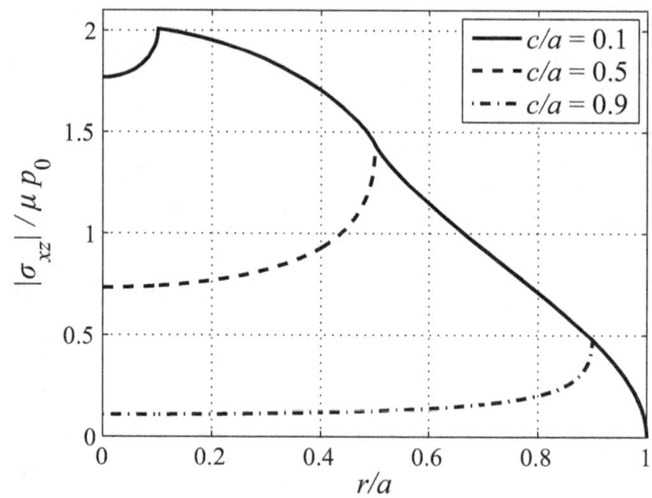

Fig. 4.30 Normalized tangential stresses in the indentation by a cone with a rounded tip for $b = 0.5a$ and different values of the normalized radius of the stick zone

4.6.9 The Paraboloid with Parabolic Cap

The method in this problem is completely analogous to that of the previous sections. The indenter profile can be described by the function:

$$f(r) = \begin{cases} \dfrac{r^2}{2R_1}, & r \le b, \\ \dfrac{r^2 - h^2}{2R_2}, & r > b. \end{cases} \tag{4.108}$$

Here, R_1 is the radius of curvature of the parabolic cap, and R_2 is the radius of the parabolic base body (see Fig. 4.31). The continuity of f implies

$$h^2 = b^2 \left(1 - \frac{R_2}{R_1}\right) \tag{4.109}$$

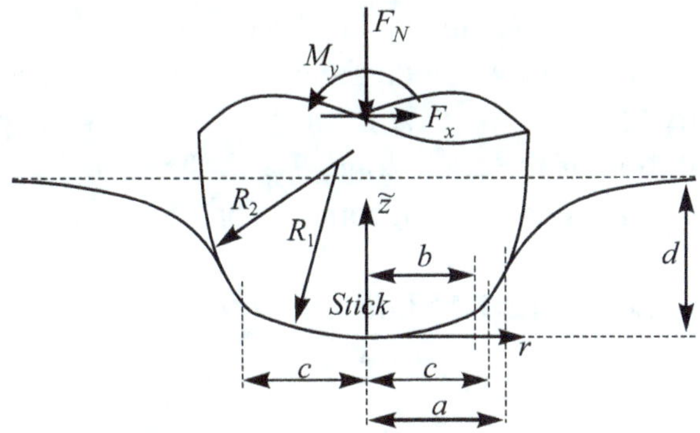

Fig. 4.31 Tangential contact between a rigid paraboloid with parabolic cap and an elastic half-space

for the length h. The normal contact problem with a contact radius $a \geq b$, the indentation depth d, the normal force F_N, and the normal stresses and displacements of the half-space σ_{zz} and w is described by the relations:

$$
\begin{aligned}
d(a) &= \frac{a^2}{R_1} + \frac{a}{R^*}\sqrt{a^2 - b^2}, \\
F_N(a) &= \frac{2E^*}{3}\left[\frac{2a^3}{R_1} + \frac{1}{R^*}\left(2a^2 + b^2\right)\sqrt{a^2 - b^2}\right], \\
\sigma_{zz}(r;a) &= -\frac{E^*}{\pi}\begin{cases} \dfrac{2\sqrt{a^2 - r^2}}{R_1} + \displaystyle\int_b^a \frac{\left(2x^2 - b^2\right)\mathrm{d}x}{R^*\sqrt{x^2 - b^2}\sqrt{x^2 - r^2}}, & r \leq b, \\ \dfrac{2\sqrt{a^2 - r^2}}{R_1} + \displaystyle\int_r^a \frac{\left(2x^2 - b^2\right)\mathrm{d}x}{R^*\sqrt{x^2 - b^2}\sqrt{x^2 - r^2}}, & b < r \leq a, \end{cases} \\
w(r;a) &= w_P(r;a;R = R_1) + w_{PS}(r;a;R = R^*), \quad r > a,
\end{aligned}
\tag{4.110}
$$

derived in Sect. 2.5.13. Here, w_P and w_{PS} denote the displacements in the indentation by a complete or truncated paraboloid:

$$
\begin{aligned}
w_P(r;a;R_1) &= \frac{a^2}{\pi R_1}\left[\left(2 - \frac{r^2}{a^2}\right)\arcsin\left(\frac{a}{r}\right) + \frac{\sqrt{r^2 - a^2}}{a}\right], \\
w_{PS}(r;a;R^*) &= \frac{2a}{\pi R^*}\sqrt{a^2 - b^2}\arcsin\left(\frac{a}{r}\right) \\
&\quad - \frac{1}{\pi R^*}\left[\left(r^2 - b^2\right)\arcsin\left(\frac{\sqrt{a^2 - b^2}}{\sqrt{r^2 - b^2}}\right) - \sqrt{a^2 - b^2}\sqrt{r^2 - a^2}\right].
\end{aligned}
\tag{4.111}
$$

R^* is an effective radius, which is described by the relation:

$$
R^* = \frac{R_1 R_2}{R_1 - R_2}. \tag{4.112}
$$

For $a < b$, we have a contact with a pure paraboloid with the radius of curvature R_1, for which the results can be looked up in Sect. 4.6.3. The solution of the tangential contact problem in the three cases already introduced in the previous section is as follows (all open functions in (4.113) and (4.114) can be taken from (4.108) to (4.112). $u^{(0)}$ denotes the tangential displacement of the rigid indenter, c the radius of the stick zone, F_R the tangential force, σ_{xz} the tangential stresses, and u the tangential displacements on the surface of the half-space).

Case 1: Partial Sliding with $c > b$

$$
\begin{aligned}
u^{(0)}(a,c) &= \frac{\mu E^*}{G^*}\left(\frac{a^2}{R_1} + \frac{a}{R^*}\sqrt{a^2-b^2} - \frac{c^2}{R_1} - \frac{c}{R^*}\sqrt{c^2-b^2}\right),\\
F_x(a,c) &= \frac{2\mu E^*}{3}\left[\frac{2a^3}{R_1} + \frac{1}{R^*}(2a^2+b^2)\sqrt{a^2-b^2}\right.\\
&\qquad \left. - \frac{2c^3}{R_1} - \frac{1}{R^*}(2c^2+b^2)\sqrt{c^2-b^2}\right],\\
\sigma_{xz}(r;a,c) &= \mu\begin{cases}\sigma_{zz}(r;a) - \sigma_{zz}(r;c), & r \le c,\\ \sigma_{zz}(r;a), & c < r \le a,\end{cases}\\
u(r,a,c) &= \frac{\mu E^*}{G^*}\begin{cases} d(a) - f(r) - w(r;c), & c < r \le a,\\ w(r;a) - w(r;c), & r > a.\end{cases}
\end{aligned}
\tag{4.113}
$$

Case 2: Partial Sliding with $c \le b$

$$
\begin{aligned}
u^{(0)}(a,c) &= \frac{\mu E^*}{G^*}\left(\frac{a^2}{R_1} + \frac{a}{R^*}\sqrt{a^2-b^2} - \frac{c^2}{R_1}\right),\\
F_x(a,c) &= \frac{2\mu E^*}{3}\left[\frac{2a^3}{R_1} + \frac{1}{R^*}\left(2a^2+b^2\right)\sqrt{a^2-b^2} - \frac{2c^3}{R_1}\right],\\
\sigma_{xz}(r;a,c) &= \mu\begin{cases}\sigma_{zz}(r;a) + \dfrac{2E^*}{\pi R_1}\sqrt{c^2-r^2}, & r \le c,\\ \sigma_{zz}(r;a), & c < r \le a,\end{cases}\\
u(r;a,c) &= \frac{\mu E^*}{G^*}\begin{cases} d(a) - f(r) - w_p(r;c;R_1), & c < r \le a,\\ w(r;a) - w_p(r;c;R_1), & r > a.\end{cases}
\end{aligned}
\tag{4.114}
$$

Case 3: Complete Sliding

The solution arises here by inserting $c = 0$ into (4.114).

One obtains the usual limiting cases for this indenter profile: for $R_1 = R_2$, respectively $R^* \to \infty$, the solution of Cattaneo and Mindlin from Sect. 4.6.3 can be used, for $R_1 \to \infty$ the solution of the truncated paraboloid from Sect. 4.6.6 and for $b = 0$ the solution of Cattaneo and Mindlin with radius R_2.

Now let us visualize some of the results obtained from this. For the sake of simplicity, let us choose $R_1 = R^*$. However, all of the effects characteristic of the indenter profile described in this section occur with this limitation.

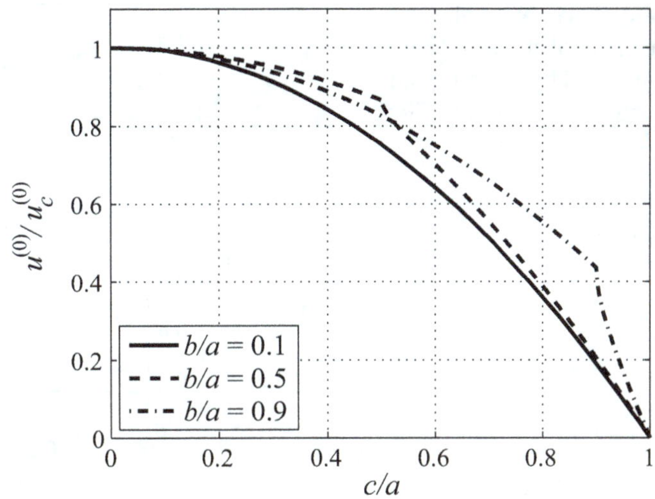

Fig. 4.32 Normalized tangential displacement as a function of the normalized radius of the stick zone for different values b/a when indented by a paraboloid with parabolic cap with $R_1 = R^*$

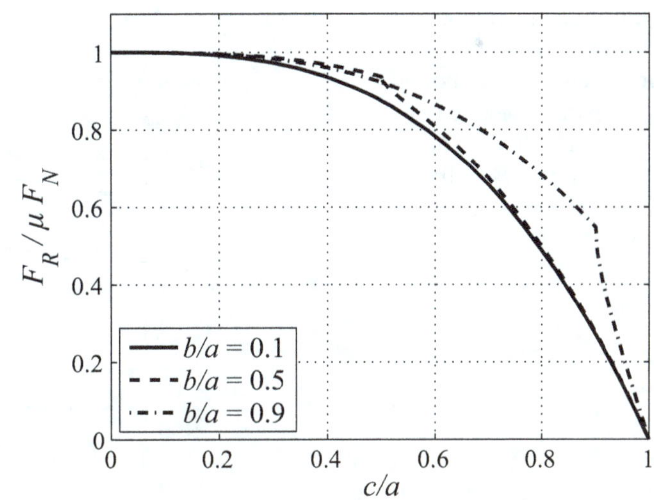

Fig. 4.33 Normalized tangential force as a function of the normalized stick radius for different values b/a when indented by a paraboloid with parabolic cap with $R_1 = R^*$

In Figs. 4.32 and 4.33 the normalized global tangential displacement $u^{(0)}$ and the normalized tangential force F_x are shown as a function of the normalized radius of the stick zone for different values of b. As always, one recognizes the characteristic limiting cases. In addition, one observes a kink at $c = b$, which has not yet appeared in the previous sections. This kink is due to the fact that the profile at $r = b$ is not continuously differentiable and yet radii of the stick zone $c < b$ are possible. This combination has not yet occurred in this chapter, or the kink corresponding to the transition to complete sliding of the contact.

Some normalized curves of the tangential stresses are shown in Figs. 4.34 and 4.35. As in the case of the cone with a rounded tip, the curves for $c > b$ barely differ from the dependencies for the ideal body; in this case the paraboloid from Sect. 4.6.3. In particular, the tangential stresses for $c > b$—in contrast to the normal stresses—are not singular at $r = b$.

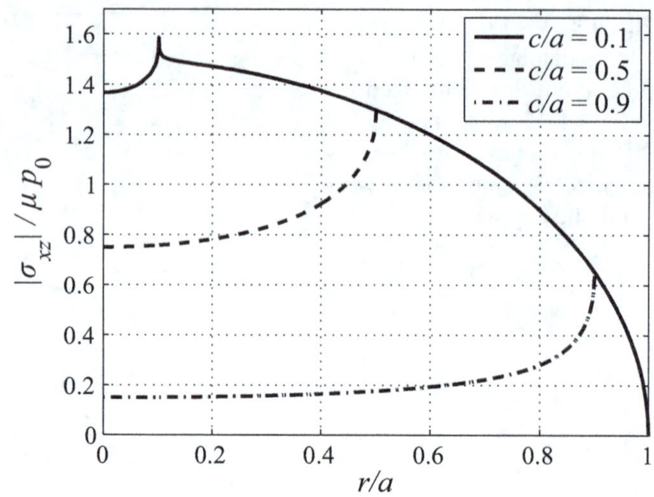

Fig. 4.34 Normalized tangential stresses when indented by a paraboloid with paraboloidal cap for $R_1 = R^*$, $b = 0.1a$ and different values of the normalized radius of the stick zone

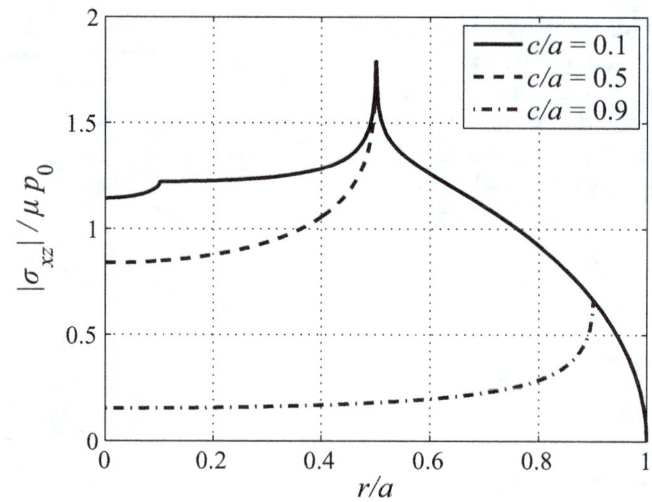

Fig. 4.35 Normalized tangential stresses when indented by a paraboloid with paraboloidal cap for $R_1 = R^*$, $b = 0.5a$ and different values of the normalized radius of the stick zone

4.6.10 The Cylindrical Flat Punch with a Rounded Edge

The tangential contact problem of a flat cylindrical punch with rounded corners and an elastic half-space (see Fig. 4.36) was first solved by Ciavarella (1999). The indenter profile has the shape

$$f(r) = \begin{cases} 0, & r \le b, \\ \dfrac{(r-b)^2}{2R}, & r > b, \end{cases} \tag{4.115}$$

with the radius of curvature R of the rounded corner and the radius b of the flat punch surface. The solution of the normal contact problem can be found in Chap. 2 (Sect. 2.5.14). As usual, a denotes the contact radius, d the indentation depth, and

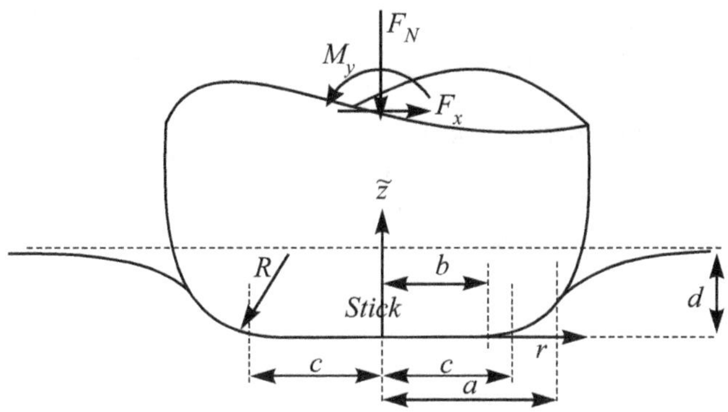

Fig. 4.36 Tangential contact between a rigid flat cylindrical punch with rounded corners and an elastic half-space

F_N the normal force. It is:

$$\begin{aligned} d(a) &= \frac{a^2}{R}(\sin\varphi_0 - \varphi_0\cos\varphi_0), \\ F_N(a) &= \frac{E^* a^3}{3R}\left[\sin\varphi_0(4-\cos^2\varphi_0) - 3\varphi_0\cos\varphi_0\right], \end{aligned} \tag{4.116}$$

with the angle:

$$\varphi_0 := \arccos\left(\frac{b}{a}\right). \tag{4.117}$$

The stresses and displacements in the normal direction are:

$$\begin{aligned} \sigma_{zz}(r;a) &= -\frac{E^*}{\pi R}\begin{cases} \displaystyle\int_b^a \left[2\sqrt{x^2-b^2} - b\arccos\left(\frac{b}{x}\right)\right]\frac{\mathrm{d}x}{\sqrt{x^2-r^2}}, & r \le b, \\ \displaystyle\int_r^a \left(2\sqrt{x^2-b^2} - b\arccos\left(\frac{b}{x}\right)\right)\frac{\mathrm{d}x}{\sqrt{x^2-r^2}}, & b < r \le a, \end{cases} \\ w(r;a) &= \frac{2d(a)}{\pi}\arcsin\left(\frac{a}{r}\right) \\ &\quad - \frac{2}{\pi}\left\{\int_b^a \frac{x}{R}\left[\sqrt{x^2-b^2} - b\arccos\left(\frac{b}{x}\right)\right]\frac{\mathrm{d}x}{\sqrt{r^2-x^2}}\right\}, \\ &\quad r > a. \end{aligned} \tag{4.118}$$

With a tangential loading of the contact, the radius of the stick zone cannot fall below the value of $c = b$, since the contact will begin to slide completely at this value. However, this makes it very easy to specify the solution of the tangential contact problem ($u^{(0)}$ denotes the tangential displacement of the rigid indenter, c the radius of the stick zone, F_x the tangential force, σ_{xz} the tangential stresses, and u the tangential displacements at the surface of the half-space). With help from the

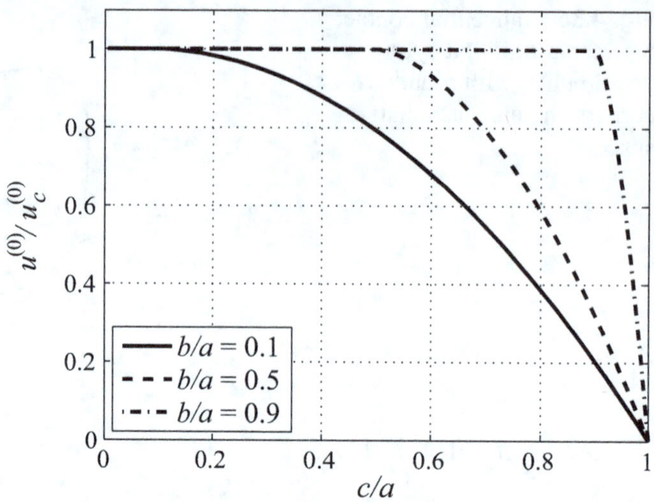

Fig. 4.37 Normalized tangential displacement as a function of the normalized radius of the stick zone for different values b/a when indenting with a flat punch with a rounded edge

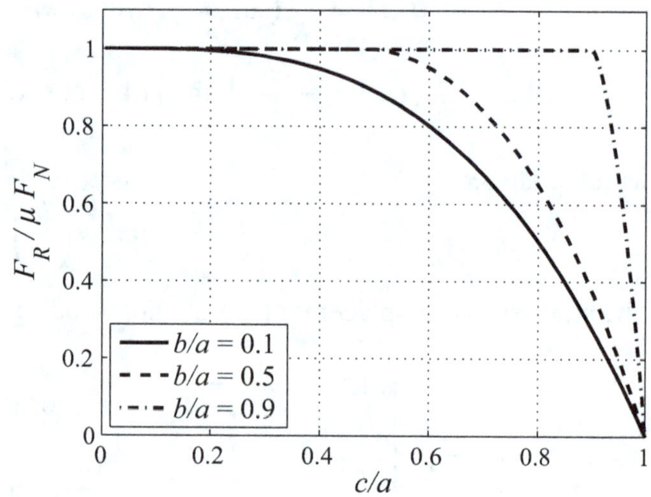

Fig. 4.38 Normalized tangential force as a function of the normalized radius of the stick zone for different values b/a when indenting with a flat punch with a rounded edge

general relations (4.38), one obtains:

$$\begin{aligned}
u^{(0)}(a,c) &= \frac{\mu E^*}{G^* R}\left[a^2(\sin\varphi_0 - \varphi_0\cos\varphi_0) - c^2(\sin\psi_0 - \psi_0\cos\psi_0)\right],\\
F_x(a,c) &= \frac{E^*}{3R}\{a^3\left[\sin\varphi_0(4-\cos^2\varphi_0) - 3\varphi_0\cos\varphi_0\right]\\
&\qquad - c^3\left[\sin\psi_0(4-\cos^2\psi_0) - 3\psi_0\cos\psi_0\right]\},\\
\sigma_{xz}(r;a,c) &= \mu\begin{cases}\sigma_{zz}(r;a) - \sigma_{zz}(r;c), & r \le c,\\ \sigma_{zz}(r;a), & c < r \le a,\end{cases}\\
u(r;a,c) &= \frac{\mu E^*}{G^*}\begin{cases} d(a) - f(r) - w(r;c), & c < r \le a,\\ w(r;a) - w(r;c), & r > a,\end{cases}
\end{aligned} \tag{4.119}$$

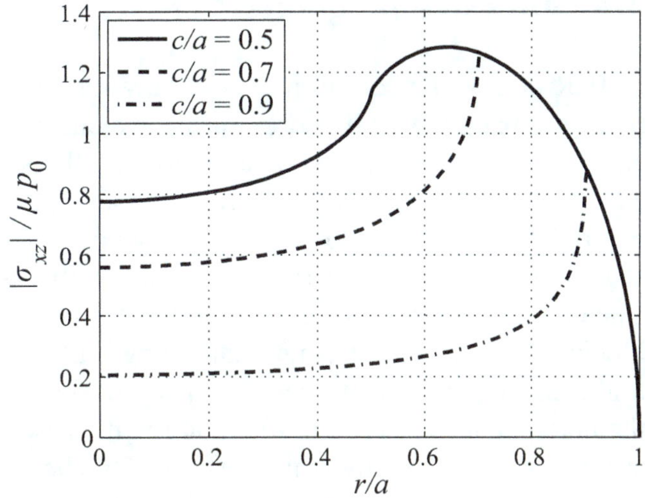

Fig. 4.39 Normalized tangential stresses in contact for different values of the normalized radius of the stick zone c/a, while indenting with a flat punch with rounded corners with $b = 0.49a$

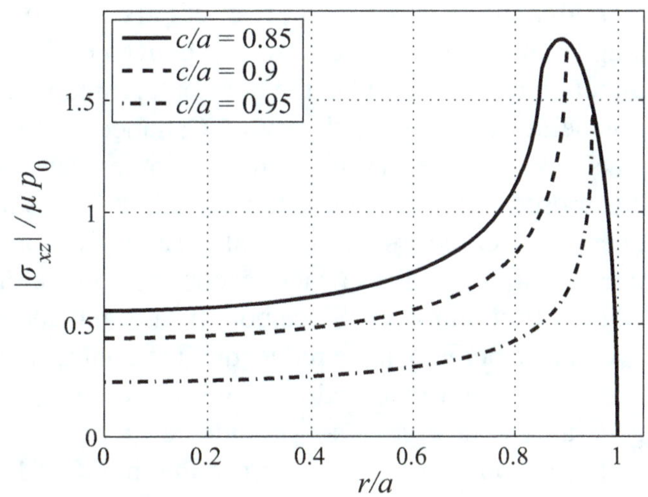

Fig. 4.40 Normalized tangential stresses in contact for different values of the normalized radius of the stick zone c/a, while indenting with a flat punch with rounded corners with $b = 0.84a$

with

$$\psi_0 := \arccos\left(\frac{b}{c}\right). \tag{4.120}$$

The solution for complete sliding is achieved by inserting $c = b$ into (4.119).

In Figs. 4.37 and 4.38 the tangential displacement and force are shown in normalized variables as functions of the normalized radius of the stick zone. Some curves of the tangential stresses are shown as an example in Figs. 4.39 and 4.40.

4.7 Adhesive Tangential Contact

In JKR theory, the equilibrium configuration of an adhesive contact is determined by minimizing the total energy of the system, which consists of the energy of the elastic deformation of the contact partners, the surface energy, and the work from external forces. Since this total energy does not depend on the tangential displacement, the JKR contact does not formally possess any "tangential strength". Strictly speaking, the absence of friction is one of the assumptions of the JKR solution, since it uses the solutions of the frictionless normal contact problem as its building blocks. Yet the lack of tangential strength of adhesive contacts is obviously contradicted by experimental results. The physical cause of this contradiction lies in the heterogeneous structure found at the microscopic scale (or at the atomic scale, at the very least) of any real interface. This heterogeneity leads to a finite contact strength (or static friction) in the tangential direction.

In this subsection, we will restrict ourselves to the simplest model of an adhesive tangential contact problem, with a conveniently defined "adhesion" in the normal direction and "friction" in the horizontal direction. This problem can be considered a generalization of the theory of Cattaneo and Mindlin to include adhesive contacts. We assume that the adhesive forces have sufficient range to be considered "macroscopic" with regards to the friction forces in the contact. In other words, we operate under the assumption that the adhesive forces create additional macroscopic pressure in the contact which, according to Coulomb's law of friction, leads to increased friction forces. Since both the normal and the tangential contact problem of two elastic bodies can be reduced to the contact between a rigid body and an elastic half-space (with modified material properties), we will consider—without loss of generality—the case of a rigid indenter in contact with an elastic half-space.

For the adhesive forces, we use the model of Dugdale where the adhesive pressure remains constant up to a certain distance h between the surfaces and abruptly drops to zero after that distance (3.147). The theory of adhesive contacts for this particular interaction was created by Maugis and is featured in Sect. 3.8 of this book. Therefore, the following theory can also be viewed as a generalization of the theory by Maugis relating to adhesive tangential contacts.

Let us consider a rigid indenter with a three-dimensional axially symmetric profile $f(r)$ and the corresponding MDR transformed profile $g(x)$. The profile $g(x)$ is initially pressed into the Winkler foundation, which are defined by the MDR rules, by d and subsequently displaced by $u^{(0)}$ in the tangential direction. The corresponding adhesive normal contact problem was solved in Sect. 3.8, from which the entire notation is inherited. For a sufficiently small rigid body displacement, the springs near the edge of the contact (contact radius a) slip while the springs in the interior zone (with the radius c) remain sticking. The radius c of the stick zone is given by the equation:

$$G^* u^{(0)} = \mu \left(2\sigma_0 \sqrt{b^2 - c^2} + E^* \left[d - g(c) \right] \right), \tag{4.121}$$

which sets the tangential force at the position c equal to the product of the normal force and the coefficient of friction. The total tangential force is calculated by integrating over all springs in the contact:

$$
\begin{aligned}
F_x &= 2\int_0^c G^* u^{(0)} \mathrm{d}x + 2\mu \int_c^a q_z(x)\mathrm{d}x \\
&= 2G^* u^{(0)} c + 2\mu \int_c^a \left(2\sigma_0\sqrt{b^2-x^2} + E^*[d-g(x)]\right)\mathrm{d}x \\
&= 2G^* u^{(0)} c + 2\mu\sigma_0 \left[x\sqrt{b^2-x^2} + b^2 \arcsin\frac{x}{b}\right]\Bigg|_c^a \\
&\quad + 2\mu E^* d(a-c) - 2\mu E^* \int_c^a g(x)\mathrm{d}x.
\end{aligned}
\tag{4.122}
$$

The first term to the right-hand side is the contribution of the inner stick zone (rigid translation $u^{(0)}$). The second term is a contribution from springs in the slip zone (distributed load in the normal direction consisting of the elastic component $E^*[d-g(x)]$ and adhesive component $2\sigma_0\sqrt{b^2-x^2}$ given by (3.148), and multiplied with the coefficient of friction). Radius b of the interaction zone is determined from the equations of the normal contact problem which are listed in Sect. 3.8.

A complete theory of the tangential contact problem under those assumptions can be found in (Popov and Dimaki 2016). Here our consideration only deals with the limiting case of very short-range adhesive interactions (i.e., the parameter h is much smaller than all other characteristic system measures). This limiting case corresponds to the JKR approximation for the adhesive normal contact. The difference between the contact radius a and the radius of the adhesive interaction $b > a$ should, therefore, remain small:

$$
\varepsilon = b - a \ll a, b. \tag{4.123}
$$

In this approximation (see (3.160)),

$$
\varepsilon = \frac{\pi h E^*}{4\sigma_0}. \tag{4.124}
$$

For the normal contact problem, expanding by the small parameter ε yields (see (3.161)):

$$
\begin{aligned}
d &\approx g(a) - \sqrt{\frac{2\pi a \Delta\gamma}{E^*}}, \\
F_{N,\mathrm{JKR}}(a) &\approx 2E^* \int_0^a [d - g(x)]\mathrm{d}x,
\end{aligned}
\tag{4.125}
$$

with

$$\Delta\gamma = \sigma_0 h, \tag{4.126}$$

which represent the exact JKR solution for an arbitrary axially symmetric profile.

Substituting $b = a + \varepsilon$ with ε from (4.124) into (4.121), and expanding to the lowest order of ε gives:

$$\begin{aligned}\frac{G^* u^{(0)}}{\mu} &= \sqrt{4\sigma_0^2(a^2 - c^2) + 2\pi a E^* \Delta\gamma} - \sqrt{2\pi a E^* \Delta\gamma} \\ &\quad + E^*[g(a) - g(c)].\end{aligned} \tag{4.127}$$

This equation determines the relationship between the radius of the stick zone c and the tangential body displacement $u^{(0)}$. It shows that even arbitrarily small tangential displacements induce partial slip in a narrow peripheral zone. This is in complete analogy to the case of the non-adhesive contact. The critical tangential displacement for the transition from partial to complete slip is calculated by setting $c = 0$:

$$\begin{aligned}u_c^{(0)} &= \frac{\mu}{G^*}\left(\sqrt{4\sigma_0^2 a^2 + 2\pi a E^* \Delta\gamma} + E^* g(a) - \sqrt{2\pi a E^* \Delta\gamma}\right) \\ &= \mu\left(\frac{1}{G^*}\sqrt{4\sigma_0^2 a^2 + 2\pi a E^* \Delta\gamma} + \frac{E^*}{G^*} d\right) \approx \mu\left(\frac{2\sigma_0}{G^*} a + \frac{E^*}{G^*} d\right).\end{aligned} \tag{4.128}$$

The approximation yields the tangential force:

$$\begin{aligned}\frac{F_x}{\mu} &= c\sqrt{4\sigma_0^2(a^2 - c^2) + 2\pi a E^* \Delta\gamma} - a\sqrt{2\pi a E^* \Delta\gamma} \\ &\quad + 2E^*[a g(a) - c g(c)] \\ &\quad + \pi\sigma_0 a^2\left(1 - \frac{2}{\pi}\arcsin\frac{c}{a}\right) - 2E^* \int_c^a g(x)\mathrm{d}x.\end{aligned} \tag{4.129}$$

At the onset of complete slip ($c = 0$) it is:

$$\begin{aligned}\frac{F_x}{\mu} &= \pi\sigma_0 a^2 + 2E^* a g(a) - a\sqrt{2\pi a E^* \Delta\gamma} - 2E^* \int_0^a g(x)\mathrm{d}x \\ &= \sigma_0 \pi a^2 + 2E^* \int_0^a [d - g(x)]\,\mathrm{d}x = \pi a^2 \sigma_0 + F_{N,\mathrm{JKR}}(a),\end{aligned} \tag{4.130}$$

where $F_{N,\mathrm{JKR}}(a)$ represents the solution of the adhesive normal contact problem in JKR approximation.

The final (4.130) could be written directly without the preceding calculations since the expression in parentheses ($\pi a^2 \sigma_0 + F_{N,\mathrm{JKR}}(a)$) is simply the total compressive force in the contact (sum of the elastic force and the additional adhesive compressive force). According to Coulomb's law of friction, the tangential force for complete slip is equal to the total compressive force of the contacting surfaces multiplied with the coefficient of friction (independent of the particular pressure distribution in the contact).

4.7.1 The Paraboloid

A parabolic profile $\tilde{z} = f(r) = r^2/(2R)$ implies $g(x) = x^2/R$, and (4.127) and (4.129) take on the following form:

$$\begin{aligned}\frac{G^* u^{(0)}}{\mu} &= \sqrt{4\sigma_0^2(a^2 - c^2) + 2\pi a E^* \Delta\gamma} - \sqrt{2\pi a E^* \Delta\gamma} \\ &\quad + \frac{E^*}{R}(a^2 - c^2), \\ \frac{F_x}{\mu} &= c\sqrt{4\sigma_0^2(a^2 - c^2) + 2\pi a E^* \Delta\gamma} - a\sqrt{2\pi a E^* \Delta\gamma} \\ &\quad + \pi\sigma_0 a^2 \left(1 - \frac{2}{\pi}\arcsin\frac{c}{a}\right) + \frac{4}{3}\frac{E^*}{R}\left(a^3 - c^3\right).\end{aligned} \tag{4.131}$$

References

Cattaneo, C.: Sul Contatto di due Corpore Elastici: Distribuzione degli sforzi. Rendiconti Dell' Acad. Nazionale Dei Lincei **27**, 342–348, 434–436, 474–478 (1938)

Cerruti, V.: Ricerche intorno all' equilibrio de' corpi elastici isotropi. Memorie Dell'accademia Nazionale Dei Lincei **13**(3), 81–122 (1882)

Ciavarella, M.: Tangential loading of general three-dimensional contacts. J. Appl. Mech. **65**(4), 998–1003 (1998)

Ciavarella, M.: Indentation by nominally flat or conical indenters with rounded corners. Int. J. Solids Struct. **36**(27), 4149–4181 (1999)

Hamilton, G.M., Goodman, L.E.: The stress field created by a circular sliding contact. J. Appl. Mech. **33**(2), 371–376 (1966)

Hertz, H.: Über die Berührung fester elastischer Körper. J. Reine Angew. Math. **92**, 156–171 (1882)

Jäger, J.: Elastic contact of equal spheres under oblique forces. Arch. Appl. Mech. **63**(6), 402–412 (1993)

Jäger, J.: A new principle in contact mechanics. J. Tribol. **120**(4), 677–684 (1998)

Johnson, K.L.: Surface interaction between elastically loaded bodies under tangential forces. Proc. R. Soc. London Ser. A **230**, 531–549 (1955)

Johnson, K.L.: Contact mechanics. Cambridge University Press, Cambridge (1985)

Landau, L.D., Lifshitz, E.M.: Elastizitätstheorie. Lehrbuch der theoretischen Physik, vol. 7. Akademie Verlag, Berlin (1991)

Mindlin, R.D.: Compliance of elastic bodies in contact. J. Appl. Mech. **16**, 259–268 (1949)

Mindlin, R.D., Deresiewicz, H.: Elastic spheres in contact under varying oblique forces. J. Appl. Mech. **20**, 327–344 (1953)

Popov, V.L., Dimaki, A.V.: Friction in an adhesive tangential contact in the Coulomb-Dugdale approximation. J. Adhes. **93**(14), 1131–1145 (2017)

Popov, V.L., Heß, M.: Method of dimensionality reduction in contact mechanics and friction. Springer, Heidelberg (2015). ISBN 978-3-642-53875-9

Truman, C.E., Sackfield, A., Hills, D.A.: Contact mechanics of wedge and cone indenters. Int. J. Mech. Sci. **37**(3), 261–275 (1995)

Vermeulen, P.J., Johnson, K.L.: Contact of Non-spherical Elastic Bodies Transmitting Tangential Forces. J. Appl. Mech. **31**(2), 338–340 (1964)

Torsional Contact

5

This chapter is dedicated to contacts between a rigid, rotationally symmetric indenter and an elastic half-space, which are subjected to a twisting moment along the z-axis in the normal direction of the half-space. The fundamental equations of elastostatics exhibit an interesting property; that purely torsional problems are generally elastically decoupled in cases of rotational symmetry. This means that the tangential displacements u_φ in no way affects the radial and normal displacements. (Note regarding the notation in this chapter: contrary to the previous chapter, the word "tangential" refers to "circumferential direction" in this chapter. With this in mind, all tangential displacements will be denoted by u with the corresponding index of the tangential direction, i.e., u_x, u_y, u_φ, etc. Normal displacements will retain the notation w.) However, in spite of the elastic decoupling, there exists the coupling caused by friction. We initially consider contacts without slip, which accordingly are decoupled from the normal contact problem, and we subsequently examine finite coefficients of friction.

5.1 No-Slip Contacts

5.1.1 The Cylindrical Flat Punch

Twisting a rigid flat punch of radius a that is in no-slip contact with an elastic half-space (see Fig. 5.1) results in the tangential displacements caused by the rigid body rotation of the punch by the torsion angle φ:

$$u_\varphi(r) = r\varphi, \quad r \leq a. \tag{5.1}$$

V.L. Popov, M. Heß, E. Willert, *Handbook of Contact Mechanics*,
https://doi.org/10.1007/978-3-662-58709-6_5

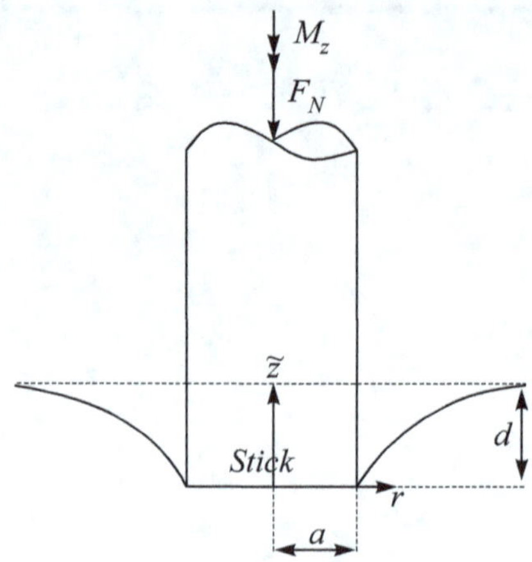

Fig. 5.1 Torsional contact between a rigid flat punch and an elastic half-space

The torsional stresses $\sigma_{\varphi z}$ in the contact area and the outer displacements u_φ are given by:

$$\sigma_{\varphi z}(r) = -\frac{4G\varphi}{\pi}\frac{r}{\sqrt{a^2 - r^2}}, \quad r \le a,$$
$$u_\varphi(r) = \frac{2}{\pi}\varphi\left[r \arcsin\left(\frac{a}{r}\right) - a\sqrt{1 - \frac{a^2}{r^2}}\right], \quad r > a, \tag{5.2}$$

where G is the shear modulus. The stresses and displacements are represented in a normalized form in Figs. 5.2 and 5.3. The total torsional moment is:

$$M_z = \frac{16}{3}Ga^3\varphi. \tag{5.3}$$

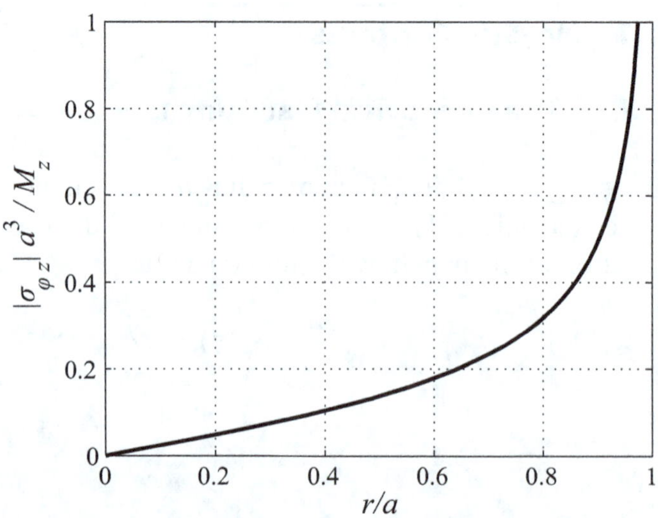

Fig. 5.2 Normalized torsional stresses as a function of the radial coordinate for torsion by a flat cylindrical punch

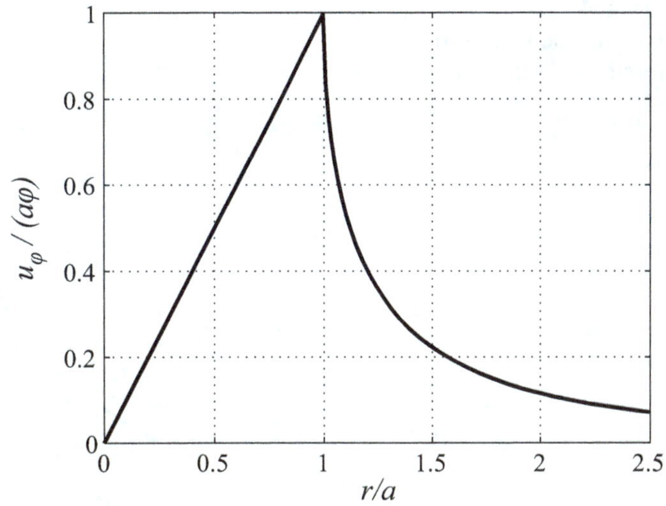

Fig. 5.3 Normalized tangential displacements as a function of the radial coordinate for torsion by a flat cylindrical punch

5.1.2 Displacement from Torsion by a Thin Circular Ring

We will now consider the torsion of the elastic half-space induced by a moment M_z that is distributed over a thin circular ring of radius a. Let the ring be sufficiently thin for the stress distribution to be described by a Dirac function:

$$\sigma_{\varphi z}(r) = -\frac{M_z}{2\pi a^2}\delta(r-a). \tag{5.4}$$

The resulting displacement of the half-space can be gained from the superposition of the fundamental solutions of elasticity theory. A point force F_x acting in the x-direction on the origin causes the tangential displacements at the half-space surface (Johnson 1985) of:

$$\begin{aligned} u_x &= \frac{F_x}{2\pi G}\left[\frac{1}{s}(1-\nu)+\nu\frac{x^2}{s^3}\right], \\ u_y &= \frac{\nu F_x}{2\pi G}\frac{xy}{s^3}, \end{aligned} \tag{5.5}$$

with s being the distance from the point of the force application. A force in the y-direction results in correspondingly identical expressions for the opposite coordinates. A slightly involved yet elementary calculation yields the displacements caused by the stress distribution (5.4):

$$\begin{aligned} u_\varphi(r;a) &= \frac{1}{2\pi G}\int_0^{2\pi}\frac{M_z}{2\pi a}\frac{\cos\varphi\,\mathrm{d}\varphi}{\sqrt{a^2+r^2-2ar\cos\varphi}} \\ &= \frac{M_z}{2\pi^2 G a^2}\left[\frac{r^2+a^2}{r^2+ar}\mathrm{K}\left(\frac{2\sqrt{ra}}{r+a}\right)-\frac{(r+a)^2}{r^2+ar}\mathrm{E}\left(\frac{2\sqrt{ra}}{r+a}\right)\right], \end{aligned} \tag{5.6}$$

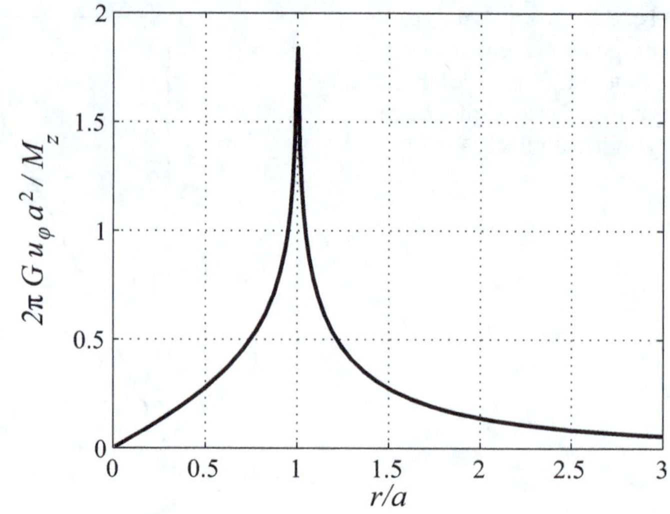

Fig. 5.4 Tangential displacement of the surface from torsion by a thin circular ring

with the complete elliptical integrals of the first and second kind:

$$\mathrm{K}(k) := \int_0^{\pi/2} \frac{\mathrm{d}\varphi}{\sqrt{1-k^2\sin^2\varphi}},$$

$$\mathrm{E}(k) := \int_0^{\pi/2} \sqrt{1-k^2\sin^2\varphi}\,\mathrm{d}\varphi. \tag{5.7}$$

These displacements are displayed in Fig. 5.4 and allow the direct calculation of the displacements from a given rotationally symmetric torsional stress distribution.

5.2 Contacts with Slip

We now consider contacts which are simultaneously loaded in the z-direction by a normal force F_N and a twisting moment M_z. Once again, the problem can be reduced to the contact between a rigid indenter and an elastic half-space by introducing the effective modulus of elasticity:

$$\frac{1}{E^*} = \frac{1-\nu_1}{2G_1} + \frac{1-\nu_2}{2G_2} = \frac{1-\nu}{2G}, \tag{5.8}$$

with the shear moduli G_i and Poisson's-ratios ν_i. The index "1" denotes the indenter and "2" the half-space. Many statements from Chap. 4 concerning tangential contacts with slip also hold true for torsional contacts with slip: the contacts exhibit *hysteresis* and *memory*, i.e., the solution of the contact problem is dependent on the loading history. Once again, we restrict ourselves to contacts with a constant normal force and a subsequently applied, increasing twisting moment. This induces a slip zone of radius a, which gradually expands inwards from the boundary of the contact. The inner stick zone is characterized by the radius c.

Contrary to tangential contacts, there exists no theorem for torsional contacts permitting the reduction to the solution of the frictionless normal contact. Nevertheless, Jäger (1995) published a general solution for arbitrary rotationally symmetric indenters with a profile shape $\tilde{z} := f(r)$. The boundary conditions for the normal and tangential stresses σ_{zz} and $\sigma_{\varphi z}$ as well as the normal and tangential displacements w and u_φ at the surface of the half-space are:

$$
\begin{aligned}
w(r) &= d - f(r), && r \le a,\\
u_\varphi(r) &= r\varphi, && r \le c,\\
\sigma_{\varphi z}(r) &= \mu\sigma_{zz}(r), && c < r \le a,\\
\sigma_{zz}(r) &= 0, && r > a,\\
\sigma_{\varphi z}(r) &= 0, && r > a,
\end{aligned}
\tag{5.9}
$$

with the indentation depth d, the torsion angle φ, and the coefficient of friction μ. We assume that the pure normal contact problem has been solved and the corresponding normal stresses σ_{zz} are known (refer to Chap. 2 for details). The solution of the torsion problem then requires just a single function $\tilde{\phi}$, which can be determined from the condition:

$$
\tilde{\phi}(x; a) = -\frac{\mu}{2G}\int_x^a \sigma_{zz}(r)\frac{\mathrm{d}r}{\sqrt{r^2 - x^2}}. \tag{5.10}
$$

The relationship between the torsion angle and the two characteristic contact radii c and a is then given by:

$$
\varphi = \tilde{\phi}(c; a). \tag{5.11}
$$

Moreover, the twisting moment can be calculated from the equation:

$$
\begin{aligned}
M_z(c, a) &= 16G\left(\varphi\frac{c^3}{3} + \int_c^a x^2\tilde{\phi}(x, a)\mathrm{d}x\right)\\
&= 16G\varphi\frac{c^3}{3} - 4\mu\int_c^a \left[c\sqrt{r^2 - c^2} + r^2 \arccos\left(\frac{c}{r}\right)\right]\sigma_{zz}(r)\mathrm{d}r,
\end{aligned}
\tag{5.12}
$$

and the torsional stresses in the stick zone are determined by the relationship:

$$
\sigma_{\varphi z}(r) = -\frac{4Gr}{\pi}\int_c^a \frac{\mathrm{d}\tilde{\phi}(x, a)}{\mathrm{d}x}\frac{\mathrm{d}x}{\sqrt{x^2 - r^2}}, \quad r \le c. \tag{5.13}
$$

The tangential displacements in the slip zone are given by the relationship:

$$u_{\varphi}(r)=r\left[\varphi-\frac{4}{\pi r^{2}}\int_{c}^{r}x^{2}\left[\varphi-\tilde{\phi}(x;a)\right]\frac{\mathrm{d}x}{\sqrt{r^{2}-x^{2}}}\right],\quad c<r\leq a. \tag{5.14}$$

Equations (5.10) to (5.14) completely solve the described torsion problem. The problem is fully defined by providing the indenter profile and one quantity of each trio $\{d, a, F_N\}$ and $\{\varphi, c, M_z\}$. We assume in both cases that the two radii are the given quantities. Regrettably, these relationships very rarely permit an analytical solution. Therefore, we will limit the scope of detail provided in this book to the indentation by a flat cylindrical punch and by a paraboloid. The following contact problems occur, for example, in stick-slip (purely mechanical) rotary drive systems. There, the most commonly used indenter shape is the sphere (see Sect. 5.2.2).

5.2.1 The Cylindrical Flat Punch

In Chap. 2 (see Sect. 2.5.1) we considered the normal indentation of an elastic half-space to the depth d by a rigid, flat cylindrical indenter of radius a. The following stress distribution was found:

$$\sigma_{zz}(r;d)=-\frac{E^{*}d}{\pi\sqrt{a^{2}-r^{2}}},\quad r\leq a. \tag{5.15}$$

Using (5.10) we obtain:

$$\tilde{\phi}(x;a)=\frac{\mu E^{*}d}{2\pi G}\int_{x}^{a}\frac{1}{\sqrt{a^{2}-r^{2}}}\frac{\mathrm{d}r}{\sqrt{r^{2}-x^{2}}}=\frac{\mu E^{*}d}{2\pi Ga}\mathrm{K}\left(\sqrt{1-\frac{x^{2}}{a^{2}}}\right), \tag{5.16}$$

with the complete elliptical integral of the first kind:

$$\mathrm{K}(k):=\int_{0}^{\pi/2}\frac{\mathrm{d}\varphi}{\sqrt{1-k^{2}\sin^{2}\varphi}}. \tag{5.17}$$

The torsion problem is solved through the following relationships between the global torsion angle φ, the radius of the stick zone c, and the torsion moment M_z as well as the tangential stresses $\sigma_{\varphi z}$:

$$\varphi=\frac{\mu E^{*}d}{2\pi Ga}\mathrm{K}\left(\sqrt{1-\frac{c^{2}}{a^{2}}}\right),$$

$$M_{z}=16G\varphi\frac{c^{3}}{3}+\frac{8\mu E^{*}d}{\pi a}\int_{c}^{a}x^{2}\mathrm{K}\left(\sqrt{1-\frac{x^{2}}{a^{2}}}\right)\mathrm{d}x,$$

$$\sigma_{\varphi z}(r) = -\frac{4Gr}{\pi}\left[\int_c^a \frac{\mathrm{d}\tilde{\phi}(x,a)}{\mathrm{d}x}\frac{\mathrm{d}x}{\sqrt{x^2-r^2}} + \frac{\tilde{\phi}(a,a)}{\sqrt{a^2-r^2}}\right], \quad r \le c$$

$$= -\frac{2\mu E^* dr}{\pi^2 a} \cdot \left\{\int_c^a \left[x\mathrm{K}\left(\sqrt{1-\frac{x^2}{a^2}}\right) - \frac{a^2}{x}\mathrm{E}\left(\sqrt{1-\frac{x^2}{a^2}}\right)\right]\frac{\mathrm{d}x}{(a^2-x^2)\sqrt{x^2-r^2}} + \frac{\pi}{2\sqrt{a^2-r^2}}\right\}, \tag{5.18}$$

with the complete elliptical integral of the second kind:

$$\mathrm{E}(k) := \int_0^{\pi/2} \sqrt{1-k^2\sin^2\varphi}\,\mathrm{d}\varphi. \tag{5.19}$$

For the sake of brevity, the tangential displacements in their general form will not be repeated here as they have already been demonstrated in (5.14). Complete stick in the contact is possible even for a non-vanishing torsion angle and a corresponding torque. The respective limiting values are:

$$\varphi(c=a) = \varphi_c = \frac{\mu E^* d}{4Ga} = \frac{\mu d}{2(1-\nu)a},$$
$$M_z(c=a) = M_c = 16G\varphi_c\frac{a^3}{3} = \frac{4}{3}\mu E^* da^2 = \frac{2}{3}\mu F_N a. \tag{5.20}$$

The torsional moment and torsion angle, both normalized to these critical values, are displayed in Figs. 5.5 and 5.6 as functions of the normalized radius of the stick

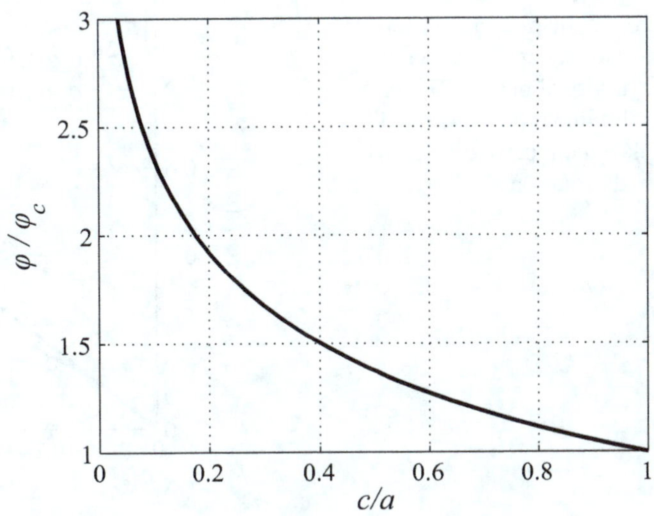

Fig. 5.5 Torsion angle, normalized to the critical value for complete stick, as a function of the normalized radius of the stick zone for the torsional contact with a flat punch

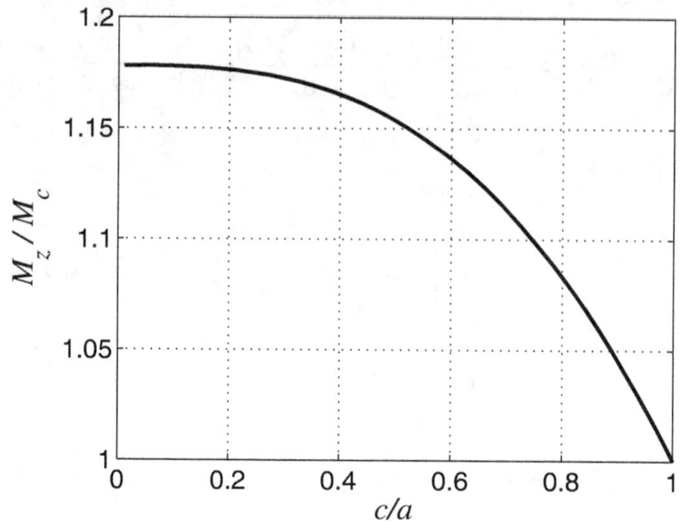

Fig. 5.6 Torsional moment, normalized to the critical value for complete stick, as a function of the normalized radius of the stick zone for the torsional contact with a flat punch

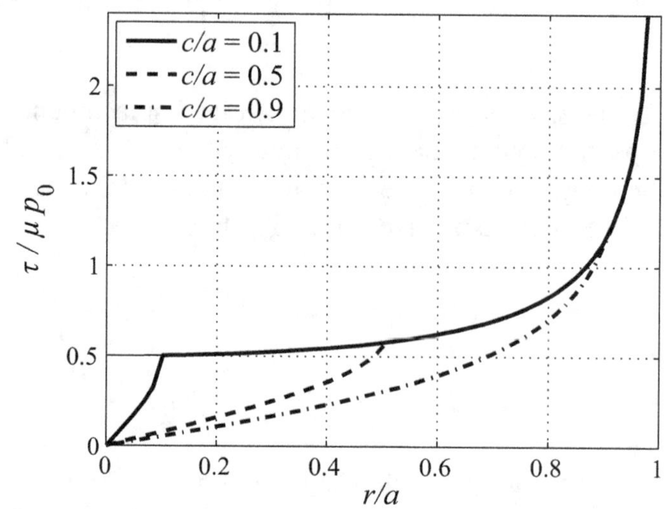

Fig. 5.7 Torsional stresses normalized to the average pressure p_0 multiplied with the coefficient of friction μ for the torsional contact with a flat cylindrical punch. The thin solid line represents the stress distribution for complete slip

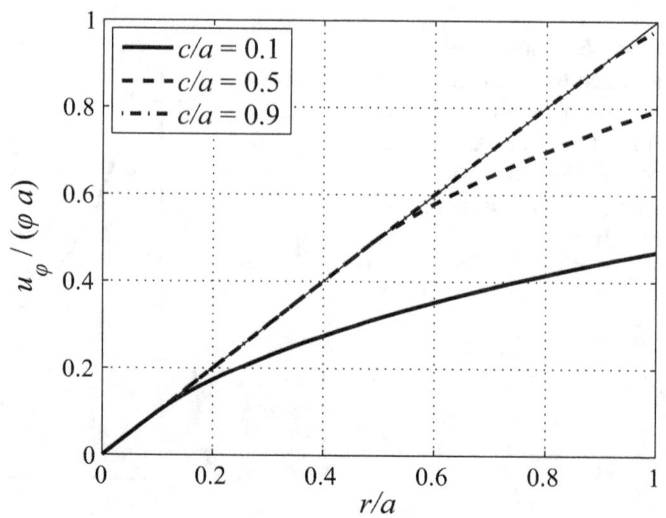

Fig. 5.8 Normalized tangential displacements for the torsional contact with a flat cylindrical punch. The thin solid line represents the displacement caused by the rigid body rotation

zone. The variation of the torsion moment during the transition from complete stick to complete slip is less than 18%. The normalized distribution of the torsional stresses is given in Fig. 5.7 and tangential displacements is provided in Fig. 5.8.

5.2.2 The Paraboloid

In Chap. 2 (see Sect. 2.5.3), for a paraboloid with the curvature radius R and the corresponding profile shape

$$f(r) = \frac{r^2}{2R}, \tag{5.21}$$

the following solution of the normal contact problem was derived:

$$\begin{aligned} d(a) &= \frac{a^2}{R}, \\ F_N(a) &= \frac{4}{3}\frac{E^* a^3}{R}, \\ \sigma_{zz}(r;a) &= -\frac{2E^*}{\pi R}\sqrt{a^2 - r^2}, \quad r \le a. \end{aligned} \tag{5.22}$$

As usual, a denotes the contact radius, d the indentation depth, F_N the normal force, and σ_{zz} the normal stress distribution in the contact.

Taking (5.10) into account gives us (for $\tilde{\phi}$):

$$\tilde{\phi}(x;a) = \frac{\mu E^* a}{\pi G R}\left[\mathrm{K}\left(\sqrt{1 - \frac{x^2}{a^2}}\right) - \mathrm{E}\left(\sqrt{1 - \frac{x^2}{a^2}}\right)\right], \tag{5.23}$$

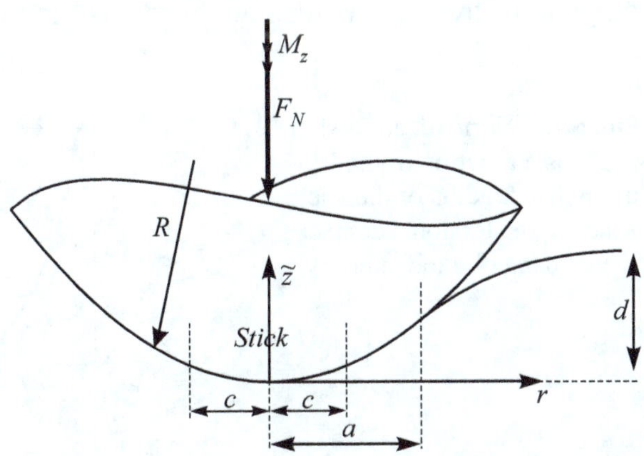

Fig. 5.9 Torsional contact between a rigid paraboloid and an elastic half-space

with the complete elliptical integrals of the first and second kind:

$$\mathrm{K}(k) := \int_0^{\pi/2} \frac{\mathrm{d}\varphi}{\sqrt{1-k^2\sin^2\varphi}},$$

$$\mathrm{E}(k) := \int_0^{\pi/2} \sqrt{1-k^2\sin^2\varphi}\,\mathrm{d}\varphi. \tag{5.24}$$

The solution for the torsion problem (see Fig. 5.9), initially found by Lubkin (1951), is then given by:

$$\begin{aligned}
\varphi &= \frac{\mu E^* a}{\pi G R}\left[\mathrm{K}\left(\sqrt{1-\frac{c^2}{a^2}}\right) - \mathrm{E}\left(\sqrt{1-\frac{c^2}{a^2}}\right)\right],\\
M_z &= 16 G\varphi \frac{c^3}{3}\\
&\quad + \frac{16\mu E^* a}{\pi R}\int_c^a x^2\left[\mathrm{K}\left(\sqrt{1-\frac{x^2}{a^2}}\right) - \mathrm{E}\left(\sqrt{1-\frac{x^2}{a^2}}\right)\right]\mathrm{d}x,\\
\sigma_{\varphi z}(r) &= -\frac{4\mu E^* a r}{\pi^2 R}\int_c^a \mathrm{E}\left(\sqrt{1-\frac{x^2}{a^2}}\right)\frac{\mathrm{d}x}{x\sqrt{x^2-r^2}}, \quad r \le c.
\end{aligned} \tag{5.25}$$

Here, φ represents the global torsion angle of the rigid paraboloid, M_z the torsional moment, and $\sigma_{\varphi z}$ the torsional stresses. The torsion angle and the torsional moment as functions of the radius of the stick zone are represented in a normalized form in Figs. 5.10 and 5.11. Furthermore, Figs. 5.12 and 5.13 display the normalized tangential stresses and displacements as functions of the radial coordinate.

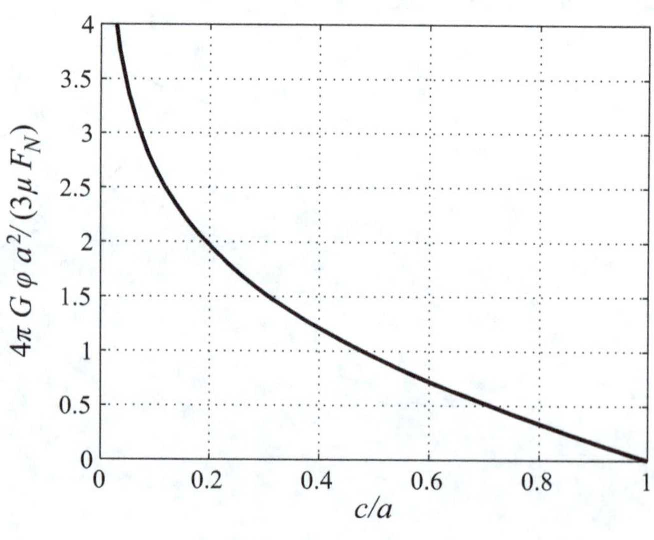

Fig. 5.10 Normalized torsion angle as a function of the normalized radius of the stick zone for the torsional contact with a parabolic indenter

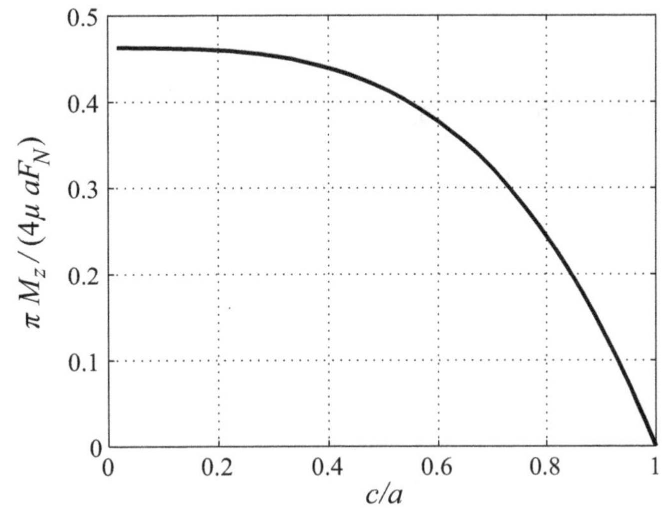

Fig. 5.11 Normalized torsional moment as a function of the normalized radius of the stick zone for the torsional contact with a parabolic indenter

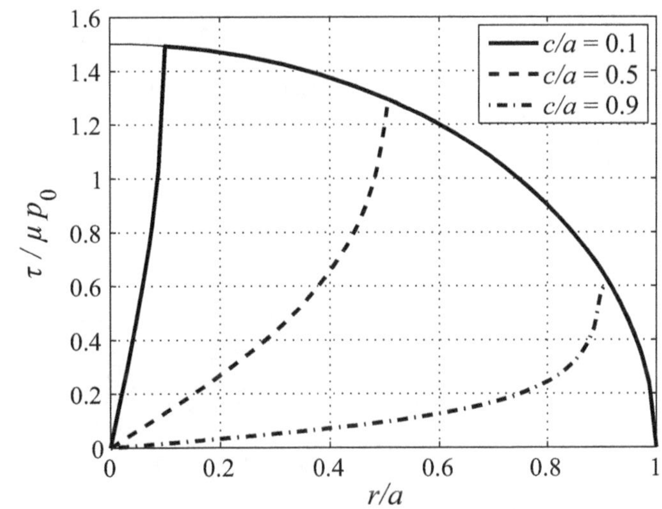

Fig. 5.12 Torsional stresses normalized to the average pressure p_0 multiplied by the coefficient of friction μ for the torsional contact with a flat cylindrical punch. The thin solid line represents the stress distribution for complete slip

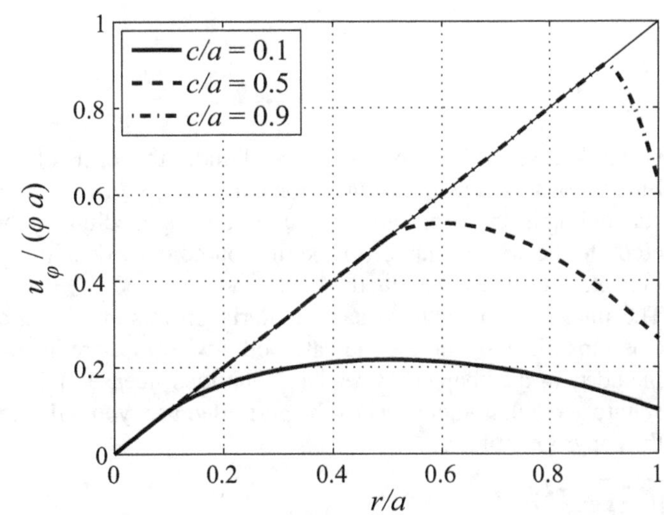

Fig. 5.13 Normalized tangential displacements for the torsional contact with a parabolic indenter. The thin solid line represents the displacement caused by the rigid body rotation

References

Jäger, J.: Axi-symmetric bodies of equal material in contact under torsion or shift. Arch. Appl. Mech. **65**(7), 478–487 (1995)

Johnson, K.L.: Contact mechanics. Cambridge University Press, Cambridge (1985)

Lubkin, J.L.: The torsion of elastic spheres in contact. J. Appl. Mech. **73**, 183–187 (1951)

6 Wear

Wear is the mechanical or chemical degradation of surfaces. Along with the closely related phenomenon of fatigue, wear is a central aspect in estimating the service life of any technical system. As is the case with friction, the microscopic and mesoscopic mechanisms causing the macroscopically observable phenomenon of wear are extremely varied and range from abrasive or adhesive debris formation, to the reintegration of previously removed material, to oxidation, and to chemical or mechanical intermixing of the involved surfaces. Accordingly, the formulation of a general wear law is quite difficult. Analogous to the Amontons–Coulomb law of dry friction, a common approach involves an elementary, linear relationship introduced by Reye (1860), Archard and Hirst (1956), as well as Khrushchov and Babichev (1960). When using this law, one must bear in mind that it is a very rough approximation.

In this chapter, we consider rotationally symmetric profiles that wear while retaining rotational symmetry. Trivially, wear is axially-symmetric for an axially symmetric load (such as a torsional load around the center line of the profile, which was studied by Galin and Goryacheva (1977)). However, the retention of axial symmetry of the profile does not necessarily require axial symmetry of the load condition.

Wear can even be rotationally-symmetric during movements in directions which violate the symmetry of the system. This is the case, for example, when a rotationally symmetric profile is in a state of gross slip at a constant speed relative to an elastic base, where only the profile yet not the elastic base is the subject of wear. Since (in this case) the pressure is axially-symmetric and the relative slip speed of both surfaces is constant in all points of contact, the wear intensity is also axially-symmetric. At least, this is valid for all local wear laws where the wear intensity is only dependent on the pressure and relative slip rate. If the counterpart exhibits wear, the system loses its axial symmetry.

Approximate axial symmetry is also retained in the practically relevant case of wear due to low-amplitude oscillations (fretting). In this case, the stress tensor is not isotropic. For a unidirectional oscillation in the Cattaneo–Mindlin approximation, the stress tensor only has a tangential component which, however, is solely depen-

V.L. Popov, M. Heß, E. Willert, *Handbook of Contact Mechanics*,
https://doi.org/10.1007/978-3-662-58709-6_6

dent on the polar radius. In the same Cattaneo–Mindlin approximation, the relative displacements of the surfaces also only depend on the radius. Wear determined by the local pressure and the relative displacements is thus axially-symmetric, regardless of the particular form of the law of wear. This leads to the common phenomenon of ring-shaped wear patterns.

Many of the results in this chapter do not depend on the exact form of the law of wear. The only assumption held is that the wear occurs continually to prevent the generation of wear particles of the same characteristic length as the contact problem. We further assume that the law of wear is local, i.e., the wear intensity at a location depends only on the pressure and the velocity occurring at this location. In the rare cases demanding a concrete law of wear, we will assume the simplest law by Reye–Archard–Khrushchov, which says that the local wear intensity is proportional to the local pressure and the local relative velocity. With the profile shape of the wearing body at the point in time t denoted by $f(r,t)$, the law can be written as:

$$\dot{f}(r,t) \sim p(r,t)\,|v_{\mathrm{rel}}(r,t)|\,. \tag{6.1}$$

Here, v_{rel} is the relative velocity between the contacting surfaces and $p(r,t)$ is the local pressure. The dot above the quantity signifies the differentiation with respect to time.

In this chapter, we initially examine wear for gross slip of the contacting surfaces and then wear due to fretting.

6.1 Wear Caused by Gross Slip

We consider an axially symmetric punch with the profile shape $f(r,t)$ and characterize its normal load by the (generally time-dependent) indentation depth $d(t)$. The profile shape and the indentation depth uniquely define the entire solution of the normal contact problem, including the contact radius a, the pressure distribution $p(r,t)$, and the normal force:

$$F_N(t) = 2\pi \int_0^a p(r,t) r \mathrm{d}r. \tag{6.2}$$

The two common load types we consider in the following sections are the force-controlled loading (i.e., F_N is given) and displacement-controlled loading (i.e., the "height" d is given).

In the context of wear problems, the "indentation depth" in a contact mechanical sense is measured from the lowest point of the current (i.e., worn) profile. Therefore, its definition changes with progressing wear.

6.1.1 Wear at Constant Height

The simplest case of wear is realized when an indenter with the original profile $f_0(r)$ is initially pressed to the indentation depth d_0, and subsequently displaced tangentially at this *height*. While the calculation of the *process* of wear, in this case, remains a relatively complicated contact mechanical task (see the work of Dimaki et al. (2016)) with the process also depending on the explicit type of the law of wear, the *final state* is simply defined by the condition that the profile just barely remains in contact at all points with the base. This state is reached when the original profile is "cut off" at the height d_0. This conclusion is not bound to the axial symmetry, nor to the type of movement (as long as a constant height is maintained): any profile displaced arbitrarily at a constant height (ensuring a relative displacement in all contact points) tends to assume the shape of the original profile cut off at the height d_0.

6.1.2 Wear at Constant Normal Force

During force-controlled processes, the normal force remains constant and the wear process never ceases. The case of cylindrical indenting bodies, where the contact area remains constant despite wear, is particularly simple. Pressing a cylindrical punch into the base with subsequent tangential displacement at a constant speed leads into a state of steady wear after a transient period, in which all points of the punch wear at the same linear rate. Under the assumption of any local law of wear in which the wear intensity is proportional to the product of the pressure and the sliding velocity, it follows that in the state of steady wear, the pressure is constant in the entire contact area and equal to $p_0 = F_N/A$, where A is the (constant) contact area. The profile shape that generates a constant pressure was previously determined in Sect. 2.5.6:

$$w(r; a, p_0) = \frac{4p_0 a}{\pi E^*}\mathrm{E}\left(\frac{r}{a}\right), \quad r \le a. \tag{6.3}$$

Here, $\mathrm{E}(\cdot)$ refers to the complete elliptical integral of the second kind:

$$\mathrm{E}(k) := \int_0^{\pi/2} \sqrt{1 - k^2 \sin^2 \varphi}\,\mathrm{d}\varphi. \tag{6.4}$$

The expression (6.3) is shown in Fig. 6.1 in normalized variables.

For general cylindrical punches (with any arbitrary, not necessarily circular face), this analysis can be generalized to heterogeneous punches with wear coefficients $k = k(x, y)$ that depend on the position in the contact area (characterized by the coordinates x and y), but not on the z-coordinate. One such example is a fiber-composite pin, whose fibers run parallel to the longitudinal axis. Once the system

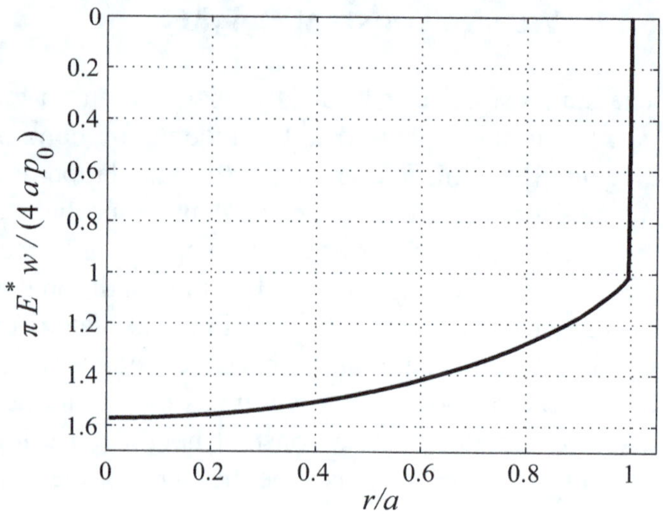

Fig. 6.1 Shape of the punch in the phase of steady wear during unidirectional slip at a constant rate

settles into a steady state, the wear intensity is constant at all points of the contact area:

$$\dot{f} = k(x, y) \cdot v_0 \cdot p(x, y) = C. \tag{6.5}$$

It follows that

$$p(x, y) = \frac{C}{k(x, y) \cdot v_0}. \tag{6.6}$$

The constant C is obtained from the condition:

$$F_N = \int p(x, y)\mathrm{d}x\mathrm{d}y. \tag{6.7}$$

The pressure distribution is then determined from (6.5) to be:

$$p(x, y) = \frac{F_N}{k(x, y) \int k(x, y)^{-1}\mathrm{d}x\mathrm{d}y}. \tag{6.8}$$

Thus, the pressure distribution for a given normal force is explicitly determined by the heterogeneity of the wear coefficient. The shape of the wear surface in the steady state is given by the shape of the elastic continuum under the effect of the pressure distribution given in (6.8).

The steady state does not exist for non-cylindrical indenters. Yet qualitatively, the situation is unchanged. In this case, the stress distribution in the worn contact area approaches a constant value as well, determined by the normal force and the current contact area (Dimaki et al. 2016). However, the state of constant pressure is only reached in approximation.

A special case of gross slip wear is the wear of a rotationally symmetric profile under the effect of axial twisting (torsion). A cylindrical indenter of radius a twisting with the angular velocity Ω has a local slip speed given by $v_{\mathrm{rel}}(r, t) = \Omega r$. Accordingly, the linear wear rate is:

$$\dot{f} = k\Omega r \cdot p(r) = C, \tag{6.9}$$

and it is constant in all points of the contact area in the steady state. The constant C is obtained from the condition:

$$F_N = 2\pi \int_0^a p(r) r \mathrm{d}r. \tag{6.10}$$

The pressure distribution is determined from (6.9) and (6.10):

$$p(r) = \frac{F_N}{2\pi r a}. \tag{6.11}$$

The profile corresponding to this stress distribution can be determined using the MDR. In the first step, the distributed load in the MDR space is calculated with the following transformation:

$$\begin{aligned} q_z(x) &= 2\int_x^\infty \frac{r p(r)}{\sqrt{r^2 - x^2}} \mathrm{d}r = \frac{F_N}{\pi a} \int_x^a \frac{1}{\sqrt{r^2 - x^2}} \mathrm{d}r \\ &= \frac{F_N}{\pi a} \ln \left(\frac{a + \sqrt{a^2 - x^2}}{x} \right). \end{aligned} \tag{6.12}$$

This gives the result for the vertical displacement in the one-dimensional MDR model:

$$w_{1D}(x) = \frac{q_z(x)}{E^*} = \frac{F_N}{\pi a E^*} \ln \left(\frac{a + \sqrt{a^2 - x^2}}{x} \right), \tag{6.13}$$

and subsequently the displacements in the original three-dimensional space (Fig. 6.2):

$$\begin{aligned} w(r) &= \frac{2}{\pi} \int_0^r \frac{u_z(x)}{\sqrt{r^2 - x^2}} \mathrm{d}x \\ &= \frac{F_N}{\pi a E^*} \frac{2}{\pi} \int_0^r \ln \left(\frac{a + \sqrt{a^2 - x^2}}{x} \right) \frac{\mathrm{d}x}{\sqrt{r^2 - x^2}} \\ &= \frac{F_N}{\pi a E^*} \Psi(r/a), \end{aligned} \tag{6.14}$$

where

$$\Psi(\zeta) = \frac{2}{\pi} \int_0^\zeta \ln \left(\frac{1 + \sqrt{1^2 - \xi^2}}{\xi} \right) \frac{\mathrm{d}\xi}{\sqrt{\zeta^2 - \xi^2}}. \tag{6.15}$$

For small radii, this function exhibits a logarithmic singularity in the form

$$\Psi(\zeta) \approx 3.420544234 - \ln(\zeta), \quad \text{for small } \zeta. \tag{6.16}$$

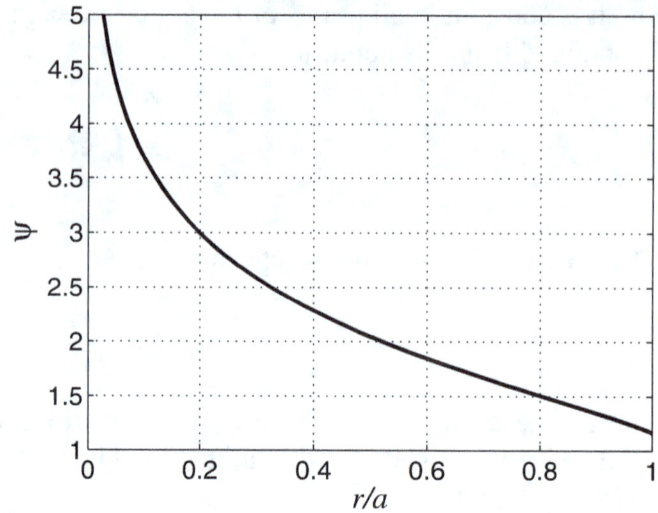

Fig. 6.2 Shape of the profile generated by wear due to rotation around the vertical axis at a constant angular velocity (see (6.14))

Galin and Goryacheva (1977) also investigated non-cylindrical, rotationally symmetric profiles in their study of torsional wear. Soldatenkov (2010) examined various rotationally symmetric wear problems of technical relevance (including stochastic ones), such as ball bearings or the wheel-rail contact.

6.2 Fretting Wear

Many technical or biological systems are part of a periodically working mechanism or are subjected to vibrations. The tribological contacts in such systems are loaded in an oscillating manner due to the periodicity. In general, the displacement amplitudes of these oscillations are sufficiently small to avoid gross slip of the contact. However, the edge of the contact area unavoidably sees the formation of a slip zone, where the contacting surfaces wear and fatigue as a result of the relative displacement. This effect is called fretting and is of marked importance for the operational lifespan of tribological systems.

Just like all wear phenomena, fretting wear has very diverse mechanical and chemical mechanisms. We also distinguish between different fretting modes and fretting regimes. The fretting modes are distinguished by the type of the underlying contact problem: an oscillation normal to the contact plane is called radial fretting. The classical case of oscillations in the contact plane is called tangential fretting and the case of an oscillating rotation around the normal axis of the contact plane is called torsional fretting. An oscillating rolling contact leads to rotational fretting. However, the different modes barely differ in their qualitative behavior (Zhou and Zhu 2011).

Different *fretting regimes* are distinguished based on the behavior of the contact during an oscillation, e.g., near complete stick, partial slip, near complete gross slip, or significant gross slip. These were first systematically examined by Vingsbo and Søderberg (1988) using *fretting maps*. The authors determined that depending on the regime, either wear or fatigue was the dominant form of material degradation caused by fretting.

Even under the assumption of the simplest laws of friction and wear, the analytical calculation of the particular wear dynamics is always extremely complicated and usually impossible. As shown by Ciavarella and Hills (1999), certain cases have a final "shakedown" state where no further wear occurs. Contrary to the wear process, the worn profile in the shakedown state does not depend on the particular law of wear nor on the fretting mode and can often be determined analytically. A qualitative explanation for the existence of such a limiting profile is easily given: if within one oscillation period a part of the contact area fulfills the no-slip condition while other areas experience at least transient slip, the stick zone will not wear while the slip zone will experience progressing wear. Intuitively, this leads to a lower pressure in the slip zone, with the stick zone taking on the additional load. Over the course of further oscillations, the stick zone continues sticking while the pressure in the slip zone decreases continually until vanishing completely, at which point the surfaces are in incipient contact with no load. Based on this fact, Popov (2014) was able to determine the general solution for this limiting profile in the case of a rotationally symmetric indenter. These results were experimentally confirmed by Dmitriev et al. (2016). The existence of such a limiting profile is a universal conclusion and is not bound to the assumption of axial symmetry.

The following section is dedicated to the calculation of the limiting profiles for various initial indenter shapes, as shown by Popov (2014). Let $f(r)$ be the rotationally symmetric profile of a rigid indenter which is pressed by d into an elastic half-space with an effective elastic modulus of E^*. Moreover, let the contact area have the radius a and the pressure distribution in the contact be $p(r)$. The (oscillating) tangential or torsional load causes the formation of a periodically changing slip zone in the contact. The radius of the area of permanent stick is referred to as c. The aforementioned limiting state is determined by the conditions:

$$\begin{aligned} f_\infty(r) &= f_0(r), \quad r \leq c \\ p_\infty(r) &= 0, \quad\;\;\; r > c. \end{aligned} \tag{6.17}$$

The index "∞" designates the shakedown state and the index "0" the unworn initial state. As previously explained in this chapter, the radius c results from the no-slip condition for the oscillation of the unworn profile and remains unchanged during the progression of wear. The calculation of the limiting profile consists of three steps.

1. Determination of the radius c for the area of permanent stick for the original profiles $f_0(r)$.
2. Determination of the limiting profile using the equations derived by Popov (2014), as follows:

$$f_\infty(r) = \begin{cases} f_0(r), & r \leq c,\ r > a, \\ \dfrac{2}{\pi}\left[\displaystyle\int_0^c \frac{g_0(x)\mathrm{d}x}{\sqrt{r^2 - x^2}} + d \arccos\left(\frac{c}{r}\right)\right], & c < r \leq a, \end{cases} \tag{6.18}$$

where

$$g_0(x) = |x| \int_0^{|x|} \frac{f_0'(r)\mathrm{d}r}{\sqrt{x^2 - r^2}}. \tag{6.19}$$

3. Determination of the contact radius a_∞ in the limiting state from the condition:

$$f_\infty(a_\infty) = f_0(a_\infty). \tag{6.20}$$

The solution structure clearly shows that the radius c of the permanent stick zone is the only load-dependent and material-dependent parameter to contribute to the solution of the limiting profile. The stick radius is calculated for the unworn profile and is valid for the entire wear process. Solution (6.18) does not yet define the outer radius of the worn area a_∞. In the final step, a_∞ is determined using (6.20).

In the following sections, we first explain how to determine the radius of the permanently sticking contact area. Subsequently, we give the form of the punch in its final shakedown state for an assumed known radius c of the permanent stick zone. We will focus on a selection of indenter profiles from Chap. 2, which is justified by their technical relevance.

6.2.1 Determining the Radius of the Permanent Stick Zone

6.2.1.1 Horizontal Oscillations at Constant Indentation Depth

The easiest way to calculate the radius c of the permanent stick zone is by using the MDR. Following the steps of the MDR (4.21) (see Sect. 4.4 of this book), we define an elastic foundation and an MDR modified profile $g_0(x)$ according to (6.19) on which further contact mechanical calculations are performed instead of the original three-dimensional system.

When the profile is displaced in the tangential direction by $u^{(0)}$, the springs are loaded normally and tangentially. The radius of the stick zone is given by the following equation, which sets the maximum tangential force equal to μ multiplied with the normal force:

$$G^* u^{(0)} = \mu E^* [d - g(c)]. \tag{6.21}$$

For oscillations in the tangential direction in accordance with the law $u^{(0)}(t) = \Delta u^{(0)} \cos(\omega t)$, the *smallest* stick radius (and thus the radius of the permanent stick zone) is reached at the maximum horizontal displacement:

$$G^* \Delta u^{(0)} = \mu E^* [d - g(c)]. \tag{6.22}$$

The form of the function $g(c)$ was determined for a great number of profiles in Chap. 2. We will forego repeating them here.

6.2.1.2 Bimodal Oscillations

Let us now consider cases in which the punch experiences simultaneous oscillations in horizontal and vertical directions:

$$\begin{aligned} u^{(0)}(t) &= \Delta u^{(0)} \cos(\omega_1 t), \\ d(t) &= d_0 + \Delta d \cos(\omega_2 t - \varphi). \end{aligned} \tag{6.23}$$

The first thing to note is that, in this case, the limiting profile is still given by (6.18). However, for d we must now insert the *maximum indentation depth* over the course of one oscillation period:

$$f_\infty(r) = \begin{cases} f_0(r), & r \leq c,\ r > a, \\ \dfrac{2}{\pi}\left[\displaystyle\int_0^c \frac{g_0(x)\mathrm{d}x}{\sqrt{r^2 - x^2}} + d_{\max} \arccos\left(\frac{c}{r}\right)\right], & c < r \leq a, \end{cases} \tag{6.24}$$

with

$$d_{\max} = d_0 + \Delta d. \tag{6.25}$$

Let us now calculate c. In the MDR representation, the normal and tangential forces at the coordinate c are given by:

$$\begin{aligned} \Delta F_x &= G^* \Delta x \cdot u^{(0)}(t) = G^* \Delta x \cdot \Delta u^{(0)} \cos(\omega_1 t), \\ \Delta F_z &= E^* \Delta x \cdot d(t) = E^* \Delta x \cdot [d_0 + \Delta d \cos(\omega_2 t - \varphi)]. \end{aligned} \tag{6.26}$$

The radius of the permanent stick zone is determined from the condition that the absolute value of the tangential force at the location c must never exceed the product of the normal force and the coefficient of friction:

$$\left|G^* \Delta u^{(0)} \cos(\omega_1 t)\right| \leq \mu E^* [d_0 + \Delta d \cos(\omega_2 t - \varphi) - g(c)]. \tag{6.27}$$

If the frequencies ω_1 and ω_2 are incommensurable, or alternatively, if the phase φ is not locked (i.e., it is slowly changing), this condition is satisfied only when the maximum value of the left-hand side of the inequality is smaller than the minimal value of the right-hand side. The critical value is reached when these two values are equal:

$$G^* \Delta u^{(0)} = \mu E^* [d_0 - \Delta d - g(c)]. \tag{6.28}$$

The only difference in this equation compared to (6.22) is that it features the minimal indentation depth $d_{\min} = d_0 - \Delta d$ instead of simply d.

For commensurable or equal frequencies and locked phase shift, the solution is generally very complicated and it can be referenced in Mao et al. (2016).

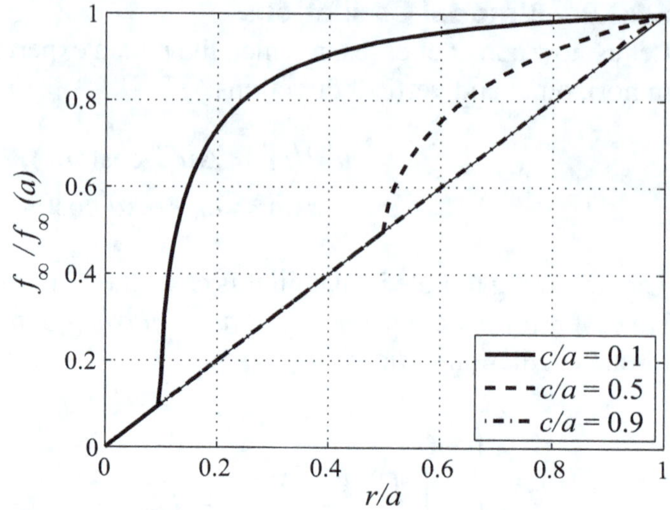

Fig. 6.3 Shakedown profile normalized to $a_\infty \tan\theta$, for different values of the permanent stick radius c for fretting wear of a conical indenter

6.2.2 The Cone

The unworn profile of a conical indenter can be written as:

$$f_0(r) = r \tan\theta. \tag{6.29}$$

Here θ is the slope of the cone. The shakedown profile is then given by:

$$f_\infty(r) = \begin{cases} r\tan\theta, & r \le c,\ r > a_\infty, \\ r\tan\theta\left(1 - \sqrt{1 - \dfrac{c^2}{r^2}}\right) + \dfrac{2d}{\pi}\arccos\left(\dfrac{c}{r}\right), & c < r \le a_\infty, \end{cases} \tag{6.30}$$

with an indentation depth d and the radius of the permanent stick zone c.

The post-shakedown contact radius a_∞ is obtained as the solution of the equation:

$$\frac{2d}{\pi}\arccos\left(\frac{c}{a_\infty}\right) = \tan\theta\sqrt{{a_\infty}^2 - c^2}. \tag{6.31}$$

The profile resulting from (6.30) is shown in Fig. 6.3.

6.2.3 The Paraboloid

For the paraboloid with the radius of curvature R and corresponding unworn profile

$$f_0(r) = \frac{r^2}{2R}, \tag{6.32}$$

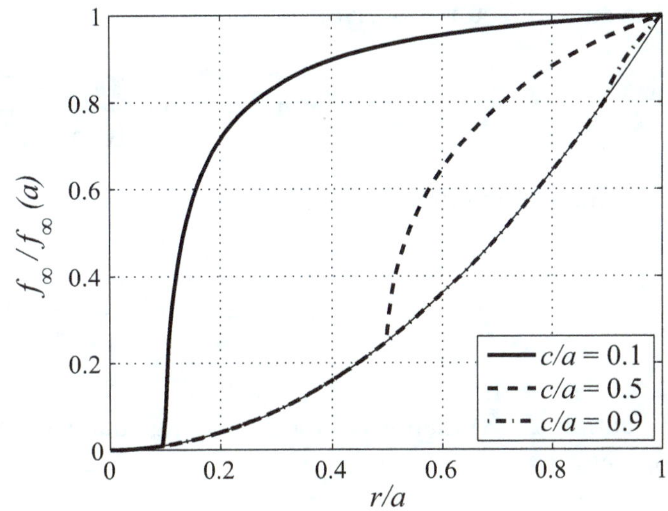

Fig. 6.4 Shakedown-profile normalized to $a_\infty^2/(2R)$, for various values of permanent stick radius c for the fretting wear of a parabolic indenter

the limiting profile is described as a function of the indentation depth d and the radius c of the permanent stick area is given by:

$$f_\infty(r) = \begin{cases} \dfrac{r^2}{2R}, & r \le c,\ r > a_\infty, \\ \dfrac{2}{\pi}\left[\dfrac{r^2}{2R}\arcsin\left(\dfrac{c}{r}\right) - \dfrac{c}{2R}\sqrt{r^2-c^2}\right. & \\ \left. \quad + d\arccos\left(\dfrac{c}{r}\right)\right], & c < r \le a_\infty. \end{cases} \tag{6.33}$$

This dependency is shown in Fig. 6.4. The contact radius in the worn state results from the solution of the equation:

$$\frac{a_\infty^2}{2R} = \frac{2}{\pi}\left[\frac{a_\infty^2}{2R}\arcsin\left(\frac{c}{a_\infty}\right) - \frac{c}{2R}\sqrt{a_\infty^2 - c^2} + d\arccos\left(\frac{c}{a_\infty}\right)\right]. \tag{6.34}$$

6.2.4 The Profile in the Form of a Lower Law

We now consider indenter profiles of the general power-law form:

$$f_0(r) = br^n, \quad n \in \mathbb{R}^+, \tag{6.35}$$

with a (dimensional) constant b and a positive real number n. In Chap. 2 (see Sect. 2.5.8) it has already been shown, that the equivalent profile in the MDR is given by a stretched power function with the same exponent n:

$$g_0(x) = \kappa(n)b|x|^n. \tag{6.36}$$

The stretch factor was given as:

$$\kappa(n) := \sqrt{\pi}\frac{\Gamma(n/2+1)}{\Gamma\left[(n+1)/2\right]}, \tag{6.37}$$

with the gamma function

$$\Gamma(z) := \int_0^\infty t^{z-1}\exp(-t)\mathrm{d}t. \tag{6.38}$$

Equation (5.18) then gives the following shakedown profile at the end of the wear process:

$$\begin{aligned} f_\infty(r) &= \begin{cases} br^n, & r \le c,\ r > a, \\ \dfrac{2}{\pi}\left[\kappa(n)b\displaystyle\int_0^c \frac{x^n\mathrm{d}x}{\sqrt{r^2-x^2}} + d\arccos\left(\dfrac{c}{r}\right)\right], & c < r \le a, \end{cases} \\ &= \begin{cases} br^n, & r \le c,\ r > a, \\ \dfrac{2}{\pi}\left[\kappa(n)b\dfrac{c^{n+1}}{(n+1)r}\,{}_2\mathrm{F}_1\left(\dfrac{1}{2},\dfrac{n+1}{2};\dfrac{n+3}{2};\dfrac{c^2}{r^2}\right)\right. & \\ \qquad \left. + d\arccos\left(\dfrac{c}{r}\right)\right], & c < r \le a, \end{cases} \end{aligned} \tag{6.39}$$

with the indentation depth d and the permanent stick radius c. The notation ${}_2\mathrm{F}_1(\cdot,\cdot;\cdot;\cdot)$ is used for the hypergeometric function

$${}_2\mathrm{F}_1(a,b;c;z) := \sum_{n=0}^{\infty}\frac{\Gamma(a+n)\Gamma(b+n)\Gamma(c)}{\Gamma(a)\Gamma(b)\Gamma(c+n)}\frac{z^n}{n!}. \tag{6.40}$$

For $n = 1$ the case of the cone is from Sect. 6.2.2 and is reproduced, and for $n = 2$ it is the solution for the paraboloid (Sect. 6.2.3).

6.2.5 The Truncated Cone

In the previous chapters, the truncated cone and paraboloid have already been discussed several times. For the truncated cone with the profile function

$$f_0(r) = \begin{cases} 0, & r \le b, \\ (r-b)\tan\theta, & r > b, \end{cases} \tag{6.41}$$

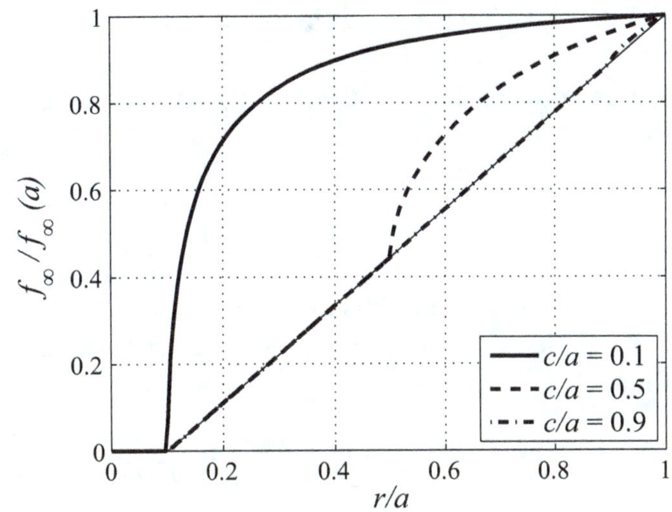

Fig. 6.5 Shakedown profile normalized to $(a_\infty - b)\tan\theta$, for $b = 0.1a_\infty$ and various values of the permanent stick radius c for fretting wear of a truncated cone

with the radius b at the blunt end and the slope angle θ, the following equivalent plane profile was determined in Chap. 2 (Sect. 2.5.9):

$$g_0(x) = \begin{cases} 0, & |x| \le b, \\ |x| \tan\theta \arccos\left(\dfrac{b}{|x|}\right), & |x| > b. \end{cases} \tag{6.42}$$

It has been shown in Chap. 4 (Sect. 4.6.5) that the radius of the stick zone c cannot fall below the value of b. Therefore, the worn boundary profile, with (6.18) and the indentation depth d, is described by

$$f_\infty(r) = \begin{cases} f_0(r), & r \le c,\ r > a_\infty, \\ \dfrac{2}{\pi}\left[\tan\theta \displaystyle\int_b^c x \arccos\left(\frac{b}{x}\right) \frac{\mathrm{d}x}{\sqrt{r^2 - x^2}} \right. & \\ \left. \quad + d \arccos\left(\dfrac{c}{r}\right)\right], & c < r \le a_\infty. \end{cases} \tag{6.43}$$

The contact radius at the end of the wear process is, as always, defined by the relationship (6.20). The profile of (6.43) is shown in normalized form in Fig. 6.5. For $b = 0$, of course, the solution of the complete cone from Sect. 6.2.2 is recovered.

6.2.6 The Truncated Paraboloid

Let us now consider the truncated parabolic, whose unworn profile is described by the rule:

$$f(r) = \begin{cases} 0, & r \leq b, \\ \dfrac{r^2 - b^2}{2R}, & r > b, \end{cases} \tag{6.44}$$

Here R denotes the radius of curvature of the parabolic base body and b the radius at the flattened tip. In Chap. 2 (Sect. 2.5.10), the equivalent profile has already been found:

$$g(x) = \begin{cases} 0, & |x| \leq b, \\ \dfrac{|x|}{R}\sqrt{x^2 - b^2}, & |x| > b. \end{cases} \tag{6.45}$$

As in the case of the truncated cone of the radius of the stick zone, c cannot fall below the value of b Equation (6.18). (d denotes the indentation depth, as always) allows the shakedown profile to be described as:

$$\begin{aligned} f_\infty(r) &= \begin{cases} f_0(r), & r \leq c,\ r > a_\infty, \\ \dfrac{2}{\pi}\left[\dfrac{1}{R}\displaystyle\int_b^c x\sqrt{x^2 - b^2}\dfrac{\mathrm{d}x}{\sqrt{r^2 - x^2}} + d\arccos\left(\dfrac{c}{r}\right)\right], & c < r \leq a_\infty, \end{cases} \\ &= \begin{cases} f_0(r), & r \leq c,\ r > a_\infty, \\ \dfrac{1}{\pi R}\left[(r^2 - b^2)\arcsin\left(\dfrac{\sqrt{c^2 - b^2}}{\sqrt{r^2 - b^2}}\right)\right. \\ \qquad \left. - \sqrt{c^2 - b^2}\sqrt{r^2 - c^2} + 2dR\arccos\left(\dfrac{c}{r}\right)\right], & c < r \leq a_\infty. \end{cases} \end{aligned} \tag{6.46}$$

This is shown in a normalized manner in Fig. 6.6. For $b = 0$ we obtain, of course, the solution of the complete paraboloid from Sect. 6.2.3.

6.2.7 Further Profiles

In Chap. 2, it was shown that the equivalent profile functions $g(x)$ of various technically relevant indenters considered in the literature can be considered as the superposition of the elementary bodies *paraboloid*, *truncated cone*, and *truncated paraboloid*. Since the integral expression in (6.18) is linear in $g_0(x)$, for the superposition the solutions of the integrals for the elementary bodies mentioned, given in the previous sections, can simply be summed up. We will, therefore, only specify the superposition rules for the functions g_0 and graph the profiles at the end of the fretting process. We will refrain from providing the complete solution in order to avoid redundancies.

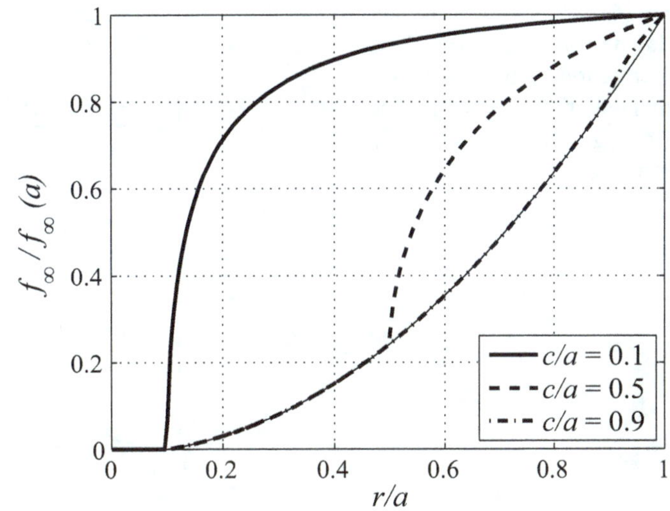

Fig. 6.6 Shakedown profile normalized to $(a_\infty^2 - b^2)/(2R)$, for $b = 0.1a_\infty$ and various values of the permanent stick radius c for the fretting wear of a truncated paraboloid

6.2.7.1 The Cone with Parabolic Cap

For a cone with the slope angle θ, which at point $r = b$ differentiably passes into a parabolic cap with the radius of curvature $R := b/\tan\theta$, the rotationally symmetric profile can be written as follows:

$$f(r) = \begin{cases} \dfrac{r^2 \tan\theta}{2b}, & r \le b, \\ r\tan\theta - \dfrac{b}{2}\tan\theta, & r > b, \end{cases} \tag{6.47}$$

For the equivalent profile in MDR, the following superposition has been shown in Chap. 2 (see Sect. 2.5.12):

$$g_0(x; b, \theta) = g_{0,P}\left(x; R = \frac{b}{\tan\theta}\right) + g_{0,KS}(x; b, \theta) - g_{0,PS}\left(x; b, R = \frac{b}{\tan\theta}\right). \tag{6.48}$$

Here, $g_{0,P}$ denotes the unworn, equivalent profile of a paraboloid (see Sect. 6.2.3), $g_{0,KS}$ that of a truncated cone (see Sect. 6.2.5), and $g_{0,PS}$ that of a truncated paraboloid (see Sect. 6.2.6). The no-slip radius can fall below the value of b, but then the limiting profile is the same as in the case of the simple paraboloid. For this reason, some variants of the shakedown profile with $c \ge b$ are shown in Fig. 6.7.

6.2.7.2 The Paraboloid with Paraboloid Cap

The rotationally symmetric profile of this body is described by the function

$$f(r) = \begin{cases} \dfrac{r^2}{2R_1}, & r \le b, \\ \dfrac{r^2 - h^2}{2R_2}, & r > b. \end{cases} \tag{6.49}$$

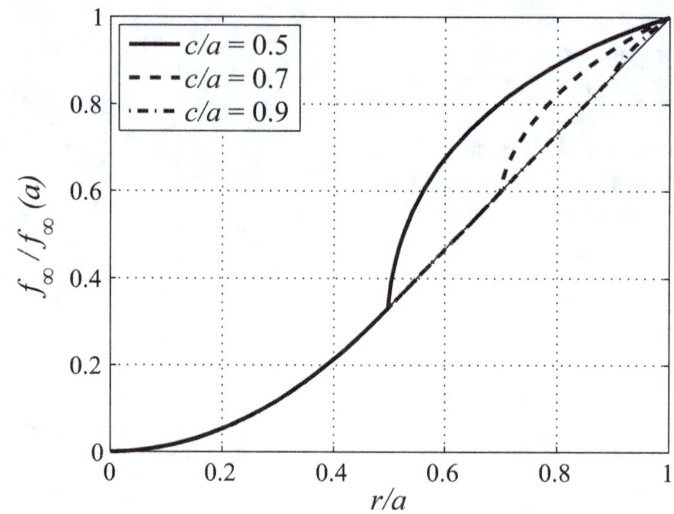

Fig. 6.7 Normalized shakedown profile for $b = 0.5a_\infty$, and various values of the permanent stick radius c for the fretting wear of a cone with a rounded tip

The radius of the cap is R_1, and that of the main body is R_2. The continuity of f at the position of $r = b$ requires

$$h^2 = b^2 \left(1 - \frac{R_2}{R_1}\right), \tag{6.50}$$

and one can introduce an effective radius of curvature:

$$R^* = \frac{R_1 R_2}{R_1 - R_2}. \tag{6.51}$$

From a contact mechanical point of view, this body can be described as a superposition:

$$g_0(x; b, R_1, R_2) = g_{0,P}(x; R = R_1) + g_{0,PS}(x; b, R = R^*), \tag{6.52}$$

as can be looked up in Chap. 2 (Sect. 2.5.13). The permanent stick radius can, again, drop below the value of b, but the limiting profile is the same as in the case of the simple paraboloid. In Fig. 6.8 some variants of the shakedown profile are shown by way of example.

6.2.7.3 The Cylindrical Flat Punch with a Rounded Edge

The indenter has the axisymmetric profile

$$f(r) = \begin{cases} 0, & r \leq b, \\ \dfrac{(r-b)^2}{2R}, & r > b, \end{cases} \tag{6.53}$$

with the radius b, for which the flat base of the punch passes into the rounded edge with the radius of curvature R. The transformed profile g can be thought of as the

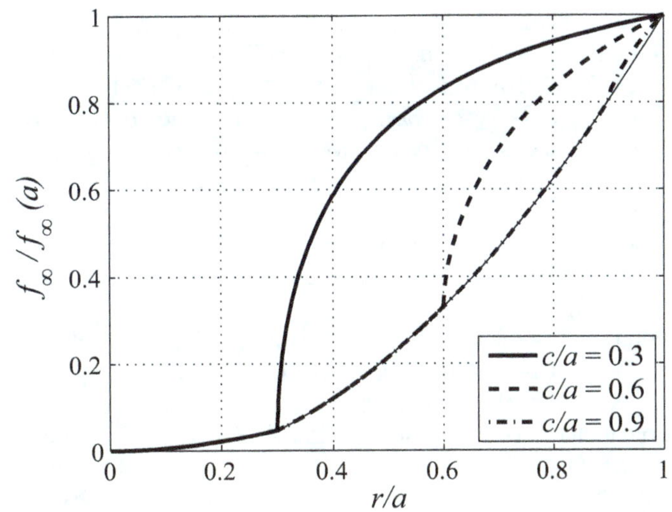

Fig. 6.8 Normalized shake-down profile for $b = 0.3a_\infty$, $R_1 = R^*$, and various values of the permanent stick radius c for the fretting wear of a paraboloid with parabolic cap

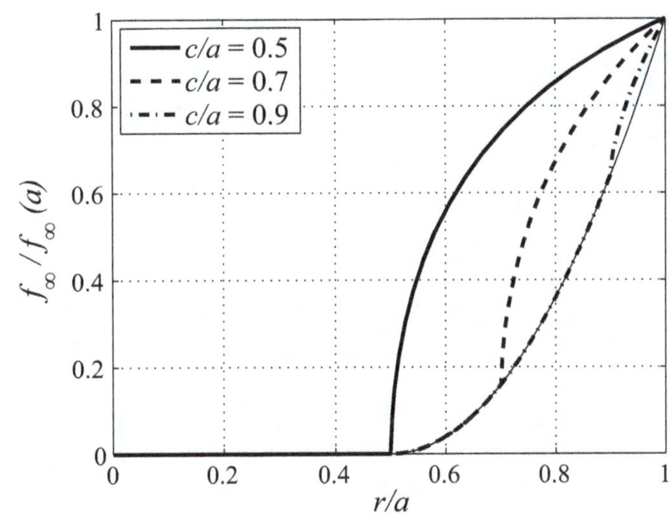

Fig. 6.9 Normalized shake-down profile for $b = 0.5a_\infty$, and various values of the permanent stick radius c for the fretting wear of a flat cylindrical punch with rounded corners

sum (see Sect. 2.5.14) $g_0(x; b, R) = g_{0,PS}(x; b, R) - g_{0,KS}(x; b, \tan\theta = b/R)$. The indices "*KS*" and "*PS*" refer to the respective results of the truncated cone and paraboloid. Figure 6.9 has been used to illustrate some profiles after the fretting process.

References

Archard, J.F., Hirst, W.: The wear of metals under unlubricated conditions. Proc. R. Soc. London Ser. A **236**, 397–410 (1956)

Ciavarella, M., Hills, D.A.: Brief note: Some observations on oscillating tangential forces and wear in general plane contacts. Eur. J. Mech. A. Solids **18**(3), 491–497 (1999)

Dimaki, A.V., Dmitriev, A.I., Menga, N., Papangelo, A., Ciavarella, M., Popov, V.L.: Fast high-resolution simulation of the gross slip wear of axially-symmetric contacts. Tribol. Trans. **59**(1), 189–194 (2016)

Dmitriev, A.I., Voll, L.B., Psakhie, S.G., Popov, V.L.: Universal limiting shape of worn profile under multiple-mode fretting conditions: theory and experimental evidence. Sci. Rep. **6**, 23231 (2016). https://doi.org/10.1038/srep23231

Galin, L.A., Goryacheva, I.G.: Axi-symmetric contact problem of the theory of elasticity in the presence of wear. J. Appl. Math. Mech. **41**(5), 826–831 (1977)

Khrushchov, M.M., Babichev, M.A.: Investigation of wear of metals. Russian Acadamy of Sciences, Moscow (1960)

Mao, X., Liu, W., Ni, Y., Popov, V.L.: Limiting shape of profile due to dual-mode fretting wear in contact with an elastomer. Pro Inst. Mech. Eng. C J. Mech. Eng. Sci. **230**(9), 1417–1423 (2016)

Popov, V.L.: Analytic solution for the limiting shape of profiles due to fretting wear. Sci. Rep. **4**, 3749 (2014). https://doi.org/10.1038/srep03749

Reye, T.: Zur Theorie der Zapfenreibung. Civilingenieur **4**, 235–255 (1860)

Soldatenkov, I.A.: Iznosokontaktnaya Zadacha. Fizmatkniga, Moscow (2010). ISBN 978-5-89-155190-9

Vingsbo, O., Søderberg, S.: On fretting maps. Wear **126**(2), 131–147 (1988)

Zhou, Z.R., Zhu, M.H.: On the mechanisms of various fretting wear modes. Tribol. Int. **44**(11), 1378–1388 (2011)

7 Transversely Isotropic Problems

7.1 Introduction

A transversely isotropic medium is a medium which has a favored direction and is isotropic in the plane perpendicular to this direction. Among crystalline media, all materials with a hexagonal crystal system belong to this class: they are elastically isotropic in the plane perpendicular to the hexagonal axis. Fiber composites with the fibers arranged in parallel in one direction also represent a transversely isotropic medium, which is isotropic in the plane perpendicular to the fiber direction (see Fig. 7.1). Many functional materials exhibiting a preferred direction can also be classified as such, e.g., some piezo-electric materials. We can find many more examples in biological media.

A linear transversely isotropic medium is fully defined by five elastic constants. For the definition of these constants using the elastic moduli and coefficients of transverse contraction, see Fig. 7.1. If we call the axis of symmetry of the medium "z", the axes "x" and "y" are "equivalent" and they can be defined arbitrarily in the plane spanned by these two axes.

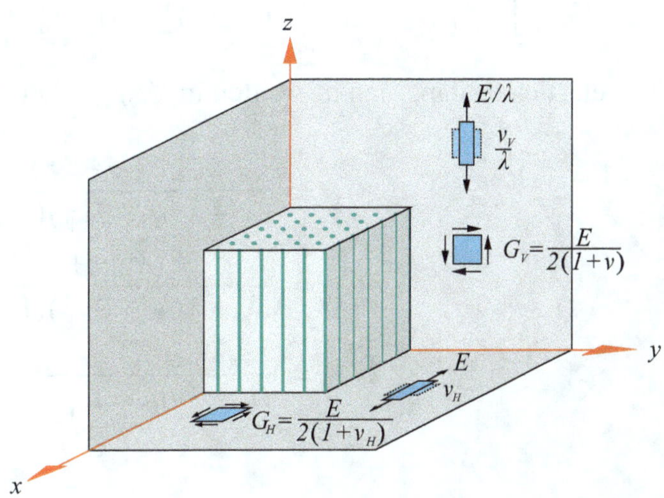

Fig. 7.1 Demonstration of the symmetry and definition of elastic constants of a transversely isotropic medium

V.L. Popov, M. Heß, E. Willert, *Handbook of Contact Mechanics*,
https://doi.org/10.1007/978-3-662-58709-6_7

Turner (1980) described the relationship between the deformation tensor and the stress tensor using the matrix of compliance coefficients:

$$\begin{bmatrix} \varepsilon_{xx} \\ \varepsilon_{yy} \\ \varepsilon_{zz} \\ 2\varepsilon_{xz} \\ 2\varepsilon_{yz} \\ 2\varepsilon_{xy} \end{bmatrix} = \frac{1}{E} \begin{bmatrix} 1 & -\nu_H & -\nu_V & 0 & 0 & 0 \\ -\nu_H & 1 & -\nu_V & 0 & 0 & 0 \\ -\nu_V & -\nu_V & \lambda & 0 & 0 & 0 \\ 0 & 0 & 0 & 2(1+\nu) & 0 & 0 \\ 0 & 0 & 0 & 0 & 2(1+\nu) & 0 \\ 0 & 0 & 0 & 0 & 0 & 2(1+\nu_H) \end{bmatrix} \begin{bmatrix} \sigma_{xx} \\ \sigma_{yy} \\ \sigma_{zz} \\ \sigma_{xz} \\ \sigma_{yz} \\ \sigma_{xy} \end{bmatrix}, \tag{7.1}$$

where

$$\varepsilon_{ij} = \frac{1}{2}\left(\frac{\partial u_i}{\partial x_j} + \frac{\partial u_j}{\partial x_i}\right) \tag{7.2}$$

is the linear symmetric deformation tensor and u_i is the displacement vector. Here, E is the elasticity modulus of the medium in the plane perpendicular to the axis of symmetry, E/λ is the elasticity modulus in the direction of the axis of symmetry, ν_H is the Poisson's ratio in the plane perpendicular to the axis of symmetry, and ν_V is the Poisson's ratio when stress is applied along the symmetry axis. $G_V = E/(2+2\nu)$ is the shear modulus for shear parallel to the axis of symmetry; note that ν has no immediate physical meaning. Additionally, it should be noted that the shear modulus in the plane of symmetry is given by the usual equation $G_H = E/(2+2\nu_H)$.

Inverting the system of equations leads to the presentation via the matrix of stiffness coefficients:

$$\begin{bmatrix} \sigma_{xx} \\ \sigma_{yy} \\ \sigma_{zz} \\ \sigma_{xz} \\ \sigma_{yz} \\ \sigma_{xy} \end{bmatrix} = \begin{bmatrix} C_{11} & C_{12} & C_{13} & 0 & 0 & 0 \\ C_{12} & C_{11} & C_{13} & 0 & 0 & 0 \\ C_{13} & C_{13} & C_{33} & 0 & 0 & 0 \\ 0 & 0 & 0 & C_{44} & 0 & 0 \\ 0 & 0 & 0 & 0 & C_{44} & 0 \\ 0 & 0 & 0 & 0 & 0 & \frac{1}{2}(C_{11}-C_{12}) \end{bmatrix} \begin{bmatrix} \varepsilon_{xx} \\ \varepsilon_{yy} \\ \varepsilon_{zz} \\ 2\varepsilon_{xz} \\ 2\varepsilon_{yz} \\ 2\varepsilon_{xy} \end{bmatrix}. \tag{7.3}$$

The elastic constants can be written in Voigt notation, as follows:

$$C_{11} = \frac{E(\lambda - \nu_V^2)}{(\lambda - \lambda\nu_H - 2\nu_V^2)(1+\nu_H)},$$

$$C_{12} = \frac{E(\lambda\nu_H + \nu_V^2)}{(\lambda - \lambda\nu_H - 2\nu_V^2)(1+\nu_H)},$$

$$C_{13} = \frac{\nu_V E}{(\lambda - \lambda\nu_H - 2\nu_V^2)},$$

$$C_{33} = \frac{(1-\nu_H)E}{(\lambda - \lambda\nu_H - 2\nu_V^2)},$$
$$C_{44} = \frac{E}{2(1+\nu)},$$
$$\frac{1}{2}(C_{11} - C_{12}) = \frac{E}{2(1+\upsilon_H)}, \tag{7.4}$$

using the moduli and Poisson's ratios.

7.2 Normal Contact Without Adhesion

For the complete formulation of the contact mechanical problem in its integral form, it is sufficient to know the fundamental solution, independent of the class of symmetry of the medium. The fundamental solution for transversely isotropic media was found by Michell (1900). He demonstrated that the normal displacement w of the surface of a transversely isotropic elastic half-space under the effect of a force F_z acting on the origin is given by the equation:

$$w(r) = \frac{1}{\pi E^*}\frac{F_z}{r}, \tag{7.5}$$

where r is the distance in-plane to the acting point of the force. The equation has the same form as the corresponding fundamental solution for the case of isotropic media, as shown in (2.2). It simply requires the following definition of the effective elasticity modulus:

$$E^* = \frac{2\sqrt{C_{44}}(C_{11}C_{33} - C_{13}^2)}{\sqrt{C_{11}}\sqrt{\left(\sqrt{C_{11}C_{33}} - C_{13}\right)\left(C_{13} + 2C_{44} + \sqrt{C_{11}C_{33}}\right)}}. \tag{7.6}$$

As such, Michell (1900) concluded that:

> It appears, therefore, that the law of depression is the same as for an isotropic solid; consequently, the applications of this law, which were made by Boussinesq and Hertz to problems concerning isotropic bodies in contact, may be at once extended to the acolotropic solids here considered, with the limitation that the normal to the plane of contact must be an axis of elastic symmetry.

Of course, the effective elasticity modulus can also be expressed by the components of the compliance matrix:

$$E^* = \frac{E}{(1-\nu_H^2)} \cdot \sqrt{\frac{2}{\left(\frac{\lambda - \nu_V^2}{1-\nu_H^2}\right)^{1/2} + \frac{1+\nu-\nu_V(1+\nu_H)}{1-\nu_H^2}}}. \tag{7.7}$$

In the case of an isotropic continuum ($\lambda = 1$, $\nu_V = \nu_H = \nu$), this expression is reduced to the known equation $E^* = E/(1 - \nu^2)$.

The integral formulation of (2.3) for the frictionless contact mechanical problem is based exclusively on the fundamental solution. Therefore, all solutions from Chap. 2 are equally valid for transversely isotropic media.

When two transversely isotropic bodies are in contact, the effective modulus is used instead of (7.6):

$$\frac{1}{E^*} = \frac{1}{E_1^*} + \frac{1}{E_2^*}, \tag{7.8}$$

where E_1^* and E_2^* represent the effective elastic moduli of the two media.

Likewise, due to the identical fundamental solutions, applying the MDR to a transversely isotropic medium simply requires substituting the effective elasticity modulus by the expressions presented in (7.6) and (7.8). All other transformation rules of the MDR remain unchanged.

Therefore, the non-adhesive normal contact problem for a transversely isotropic medium is identical to the corresponding contact problem of an isotropic continuum. This applies to the displacement field of the surface of the body and the pressure distribution in the immediate surface, but not to the deformation and stress distribution in the interior of the half-space. Consequently, there is no need for special consideration to be given to all normal contact problems for transversely isotropic media. We will simply refer to the results from Chap. 2, which are equally valid for transversely isotropic media.

Further information, particularly concerning the calculation of the stresses in the interior of the transversely isotropic half-space (which, again, do not coincide with those of the isotropic case), can be found in a paper by Yu (2001).

For a historical perspective, the paper by Conway (1956) is worth mentioning. Notably, it describes how, due to the form of the fundamental solution (7.5) by Michell (1900), the calculation method for any (isotropic) axially symmetric normal contact problem by Schubert (1942) can also be applied to the corresponding contact problem of transversely isotropic media.

7.3 Normal Contact with Adhesion

As explained in the previous section, the non-adhesive, frictionless normal contact problem for a transversely isotropic medium is identical to the corresponding contact problem for an isotropic continuum. It merely requires redefining the effective elasticity modulus according to (7.6) or (7.7). Additionally, in Chap. 3 of this book, it was shown that the adhesive, frictionless normal contact problem can be reduced to the corresponding non-adhesive contact. Therefore, the adhesive normal contact problems of isotropic and transversely isotropic media are also equivalent to the respective isotropic problems, regarding both their relationships of the global contact quantities (normal force, indentation depth, and contact radius), as well as the stresses in the contact surface, and the displacements of the medium's surface.

Consequently, there is no need for special consideration to be given to adhesive normal contact problems of transversely isotropic media. Here we will simply refer to the results in Chap. 3 of this book. Taking into account the aforementioned corresponding definition of the effective elasticity modulus, the results directly apply to transversely isotropic contacts.

An overview of the history of work done in the field of adhesive contacts of transversely isotropic media can be found in an article by Borodich et al. (2014).

7.4 Tangential Contact

Turner (1980) provided a general expression for the surface displacement of a transversely isotropic elastic half-space under the effect of an arbitrarily directed force acting on the surface of the half-space at the origin. He used the matrix of compliance coefficients, as seen in (7.1).

According to Turner, the simultaneous effect of a normal force F_N and a tangential force F_x in the x-direction generates the surface displacements

$$\begin{bmatrix} u(x,y) \\ v(x,y) \\ w(x,y) \end{bmatrix} = \frac{\varepsilon}{2\pi r}\begin{bmatrix} \gamma(x/r)F_N + \left(1+\delta(x/r)^2\right)F_x \\ \gamma(y/r)F_N + \delta(xy/r^2)F_x \\ \alpha F_N - \gamma(x/r))F_x \end{bmatrix}, \quad r^2 = x^2+y^2, \tag{7.9}$$

with

$$\begin{aligned}
\alpha &= \left(\frac{\lambda - \nu_V^2}{1-\nu_H^2}\right)^{1/2}, \\
\beta &= \frac{(1+\nu) - \nu_V(1+\nu_H)}{(1-\nu_H^2)}, \\
\gamma &= \left(\frac{2}{\alpha+\beta}\right)^{1/2}\left(\frac{\alpha}{2} - \frac{\nu_V}{2(1-\nu_H)}\right), \\
\delta &= \left(\frac{2}{\alpha+\beta}\right)^{1/2}\left(\frac{1+\nu}{1+\nu_H}\right)^{1/2}\frac{1}{1-\nu_H} - 1, \\
\varepsilon &= \left(\frac{\alpha+\beta}{2}\right)^{1/2}\frac{1-\nu_H}{G_H}.
\end{aligned} \tag{7.10}$$

For an isotropic medium, $\lambda = 1, \alpha = \beta = 1, \gamma = (1-2\nu)/(2-2\nu), \delta = \nu/(1-\nu)$, and $\varepsilon = (1-\nu)/G$; (7.9) can then be reduced to the form provided by Landau and Lifshitz (1944, 1959):

$$\begin{aligned}
u(x,y) &= \frac{1+\nu}{2\pi E}\frac{1}{r}\left\{-(1-2\nu)\frac{x}{r}F_N + \left(2(1-\nu) + \frac{2\nu x^2}{r^2}\right)F_x\right\}, \\
v(x,y) &= \frac{1+\nu}{2\pi E}\frac{1}{r}\left\{-(1-2\nu)\frac{y}{r}F_N + 2\nu\frac{xy}{r^2}F_x\right\}, \quad r^2 = x^2+y^2, \\
w(x,y) &= \frac{1+\nu}{2\pi E}\frac{1}{r}\left\{2(1-\nu)F_N + (1-2\nu)\frac{x}{r}F_x\right\}.
\end{aligned} \tag{7.11}$$

The normal and tangential contact problems are independent of one another for a vanishing γ in (7.9), i.e., when

$$\frac{1-\nu_H}{1+\nu_H}\frac{\lambda-\nu_V^2}{\nu_V{}^2}=1. \tag{7.12}$$

The criterion for the decoupling of the normal and tangential contact problem in the case of an isotropic continuum is reduced to the requirement of incompressibility ($\nu = 1/2$) of the deformable contact partner. In the case that both media are linear-elastic and transversely isotropic, the quantity $\gamma_1 - \gamma_2$ must vanish:

$$\begin{aligned}&\left(\frac{2}{\alpha_1+\beta_1}\right)^{1/2}\left(\frac{\alpha_1}{2}-\frac{\nu_{V,1}}{2(1-\nu_{H,1})}\right)\\ &-\left(\frac{2}{\alpha_2+\beta_2}\right)^{1/2}\left(\frac{\alpha_2}{2}-\frac{\nu_{V,2}}{2(1-\nu_{H,2})}\right)=0.\end{aligned} \tag{7.13}$$

7.4.1 "Cattaneo–Mindlin" Approximation for the Transversely Isotropic Contact

Assuming a decoupling of the normal and tangential contact problem ($\gamma = 0$), and neglecting the surface displacement in the direction perpendicular to the direction of force action (as assumed by the solution by Cattaneo and Mindlin in Chap. 4), (7.9) and (7.11) can be simplified to:

$$\begin{bmatrix} u(r) \\ w(r) \end{bmatrix} = \frac{\varepsilon}{2\pi r}\begin{bmatrix} \left(1+\delta\,(x/r)^2\right)F_x \\ \alpha F_N \end{bmatrix}, \tag{7.14}$$

for a transversely isotropic medium and to:

$$\begin{bmatrix} u(r) \\ w(r) \end{bmatrix} = \frac{1+\nu}{2\pi E}\frac{1}{r}\begin{bmatrix} \left(2\,(1-\nu)+2\nu\,(x/r)^2\right)F_x \\ 2\,(1-\nu)\,F_N \end{bmatrix}, \tag{7.15}$$

for an isotropic medium. It is easy to see that the expression for the tangential displacements in a transversely isotropic medium exactly matches the one for an isotropic medium for the values $\varepsilon = 2(1-\nu^2)/E$ and $\delta = \nu/(1-\nu)$.

Solving for E and ν gives $\nu = \frac{\delta}{1+\delta}$ and $E = \frac{2(1-\nu^2)}{\varepsilon} = \frac{2(1+2\delta)}{\varepsilon(1+\delta)^2}$. For the effective shear modulus, we obtain:

$$G^* = \frac{4G}{2-\nu} = \frac{2E}{(1+\nu)(2-\nu)} = \frac{4}{\varepsilon(2+\delta)}. \tag{7.16}$$

Similarly, achieving identical normal displacements for both transversely isotropic and isotropic media requires the effective moduli to follow the expressions

$$E^* = \frac{2}{\alpha\varepsilon}. \tag{7.17}$$

Inserting the definitions of δ, α, and ε from (7.10) yields the following result:

$$G^* = \frac{2E}{(1+\nu_H)\left((1-\nu_H)\left(\frac{1}{2}\left(\frac{1-\nu_V^2}{1-\nu_H^2}\right)^{1/2} + \frac{1}{2}\frac{1+\nu-\nu_V(1+\nu_H)}{1-\nu_H^2}\right) + \left(\frac{1+\nu}{1+\nu_H}\right)^{1/2}\right)}, \tag{7.18}$$

which, for the isotropic continuum ($\lambda = 1$, $\nu_V = \nu_H = \nu$), takes on the usual form of:

$$G^* = \frac{2E}{(1+\nu)(2-\nu)}. \tag{7.19}$$

Inserting the definitions of α and ε from (7.10) into (7.17) leads to the expression previously formulated in (7.7).

Taking this into consideration, this proves the equivalence of the fundamental solutions for the normal contact problem and the tangential contact problem in the Cattaneo–Mindlin approximation for isotropic and transversely isotropic continua. *Therefore, using definitions (7.16), (7.17), (7.18), and (7.7) of the effective moduli, all results from Chaps. 2, 3, and 4 regarding the relationships of the macroscopical displacements, forces, contact radii, and stress distributions carry over.* Only the stresses in the interior of the medium require special consideration.

The ratio of normal to tangential stiffness of a no-slip contact is given by:

$$\frac{E^*}{G^*} = \frac{2+\delta}{2\alpha}, \tag{7.20}$$

which (in the isotropic case) is reduced to the Mindlin ratio:

$$\frac{E^*}{G^*} = \frac{2-\nu}{2-2\nu}. \tag{7.21}$$

7.5 Summary of the Calculation of Transversely Isotropic Contacts

Once again, we will provide a summary of the approach to solving contact problems of transversely isotropic media.

The quantities defined at the surface of the medium are listed below:

- Normal force
- Contact radius
- Indentation depth
- Distribution of normal stresses and normal displacements at the surface
- The quantities previously listed for the adhesive contact in the JKR approximation
- Tangential force in contact with friction in the "Cattaneo–Mindlin approximation"

- Macroscopical tangential displacement in contact with friction in the "Cattaneo–Mindlin approximation"
- Distribution of tangential stresses and tangential displacements at the surface in the "Cattaneo–Mindlin approximation"

Transversely isotropic media exhibit exactly the same behavior as isotropic media. The only required change involves inserting the effective elasticity modulus (defined in (7.6) and equivalently in (7.7)) or the effective shear modulus defined in (7.18), respectively. The MDR method is also valid without restriction, unchanged from the case of isotropic media.

References

Borodich, F.M., Galanov, B.A., Keer, L.M., Suarez-Alvarez, M.M.: The JKR-type adhesive contact problems for transversely isotropic elastic solids. Mech. Mater. **75**, 34–44 (2014)

Conway, H.D.: The indentation of a transversely isotropic half-space by a rigid punch. Z. Angew. Math. Phys. **7**(1), 80–85 (1956)

Landau, L.D., Lifshitz, E.M.: Механика сплошных сред. Гидродинамика и теория упругости. Теоретическая физика, vol. III. ОГИЗ. ГИТТЛ, Москва (1944)

Landau, L.D., Lifshitz, E.M.: Theory of elasticity. Course of theoretical physics, vol. 7. Pergamon Press, London (1959)

Michell, J.H.: The stress in an aeolotropie elastic solid with an infinite plane boundary. Proc. London Math. Soc. **32**, 247–258 (1900)

Schubert, G.: Zur Frage der Druckverteilung unter elastisch gelagerten Tragwerken. Ing. Arch. **13**(3), 132–147 (1942)

Turner, J.R.: Contact on a transversely isotropic half-space, or between two transversely isotropic bodies. Int. J. Solids Struct. **16**, 409–419 (1980)

Yu, H.Y.: A concise treatment of indentation problems in transversely isotropic half-spaces. Int. J. Solids Struct. **38**(10), 2213–2232 (2001)

8 Viscoelastic Materials

Rubber and other elastomers are highly deformable and thus conform well to surfaces, exhibit high coefficients of friction in many material pairings (rubber-rubber, rubber-asphalt, etc.), and are water and heat resistant. As a result, elastomers see widespread use in tires, belts, cables, adhesive layers, and many other areas of application.

The most important properties of elastomers are: (1) an extremely low elasticity modulus (about 1 to 10 MPa, i.e., four to five orders of magnitude lower than that of "normal" solids), (2) an extreme degree of deformability, and (3) internal energy dissipation (viscosity) during deformation. The root of these fundamental properties of elastomers lies in their structure. Elastomers consist of relatively weakly interacting polymer molecules. In the thermodynamic state of equilibrium, they are in a statistically favored coiled state. Upon application of a mechanical stress on the elastomer, the polymer molecules begin to uncoil. Removing the load from the elastomer causes the polymer molecules to relax once again into their coiled state. This structure explains both the high deformability and the internal friction, along with the associated time-dependent behavior, of elastomers.

Due to the high degree of deformations, elastomers very often exhibit non-linear mechanical behavior. However, for reasons of simplification, we will treat them here as *linear viscoelastic materials*. The treatment of non-linearities would extend beyond the scope of this book.

Chapter 8 is dedicated to contact problems between a rigid, rotationally symmetric indenter and a homogeneous, isotropic, linear viscoelastic half-space. Section 8.1 will provide some initial general information and definitions regarding the description and characterization of linear viscoelastic materials. Sections 8.2 and 8.3 are dedicated to the explicit solution of axially symmetric contact problems using the MDR and the functional equation method by Lee and Radok (1960). These solutions apply to incompressible elastomers. Compressible normal contacts are discussed in Sect. 8.4. Finally, Sect. 8.5 deals with fretting wear of elastomers.

V.L. Popov, M. Heß, E. Willert, *Handbook of Contact Mechanics*,
https://doi.org/10.1007/978-3-662-58709-6_8

8.1 General Information and Definitions on Viscoelastic Media

8.1.1 Time-Dependent Shear Modulus and Creep Function

In a first-order approximation, elastomers can be viewed as incompressible media (Poisson's ratio $\nu \approx 0.5$). Therefore, the following consideration exclusively deals with the characterization of mechanical properties of elastomers for *shear* loads. Let us consider an elastomer block that is being acted upon by shear forces (Fig. 8.1). If it is rapidly deformed by the shear angle ε_0[1], the stress initially rises to a high level $\sigma(0)$ and subsequently relaxes to a much lower value $\sigma(\infty)$ (Fig. 8.2). In elastomers, $\sigma(\infty)$ can be three to four orders of magnitudes smaller than $\sigma(0)$. The ratio

$$G(t) = \frac{\sigma(t)}{\varepsilon_0} \tag{8.1}$$

is called the *time-dependent shear modulus*. This function completely describes the mechanical properties of a material, assuming that the material exhibits a linear behavior.

Let us assume that the block is deformed according to an arbitrary function $\varepsilon(t)$. Any time-dependency $\varepsilon(t)$ can be represented as the sum of time-shifted step functions, as shown schematically in Fig. 8.3. In this diagram, an "elementary step function" at time t' has the amplitude $\mathrm{d}\varepsilon(t') = \dot{\varepsilon}(t')\mathrm{d}t'$. Accordingly, its contribution to the stress is equal to $\mathrm{d}\sigma = G(t-t')\dot{\varepsilon}(t')\mathrm{d}t'$ and the total stress at every point

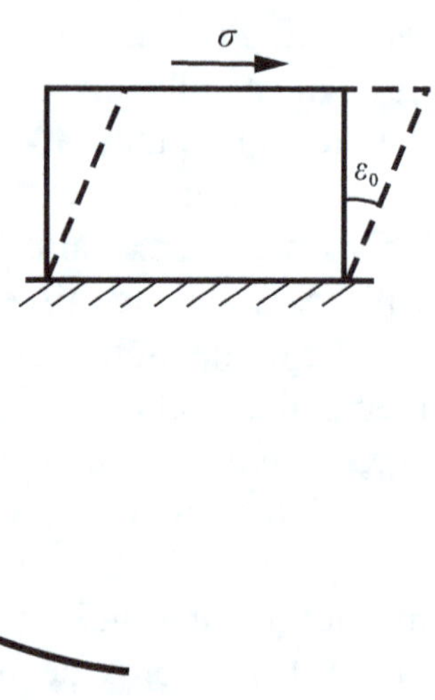

Fig. 8.1 Shear deformation of a rubber block

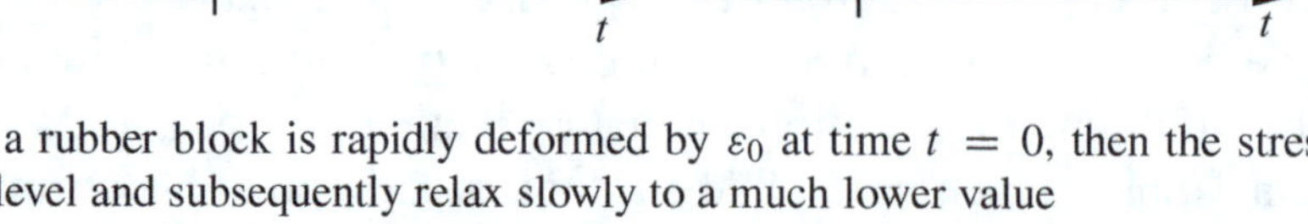

Fig. 8.2 If a rubber block is rapidly deformed by ε_0 at time $t = 0$, then the stresses rise to an initial high level and subsequently relax slowly to a much lower value

---

[1] We emphasize that the shear angle ε is equal to twice the shear component of the deformation tensor.

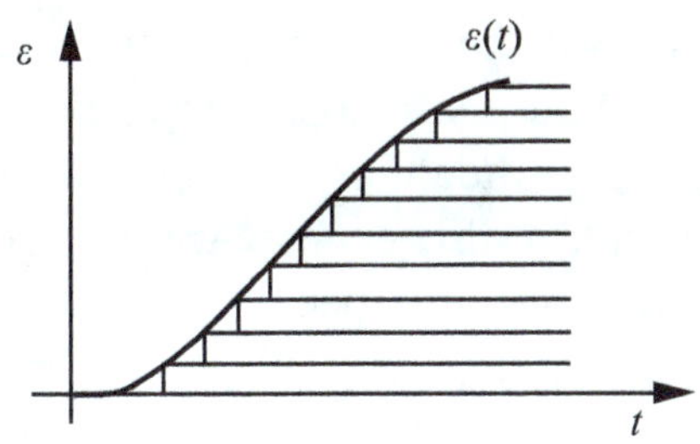

Fig. 8.3 Representation of a time-dependent function as the superposition of several shifted step functions

in time is, therefore, calculated as:

$$\sigma(t) = \int_{-\infty}^{t} G(t - t')\dot{\varepsilon}(t')\mathrm{d}t'. \tag{8.2}$$

Alternatively, the block can be effected by a sudden tangential force. After an instantaneous reaction, the shear angle will change with time. The time-dependent shear angle related to the stress is called the *creep function*, $\Phi(t)$:

$$\Phi(t) = \frac{\varepsilon(t)}{\sigma_0}. \tag{8.3}$$

Using the creep function, we can write the relationship between the stress and strain similarly to (8.2):

$$\varepsilon(t) = \int_{-\infty}^{t} \Phi(t - t')\dot{\sigma}(t')\mathrm{d}t'. \tag{8.4}$$

It can be shown that representations (8.1) and (8.4) are equivalent, i.e., for any time-dependent shear modulus a corresponding creep function can be determined and vice-versa.

From the definitions of the time-dependent shear modulus and the creep functions, the following identities can be deducted:

$$\int_{-\infty}^{t} G(t - t')\dot{\Phi}(t')\mathrm{d}t' = 1 \tag{8.5}$$

and

$$\int_{-\infty}^{t} \Phi(t - t')\dot{G}(t')\mathrm{d}t' = 1. \tag{8.6}$$

Technically, every elastomer also has a material law for the trace (hydrostatic) components of the stress and deformation tensor corresponding to a second creep function for the deformation response to hydrostatic pressure. However, as previously mentioned, elastomers can usually be considered incompressible. Therefore, this secondary creep function is usually neglected. Taking both creep functions into account also makes the analytical treatment of viscoelastic contact problems far more difficult, as demonstrated by Vandamme and Ulm (2006). With these facts in mind, in this chapter we will generally limit ourselves to solutions for an incompressible half-space. The consideration of the compressible case is touched upon in Sect. 8.4.

8.1.2 Complex, Dynamic Shear Modulus

If $\varepsilon(t)$ changes according to the harmonic function

$$\varepsilon(t) = \tilde{\varepsilon} \cos(\omega t), \tag{8.7}$$

then after the transient time, there will also be a periodic change in stress at the same frequency ω. The relationship between the change in deformation and the stress can be represented quite simply when the real function $\cos(\omega t)$ is presented as the sum of two complex exponential functions:

$$\cos(\omega t) = \frac{1}{2}(e^{i\omega t} + e^{-i\omega t}). \tag{8.8}$$

Due to the principle of superposition, one can initially calculate the stresses resulting from the complex oscillations

$$\varepsilon(t) = \tilde{\varepsilon} e^{i\omega t} \quad \text{and} \quad \varepsilon(t) = \tilde{\varepsilon} e^{-i\omega t} \tag{8.9}$$

and sum them up. If we insert $\varepsilon(t) = \tilde{\varepsilon} e^{i\omega t}$ into (8.2), then we obtain

$$\sigma(t) = \int_{-\infty}^{t} G(t - t') i\omega \tilde{\varepsilon} e^{i\omega t'} \mathrm{d}t' = i\omega \tilde{\varepsilon} e^{i\omega t} \int_{0}^{\infty} G(\xi) e^{-i\omega\xi} \mathrm{d}\xi \tag{8.10}$$

for the stress, where we substituted $\xi = t - t'$. This relation can also be written in the form:

$$\sigma(t) = \hat{G}(\omega) \tilde{\varepsilon} e^{i\omega t} = \hat{G}(\omega) \varepsilon(t). \tag{8.11}$$

The coefficient of proportionality

$$\hat{G}(\omega) = i\omega \int_{0}^{\infty} G(\xi) e^{-i\omega\xi} \mathrm{d}\xi \tag{8.12}$$

is called the *complex shear modulus*. Its real part $G'(\omega) = \operatorname{Re} \hat{G}(\omega)$ is called the *storage modulus*, and its imaginary part $G''(\omega) = \operatorname{Im} \hat{G}(\omega)$ is referred to as the *loss modulus*.

8.1.3 Rheological Models

The properties of viscoelastic media are frequently presented in the form of *rheological models*. The two fundamental elements of these models are:

(a) A *linear-elastic body*. For an ideally elastic body, the shear deformation follows Hooke's law: $\sigma = G\varepsilon$. In this case, the complex modulus only has a real part which is equal to G:

$$\hat{G} = G. \tag{8.13}$$

(b) A *linear viscous fluid*. For which the following is valid: (see Fig. 8.4):

$$\sigma = \eta \frac{\mathrm{d}v}{\mathrm{d}z}. \tag{8.14}$$

For a periodic displacement $\hat{u}(l, t) = u_0 e^{i\omega t}$ we get:

$$\hat{\sigma}(t) = \eta \left.\frac{\mathrm{d}v}{\mathrm{d}z}\right|_{z=l} = \eta \frac{\hat{v}(t)}{l} = \eta i \omega \frac{u_0}{l} e^{i\omega t} = i\omega\eta\hat{\varepsilon}(t). \tag{8.15}$$

In this case the complex modulus

$$\hat{G}(\omega) = i\omega\eta \tag{8.16}$$

only has an imaginary part: $\mathrm{Re}\,\hat{G} = 0$, $\mathrm{Im}\,\hat{G} = \omega\eta$.

These two elements enable the "construction" of different media. Although we will refer to the rheological models as "springs" and "dampers", in actuality we mean the corresponding quantities per unit volume of the medium. In this case, it should be noted that the tangential stiffness equals the shear modulus and the damping coefficient equals the dynamic viscosity. From now on we will use these continuum mechanical terms of moduli and viscosities rather than those of values of stiffness and damping coefficients.

8.1.3.1 Kelvin Medium

One of the most frequently used rheological models is the Kelvin model, consisting of a spring (shear modulus G), connected in parallel with a damper (viscosity η). The complex shear modulus of this medium is equal to

$$\hat{G} = G + i\omega\eta. \tag{8.17}$$

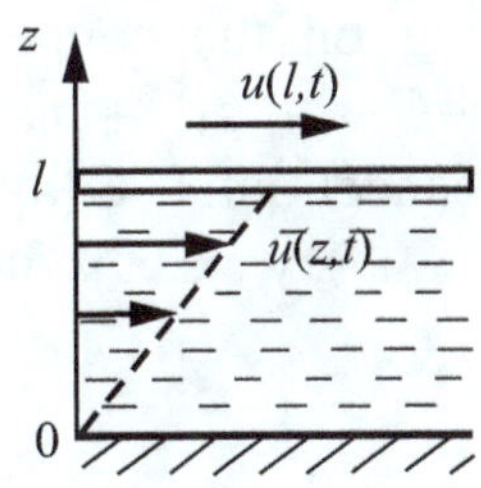

Fig. 8.4 Uniform shear flow of a linear viscous fluid

If at time $t = 0$ a constant stress σ_0 is applied to the Kelvin element, the following is valid:

$$G\varepsilon + \eta\dot{\varepsilon} = \sigma_0. \tag{8.18}$$

The shear angle is time-dependent according to

$$\varepsilon = \frac{\sigma_0}{G}(1 - e^{-t/\tau}), \tag{8.19}$$

with $\tau = \eta/G$. The creep function of this medium is, therefore, equal to:

$$\Phi(t) = \frac{\varepsilon(t)}{\sigma_0} = \frac{1}{G}(1 - e^{-t/\tau}). \tag{8.20}$$

8.1.3.2 Maxwell Medium

An important component of many rheological models is the *Maxwell element*, consisting of a spring connected in series with a linear viscous damper.

The complex moduli of the spring and damper are G and $i\eta\omega$. Due to the serial arrangement, we obtain:

$$\hat{G}_{\text{Maxwell}} = \frac{G \cdot i\eta\omega}{G + i\eta\omega} = \frac{G \cdot i\eta\omega}{(G + i\eta\omega)}\frac{(G - i\eta\omega)}{(G - i\eta\omega)} = \frac{G\left(i\eta\omega G + (\eta\omega)^2\right)}{G^2 + (\eta\omega)^2} \tag{8.21}$$

for the total modulus.

The storage and loss moduli are:

$$G'_{\text{Maxwell}} = \frac{G(\eta\omega)^2}{G^2 + (\eta\omega)^2}, \quad G''_{\text{Maxwell}} = \frac{\eta\omega G^2}{G^2 + (\eta\omega)^2}. \tag{8.22}$$

By introducing the quantity

$$\tau = \eta/G, \tag{8.23}$$

(8.22) can also be presented in the form:

$$G'_{\text{Maxwell}} = G\frac{(\omega\tau)^2}{1 + (\omega\tau)^2}, \quad G''_{\text{Maxwell}} = G\frac{\omega\tau}{1 + (\omega\tau)^2}. \tag{8.24}$$

The quantity τ has the dimension time and is called the *relaxation time*.

Let us examine the stress relaxation in a medium which is described by a Maxwell element. We use the terms introduced in Fig. 8.5. The stress acting on the connection point between the spring and the damper is equal to $-G(\varepsilon - \varepsilon_1) + \eta\dot{\varepsilon}_1$. Because the connection point is massless, the stress must cancel out: $-G(\varepsilon - \varepsilon_1) + \eta\dot{\varepsilon}_1 = 0$. By dividing this equation by G and inserting (8.23), we can write the equation as:

$$\tau\dot{\varepsilon}_1 + \varepsilon_1 = \varepsilon. \tag{8.25}$$

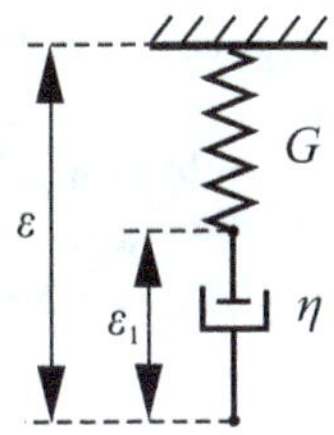

Fig. 8.5 Maxwell element

If at time $t = 0$ the material is *suddenly* deformed by ε_0, then for every point in time $t > 0$ we get

$$\tau\dot{\varepsilon}_1 + \varepsilon_1 = \varepsilon_0, \tag{8.26}$$

with the initial condition $\varepsilon_1(0) = 0$. The solution to this equation with the stated initial condition is:

$$\varepsilon_1 = \varepsilon_0(1 - e^{-t/\tau}). \tag{8.27}$$

The stress is:

$$\sigma = G(\varepsilon_0 - \varepsilon_1) = G\varepsilon_0 e^{-t/\tau}. \tag{8.28}$$

The stress decays exponentially with the characteristic time τ. Therefore, the time-dependent shear modulus in this case equals:

$$G(t) = Ge^{-t/\tau}. \tag{8.29}$$

8.1.3.3 The Standard Solid Model of Rubber

The following model (Fig. 8.6) is the simplest spring-damper model to give a qualitatively correct representation of the most important dynamic properties of rubber during periodic loading.

Since we are dealing with a parallel arrangement of a linear-elastic spring and a Maxwell element, we can immediately write:

$$G' = G_1 + G_2\frac{(\omega\tau)^2}{1 + (\omega\tau)^2}, \quad G'' = G_2\frac{\omega\tau}{1 + (\omega\tau)^2}, \tag{8.30}$$

with $\tau = \eta/G_2$. The frequency-dependency of the moduli is presented *double logarithmically* for the case $G_2/G_1 = 1000$ in Fig. 8.7.

At low frequencies $\omega < G_1/\eta$ (quasi-static loading) the modulus approaches G_1. At very high frequencies $\omega > G_2/\eta$ it approaches $G_1 + G_2 \gg G_1$. This means that for very slow loading, rubber is soft, while for rapid loading it is stiff. In the

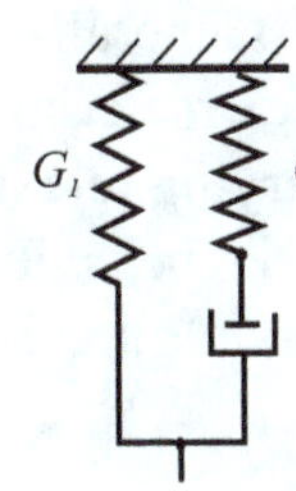

Fig. 8.6 A simple rheological model of rubber (standard solid model)

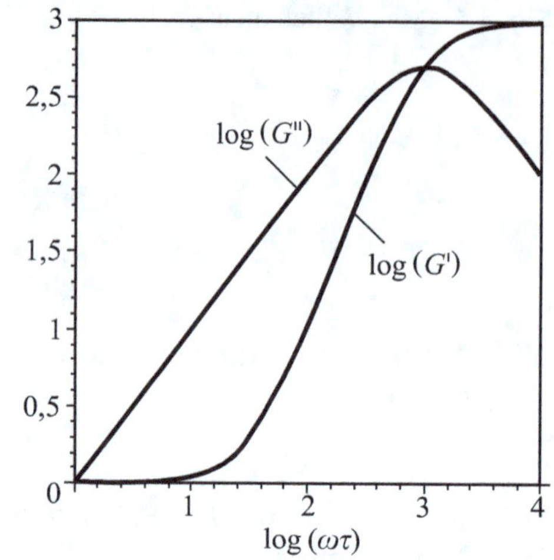

Fig. 8.7 Real and imaginary parts of the dynamic modulus for the rheological model shown in Fig. 8.6 with $G_2/G_1 = 1000$

intermediate range, the imaginary part is dominant: $G''(\omega) \approx \eta\omega$, meaning that the medium behaves like a viscous fluid during periodic loading.

Once again, since we are dealing with a parallel arrangement of a linear-elastic spring and a Maxwell element, we can immediately write:

$$\sigma(t) = \varepsilon_0(G_1 + G_2 e^{-t/\tau}). \tag{8.31}$$

The normalized stress, which we have called the *time-dependent modulus*, can be obtained by dividing this equation by ε_0:

$$G(t) = \sigma/\varepsilon_0 = (G_1 + G_2 e^{-t/\tau}). \tag{8.32}$$

It relaxes exponentially from the value $G_0 = G_1 + G_2$ for $t = 0$ to the value $G_\infty = G_1$ for $t \to \infty$.

8.1.3.4 Summary: Creep Function, Relaxation Function, and Complex Shear Modulus for the Four Most Common Viscous/Viscoelastic Material Models

The creep functions, relaxation functions, and complex shear moduli of the three most frequently used simple viscoelastic material models (standard solid, Maxwell solid, Kelvin solid) are presented in normalized form in Table 8.1. The linear viscous half-space results as the limiting case of the Maxwell element for $\tau \to 0$ and as the limiting case of the Kelvin–Voigt element for $\tau \to \infty$. Figure 8.8 shows phase diagrams of the complex shear modulus of the standard element.

Generalized Maxwell or Kelvin solids are frequently used to model more complex material behavior. These models can be given in the form of Prony series with varying relaxation times. However, since an exact analytical treatment of such material behavior is usually impossible, we will restrict our consideration to the simple models presented previously in this chapter.

Table 8.1 Creep function, relaxation function, and complex shear modulus in normalized representation for the four most frequently used viscous/viscoelastic material models. $k = G_\infty / G_0$ and τ (characteristic relaxation time) are material parameters. Note the special normalization in the case of the Kelvin element

	Standard solid model	Maxwell model	Viscous	Kelvin model
$\frac{\Phi(t)}{\Phi(t=0)}$	$\frac{1}{k}\left[1-(1-k)\exp\left(-\frac{kt}{\tau}\right)\right]$	$1+\frac{t}{\tau}$	$\frac{t}{\tau}$	$\frac{\Phi(t)}{\Phi(t\to\infty)} = 1-\exp\left(-\frac{t}{\tau}\right)$
$\frac{G(t)}{G(t=0)}$	$k\left[1-\left(1-\frac{1}{k}\right)\exp\left(-\frac{t}{\tau}\right)\right]$	$\exp\left(-\frac{t}{\tau}\right)$	$\tau\delta(t)$	$1+\tau\delta(t)$
$\frac{\hat{G}(\omega)}{\hat{G}(\omega\to\infty)}$	$\left[\frac{(\omega\tau)^2}{1+(\omega\tau)^2}+i\frac{\omega\tau}{1+(\omega\tau)^2}\right](1-k)+k$	$\frac{(\omega\tau)^2}{1+(\omega\tau)^2}+i\frac{\omega\tau}{1+(\omega\tau)^2}$	$i\omega\tau$	$\frac{\hat{G}(\omega)}{\hat{G}(\omega=0)} = 1+i\omega\tau$

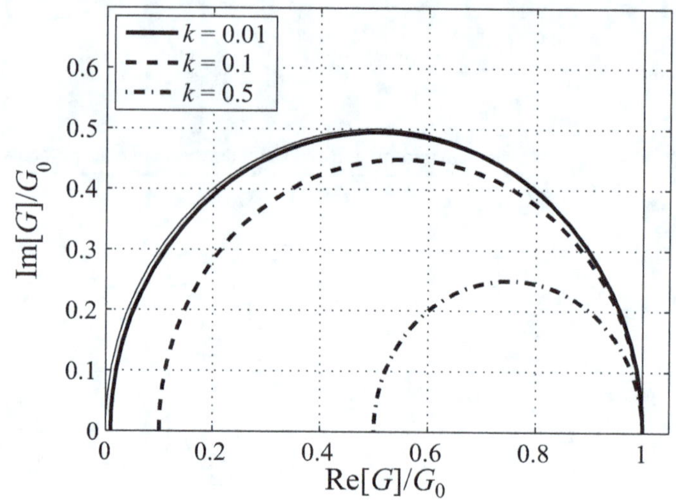

Fig. 8.8 Phase diagram of the complex shear modulus of a standard solid model for different values of $k = G_\infty/G_0$ with the parameter $\omega\tau$. The thin solid lines represent the limiting cases of the Maxwell and Kelvin models (vertical). The maximum of the normalized loss modulus lies at $\omega\tau = 1$

8.1.4 Application of the MDR to Viscoelastic Media

If the indentation velocity of an elastomer during a dynamic loading is below the lowest speed of wave propagation (which is defined by the lowest relevant modulus) then the contact can be regarded as quasi-static. If this condition is met and an area of an elastomer is excited with the angular frequency ω, there is a linear relationship between stress and deformation and, consequently, between force and displacement. The medium can be viewed as an elastic body with the effective shear modulus $G(\omega)$. All principles that are valid for a purely elastic body must also be valid for the harmonically excited viscoelastic medium. Therefore, elastomers can also be described using the MDR as presented in Chaps. 2 and 4. The only difference to the elastic contact is that the effective elasticity modulus is now a function of frequency. For incompressible media ($\nu = 1/2$), the following equation is valid for springs of the effective Winkler foundation:

$$\Delta k_z(\omega) = E^*(\omega)\Delta x = \frac{E(\omega)}{1-\nu^2}\Delta x = \frac{2G(\omega)}{1-\nu}\Delta x \approx 4G(\omega)\Delta x. \tag{8.33}$$

The stiffness of an individual "spring" of the Winkler foundation is four times the shear modulus multiplied by the discretization step size. For a harmonic excitation of the one-dimensional equivalent system, we obtain the spring force:

$$\Delta F_N(x,\omega) = \frac{2G(\omega)}{1-\nu}\Delta x \cdot w_{1D}(x,\omega) \approx 4G(\omega)\Delta x \cdot w_{1D}(x,\omega). \tag{8.34}$$

The inverse transformation into the time domain gives the force as an explicit function of time:

$$\Delta F_N(x,t) = \frac{2}{1-\nu}\Delta x \int_{-\infty}^{t} G(t-t')\dot{w}_{1D}(x,t')\mathrm{d}t'$$
$$\approx 4\Delta x \int_{-\infty}^{t} G(t-t')\dot{w}_{1D}(x,t')\mathrm{d}t'. \tag{8.35}$$

For tangential contacts, the tangential stiffness of the springs of the equivalent one-dimensional MDR foundation must be defined in accordance with (4.21):

$$\Delta k_x = G^*(\omega)\Delta x = \frac{4G(\omega)}{2-\nu}\Delta x \approx \frac{8}{3}G(\omega)\Delta x. \tag{8.36}$$

The corresponding force in the time domain is:

$$\Delta F_x(t) = \frac{4}{2-\nu}\Delta x \int_{-\infty}^{t} G(t-t')\dot{u}_{1D}(t')\mathrm{d}t'$$
$$\approx \frac{8}{3}\Delta x \int_{-\infty}^{t} G(t-t')\dot{u}_{1D}(t')\mathrm{d}t'. \tag{8.37}$$

The formal mathematical proof of this method is based on the method of functional equations by Radok (1957) and was presented by Popov and Heß (2015).

Using rheological models instead of the integral representations (8.35) and (8.37), the forces in the two fundamental elements of elastic bodies and fluids are then given by

$$\Delta F_N = 4Gw_{1D}\Delta x \quad \text{and} \quad \Delta F_N = 4\eta\dot{w}_{1D}\Delta x \tag{8.38}$$

for the normal force, and by

$$\Delta F_x = \frac{8}{3}Gu_{1D}\Delta x \quad \text{and} \quad \Delta F_x = \frac{8}{3}\eta\dot{u}_{1D}\Delta x \tag{8.39}$$

for the tangential force.

In summary, here are the most important steps of applying the MDR to viscoelastic contacts of non-compressible media:

I. The three-dimensional viscoelastic body is substituted by a one-dimensional Winkler foundation consisting of rheological models, which is defined by the following functions of force:

$$\Delta F_N(x,t) = \frac{2}{1-\nu}\Delta x \int_{-\infty}^{t} G(t-t')\dot{w}_{1D}(x,t')\mathrm{d}t'$$

$$\approx 4\Delta x \int_{-\infty}^{t} G(t-t')\dot{w}_{1D}(x,t')\mathrm{d}t', \tag{8.40}$$

$$\Delta F_x(t) = \frac{4}{2-\nu}\Delta x \int_{-\infty}^{t} G(t-t')\dot{u}_{1D}(t')\mathrm{d}t'$$

$$\approx \frac{8}{3}\Delta x \int_{-\infty}^{t} G(t-t')\dot{u}_{1D}(t')\mathrm{d}t', \tag{8.41}$$

or alternatively, is constructed from the usual rheological models of springs (stiffness k) and dampers (damping coefficients α) in accordance with:

$$\Delta k_z = 4G\Delta x \quad \text{and} \quad \Delta\alpha = 4\eta\Delta x \tag{8.42}$$

$$\Delta k_x = \frac{8}{3}G\Delta x \quad \text{and} \quad \Delta\alpha_x = \frac{8}{3}\eta\Delta x. \tag{8.43}$$

II. The three-dimensional profile $\tilde{z} = f(r)$ is transformed into a one-dimensional profile $g(x)$ according to:

$$g(x) = |x| \int_{0}^{|x|} \frac{f'(r)}{\sqrt{x^2 - r^2}}\mathrm{d}r. \tag{8.44}$$

III. The one-dimensional profile according to (8.44) is now pressed into the viscoelastic foundation according to (8.40)–(8.43). The relationships between the normal force, the indentation depth, and the contact radius resulting from the one-dimensional model corresponds exactly to the ones of the original three-dimensional problem at every point in time and independent of the loading history.

IV. The stress distribution can, if necessary, also be calculated using an equation given in Chap. 2 (2.13) and the deformation of the medium (including the area outside the contact) also with an (2.14) from Chap. 2.

8.1.5 Description of Elastomers by Radok (1957)

By substituting $t' = t - \xi$ and expanding (8.2) to powers of ξ, (8.2) can be rewritten in the following differential form:

$$\sigma(t) = \int_0^\infty G(\xi)\dot{\varepsilon}(t-\xi)\mathrm{d}\xi = \sum_{n=1}^{\infty} \frac{\mathrm{d}^n \varepsilon(t)}{\mathrm{d}t^n} \int_0^\infty \frac{(-1)^n}{n!}\xi^n G(\xi)\mathrm{d}\xi. \tag{8.45}$$

Similarly, (8.4) can be written in the form:

$$\varepsilon(t) = \sum_{n=1}^{\infty} \frac{\mathrm{d}^n \sigma(t)}{\mathrm{d}t_n} \int_0^\infty \frac{(-1)^n}{n!}\xi^n \Phi(\xi)\mathrm{d}\xi. \tag{8.46}$$

Equation (8.18) for the Kelvin model is a simple example of the series expansion (8.45). For the case of the Kelvin model, only the terms of the order $n = 0$ and $n = 1$ appear.

Radok (1957) noted that representations (8.45) and (8.46) were merely special cases of a more general representation

$$P[s_{ij}(t)] = Q[e_{ij}(t)], \tag{8.47}$$

where s_{ij} and e_{ij} denote the traceless components of the stress tensor and deformation tensor in Cartesian coordinates. P and Q are the linear differential operators defined as:

$$P := \sum_{k=0}^{\infty} p_k \frac{\mathrm{d}^k}{\mathrm{d}t^k},$$
$$Q := \sum_{k=0}^{\infty} q_k \frac{\mathrm{d}^k}{\mathrm{d}t^k}. \tag{8.48}$$

In the purely elastic case, these are zero-order operators with

$$\frac{q_0}{p_0} = 2G,$$
$$q_k = 0, \quad p_k = 0, \quad k > 0, \tag{8.49}$$

and the shear modulus G. It is easy to see that the two operators are simply indeterminate, yet this can be easily resolved by an appropriate normalization. The identities (8.5) and (8.6) can be written in the form

$$P[G(t)] = \text{const},$$
$$Q[\Phi(t)] = \text{const}. \tag{8.50}$$

The creep and relaxation functions constitute a transformation pair. If we designate the respective Laplace transforms by $\hat{\Phi}(s)$ and $\hat{G}(s)$, the following identity is valid:

$$s^2\hat{\Phi}(s)\hat{G}(s) \equiv 1. \tag{8.51}$$

8.1.6 General Solution Procedure by Lee and Radok (1960)

The material behavior of viscoelastic materials is time-dependent. The state of a viscoelastic contact is, therefore, dependent on its loading history. In the case of the normal contact problem, the sole relationship that is independent of the loading history is the contact configuration, i.e., the relationship

$$d = g(a) \tag{8.52}$$

between the indentation depth d and the contact radius a. Even this is valid only when the contact radius monotonically increases over time. In this case, the relationship is defined entirely by the form of the indenter but is independent of rheology. On the basis of this idea and the form of the fundamental field equations of elasticity and viscoelasticity theory, Lee (1955) proposed a method to derive the viscoelastic solution from the solution of the elastic problem. This method, based on the Laplace transform, was generalized by Radok (1957). In a joint effort, Lee and Radok (1960) found the solution for the contact of a parabolic indenter, provided that the contact radius is monotonically increasing over time. Hunter (1960) extended this to cases where the contact radius a exhibits a single maximum, facilitating the consideration of, for example, the Hertzian impact problem with a viscoelastic half-space. The solutions for arbitrary loading histories stem from Ting (1966, 1968) and Graham (1965, 1967). However, it should be noted that the calculation increases in complexity for every additional extremum of the contact radius.

Normal contacts with elastomers frequently appear in testing procedures to determine the material properties of the viscoelastic material. The most frequently used ones are the Shore hardness test (see Sect. 8.3.2), the rebound-indentation test (see Sect. 8.3.3.1), and the impact test (see Sect. 8.3.3.2).

8.1.6.1 Contact Radius Increasing Monotonically Over Time

The solution by Lee and Radok (1960) can be generalized without great difficulty to any rotationally symmetric indenter with the profile $\tilde{z} := f(r)$. It merely requires the solution of the corresponding elastic problem, which is detailed in Chap. 2 and summarized in the form:

$$\begin{aligned} d^{\mathrm{el}} &= d^{\mathrm{el}}(a), \\ F_N^{\mathrm{el}} &= F_N^{\mathrm{el}}(a), \\ \sigma_{zz}^{\mathrm{el}} &= \sigma_{zz}^{\mathrm{el}}(r;a) \end{aligned} \tag{8.53}$$

Here, r denotes the radial coordinate, d the indentation depth, a the contact radius, F_N the normal force, and σ_{zz} the normal stress at the surface of the half-space. The superscript "el" indicates the elastic solution. As previously mentioned, for a monotonically increasing contact radius, the relationship $d = d(a)$ is independent of the material law of the half-space. Using the method by Lee and Radok yields the following solution to the viscoelastic problem:

$$\begin{aligned} d(t) &= d(a(t)), \\ P[\sigma_{zz}(r,t)] &= \frac{1}{2G_0} Q\left[\sigma_{zz}^{\text{el}}(r;a(t))\right], \\ P[F_N(t)] &= \frac{1}{2G_0} Q\left[F_N^{\text{el}}(a(t))\right], \end{aligned} \tag{8.54}$$

with the operators P and Q introduced in (8.47) and (8.48). G_0 is the shear modulus assumed for the elastic solution. Since all elastic stresses are linear in G_0, the specific value of G_0 is irrelevant. The solution can also be expressed using the relaxation function:

$$\begin{aligned} \sigma_{zz}(r,t) &= \frac{1}{G_0} \int_{-\infty}^{t} G(t-\tau) \frac{\partial}{\partial \tau} \left\{\sigma_{zz}^{\text{el}}[r;a(\tau)]\right\} \mathrm{d}\tau, \\ F_N(t) &= \frac{1}{G_0} \int_{-\infty}^{t} G(t-\tau) \frac{\mathrm{d}}{\mathrm{d}\tau} \left\{F_N^{\text{el}}[a(\tau)]\right\} \mathrm{d}\tau. \end{aligned} \tag{8.55}$$

The stress vanishes outside the contact, i.e., for $r > a(t)$. This must be taken into account for the integrations. Two particularly elegant and simple cases occur when either the contact radius or the entire normal force is kept constant (see Sects. 8.3.1 and 8.3.2).

8.2 Explicit Solutions for Contacts with Viscoelastic Media Using the MDR

8.2.1 Indentation of a Cylindrical Punch in a Linear Viscous Fluid

In this section, we will consider the indentation of a rigid cylindrical punch of radius a with the constant force F_N in a linear viscous half-space (viscosity η, no gravity, no capillarity), as demonstrated in Fig. 8.9. We will calculate the indentation velocity and the indentation depth as a function of time.

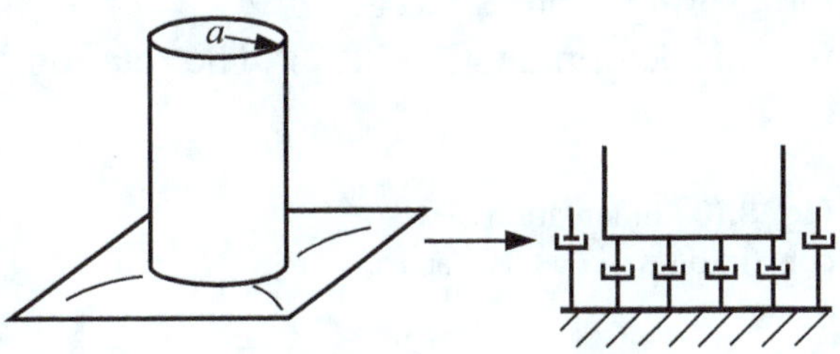

Fig. 8.9 Indentation of a cylindrical punch in a viscous half-space

The first step is to determine the equivalent one-dimensional profile using (8.44):

$$g(x) = \begin{cases} 0, & |x| < a, \\ \infty, & |x| \geq a. \end{cases} \tag{8.56}$$

The second step is to define the Winkler foundation in accordance with (8.42). For the "spring" force, we obtained $\Delta F_N = 4\eta \dot{w}_{1D} \Delta x = 4\eta \dot{d}(t) \Delta x$. At a contact radius $a(t)$, the vertical force is equal to the individual spring force multiplied by $2a/\Delta x$, which is the number of springs in the contact:

$$F_N = 8\eta a(t) \dot{d}(t). \tag{8.57}$$

In the case of a cylindrical punch, the contact radius remains constant and equal to a. The force is given by:

$$F_N = 8\eta a \dot{d}. \tag{8.58}$$

If the force is constant, integration of (8.58) with the initial condition $d(0) = 0$ yields:

$$F_N t = 8\eta a d(t). \tag{8.59}$$

The indentation depth as a function of time is then given by:

$$d(t) = \frac{F_N t}{8\eta a}. \tag{8.60}$$

8.2.2 Indentation of a Cone in a Linear Viscous Fluid

In this section, we will consider the indentation of a rigid cone $f(r) = \tan\theta \cdot |r|$ with the constant force F_N into a linear viscous half-space (viscosity η, no gravity, no capillarity), as demonstrated in Fig. 8.10. We will calculate the indentation speed and indentation depth as a function of time.

First the equivalent one-dimensional profile is determined using (8.44):

$$g(x) = \frac{\pi}{2} |x| \tan\theta. \tag{8.61}$$

The second step is to define the Winkler foundation in accordance with (8.42). For the "spring" forces we get $\Delta F_N = 4\eta \dot{w}_{1D} \Delta x = 4\eta \dot{d}(t) \Delta x$. The total normal force is determined by (8.57). The relationship between the current contact radius

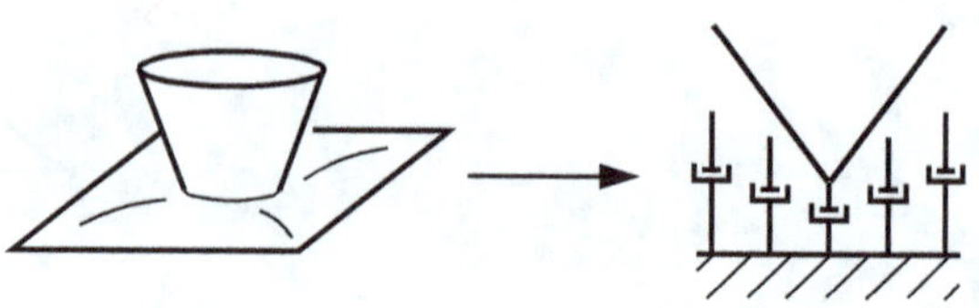

Fig. 8.10 Indentation of a cone into a viscous half-space

and the indentation depth does not depend on the rheology and follows from the equation $d(t) = g(a(t))$:

$$a(t) = \frac{2}{\pi}\frac{d(t)}{\tan\theta}. \quad (8.62)$$

Substituting this equation into (8.57) for the force yields:

$$F_N = \frac{16}{\pi \tan\theta}\eta d(t)\dot{d}(t). \quad (8.63)$$

If the force is constant, the integration with the initial condition $d(0) = 0$ results in:

$$F_N t = \frac{8}{\pi \tan\theta}\eta d(t)^2. \quad (8.64)$$

The indentation depth as a function of time is then given by:

$$d(t) = \left(\frac{\pi \tan\theta \cdot F_N t}{8\eta}\right)^{1/2}. \quad (8.65)$$

8.2.3 Indentation of a Parabolic Indenter into a Linear Viscous Fluid

In this section, we will look at the indentation of a rigid paraboloid of rotation $f(r) = r^2/(2R)$ (see Fig. 8.11) with the constant force F_N in a linear viscous half-space (viscosity η, no gravity, no capillarity). We will calculate the indentation speed and indentation depth as a function of time.

First the equivalent one-dimensional profile is determined using (8.44):

$$g(x) = x^2/R. \quad (8.66)$$

For the second step, we will define the Winkler foundation in accordance with (8.42). For the "spring" forces we get $\Delta F_N = 4\eta\dot{w}_{1D}\Delta x = 4\eta\dot{d}(t)\Delta x$. The total normal force is determined by (8.57). The relationship between the current contact radius and the indentation depth does not depend on the rheology and follows from the equation $d(t) = g(a(t))$:

$$a(t) = \sqrt{Rd(t)}. \quad (8.67)$$

Substituting the contact radius into the equation for the force yields:

$$F_N = 8\eta R^{1/2}\sqrt{d(t)}\dot{d}(t). \quad (8.68)$$

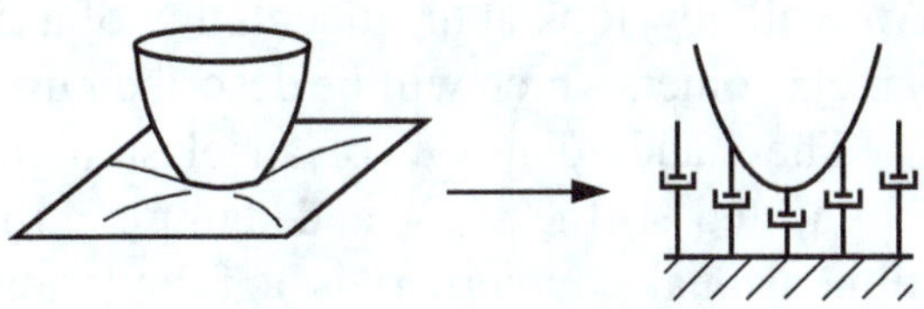

Fig. 8.11 Indentation of a paraboloid in a viscous half-space

If the force is constant, integration with the initial condition $d(0) = 0$ yields:

$$F_N t = \frac{16}{3}\eta R^{1/2} d(t)^{3/2}. \tag{8.69}$$

The indentation depth as a function of time is then given by:

$$d(t) = \left(\frac{3F_N t}{16\eta R^{1/2}}\right)^{2/3}. \tag{8.70}$$

8.2.4 Indentation of a Cone in a Kelvin Medium

Let us now consider the indentation of a cone in a Kelvin medium with shear modulus G and dynamic viscosity η. In the first step, the equivalent one-dimensional profile is determined by (8.44): $g(x) = \tan\theta \cdot |x| \cdot \pi/2$. The contact radius is determined by the equation $d = g(a)$, thus $a = (2/\pi)(d/\tan\theta)$. We now have to use a superposition of the elastic contribution for the force (see Sect. 2.5.2 in Chap. 2):

$$F_{N,\text{el}} = \frac{8G}{\pi}\frac{d^2}{\tan\theta} \tag{8.71}$$

and the viscous contribution (see (8.63)):

$$F_N = \frac{8G}{\pi}\frac{d^2}{\tan\theta} + \frac{16\eta}{\pi\tan\theta} d\dot{d}. \tag{8.72}$$

This equation can be written in the form

$$\frac{\pi\tan\theta \cdot F_N}{8G} = d^2 + 2\tau d\dot{d} = d^2 + \tau\frac{\mathrm{d}\left(d^2\right)}{\mathrm{d}t}, \tag{8.73}$$

where $\tau = \eta/G$ is the relaxation time of the medium. Integration of this equation with the initial condition $d(0) = 0$ yields:

$$d^2(t) = \frac{\pi\tan\theta \cdot F_N}{8G}\left(1 - e^{-t/\tau}\right). \tag{8.74}$$

8.2.5 Indentation of a Rigid Cylindrical Indenter into a "Standard Medium"

We will now look at the indentation of a rigid cylindrical indenter with radius a into an elastomer, which will be described using the "standard model" (see Fig. 8.6).

The standard model of an elastomer consists of a Maxwell element (series-connected stiffness G_2 and damping η) and a parallel stiffness G_1. The one-dimensional counterpart is a foundation of elements at a distance Δx, whose individual components are characterized by the parameters $4G_1\Delta x$, $4G_2\Delta x$, and $4\eta\Delta x$. The equivalent one-dimensional indenter is a rectangle whose side has the length $2a$. For the normal force the following applies:

$$F_N = 8G_1ad + 8G_2a(d - w_{1,1D}), \tag{8.75}$$

where $w_{1,1D}$ satisfies the following equation:

$$d = w_{1,1D} + \tau\dot{w}_{1,1D}, \tag{8.76}$$

with $\tau = \eta/G_2$. The solution of these equations with the initial conditions $d(0) = 0$ and $w_{1,1D}(0) = 0$ yields to:

$$\begin{aligned} w_{1,1D}(t) &= \frac{F_N}{8G_1a}\left(1 - \exp\left(-\frac{G_1 t}{\tau(G_1 + G_2)}\right)\right), \\ d(t) &= \frac{F_N}{8G_1a}\left(1 - \frac{G_2}{G_1 + G_2}\exp\left(-\frac{G_1 t}{\tau(G_1 + G_2)}\right)\right). \end{aligned} \tag{8.77}$$

In the limiting case $G_2 \gg G_1$ we recover the result for a Kelvin body:

$$d(t) = \frac{F_N}{8G_1a}\left(1 - \exp\left(-\frac{G_1 t}{\eta}\right)\right). \tag{8.78}$$

8.3 Explicit Solutions for Contacts with Viscoelastic Media by Lee and Radok (1960)

8.3.1 Constant Contact Radius

For a constant contact radius, i.e., the limiting case of the requisite monotonic behavior, the right-hand side of (8.54) is populated by time-invariant terms. The comparison to (8.50) immediately yields:

$$\begin{aligned} \sigma_{zz}(r,t) &= \frac{1}{G(t=0)}\sigma_{zz}^{\mathrm{el}}(r;a)G(t), \\ F_N(t) &= \frac{1}{G(t=0)}F_N^{\mathrm{el}}(a)G(t), \end{aligned} \tag{8.79}$$

with the relaxation function $G(t)$.

8.3.2 Constant Normal Force (Shore Hardness Test, DIN EN ISO 868)

According to (8.54), for a constant normal force $F_N(t) = \text{const} = F_0$, it is valid that $P[F_N(t)] = \text{const}$. The comparison to (8.50) yields the relationship:

$$F_N^{\mathrm{el}}(a(t)) = F_0\Phi(t), \tag{8.80}$$

with $\Phi(t)$ denoting the (normalized) creep function. A few examples for (8.80) were explicitly considered in Sect. 8.2.

From (8.80) we can design a very simple measurement procedure of the creep function: an indenter can be placed on the viscoelastic half-space and the indentation depth resulting from its own weight can be measured as a function of time.

Another important application of (8.80) is measuring the hardness of viscoelastic media according to Shore (DIN EN ISO 868). It involves indenting a test piece with a rigid indenter by applying a constant force F_0 for a predetermined amount of time and measuring the achieved indentation depth at the end of the test. This indentation depth is then the determining measure for the Shore hardness. The values of the normal force and indentation time, as well as the particular indenter geometry vary according to the specific test form (Shore-A, Shore-D, etc.). If the creep function of the material to be inspected is known, the result of the hardness test can be predicted by (8.80) (within the scope of the fundamental assumptions of the equation, i.e., linear material behavior and validity of the half-space hypothesis). As indenting bodies: a cone for the Shore-D test (see Sect. 8.3.2.1) and a truncated cone for Shore-A test (see Sect. 8.3.2.2) are utilized. We will also consider the indenter profile in the form of a power-law (see Sect. 8.3.2.3) as, through a Taylor series expansion, it can function as a building block of a generalized Shore test for any differentiable indenter profile.

8.3.2.1 Shore-D: The Cone in Contact with a Standard Solid

For illustration purposes, we will restrict the consideration to the viscoelastic material model of the standard solid, for which we determined the normalized creep function in Table 8.1 (Sect. 8.1.3.4) to be:

$$\Phi(t) = \frac{1}{k}\left[1 - (1-k)\exp\left(-\frac{kt}{\tau}\right)\right]. \tag{8.81}$$

Here, k is the ratio between the static modulus and glass modulus, while τ is the characteristic relaxation time of the standard solid model.

In Chap. 2 of this book (see Sect. 2.5.2) it was demonstrated that, for the conical indenter with the slope angle θ, which was first considered in viscoelastic contact by Graham (1965), the normal force in the elastic case is proportional to the square of the indentation depth d:

$$F_N^{\mathrm{el}} \sim d^2. \tag{8.82}$$

Equation (8.80) and the creep function (8.81) of the standard body give the following relationship for the indentation depth as a function of time:

$$\frac{d(t)}{d(t=0)} = \sqrt{\frac{1}{k}\left[1 - (1-k)\exp\left(-\frac{kt}{\tau}\right)\right]}. \tag{8.83}$$

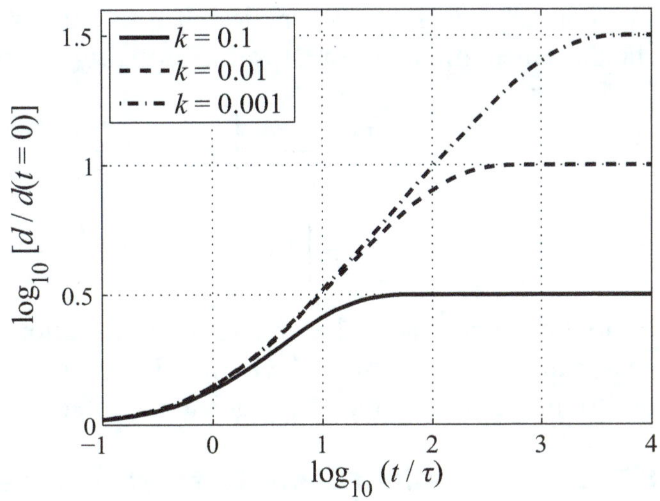

Fig. 8.12 Dependency of the normalized indentation depth on the normalized time in logarithmic representation for the Shore hardness test of a standard solid medium with a conical indenter

The indentation depth at the beginning of the test $d(t = 0)$ is then the instantaneous elastic solution with the glass modulus G_0 (see Sect. 2.5.2):

$$d(t = 0) = \sqrt{\frac{\pi F \tan \theta_0}{8G_0}}. \tag{8.84}$$

Solution (8.83) is shown in normalized logarithmic representation in Fig. 8.12.

8.3.2.2 Shore-A: The Truncated Cone in Contact with a Standard Solid

The Shore A test, which is used for softer materials, is carried out with a truncated cone as the indenting body. This body has the rotationally symmetric profile

$$f(r) = \begin{cases} 0, & r \leq b, \\ (r - b) \tan \theta, & r > b, \end{cases} \tag{8.85}$$

with the radius b at the blunt end and the conical inclination angle θ (note that the complementary angle $\theta^* = \pi/2 - \theta$ is used in the Shore hardness measurement standard). The elastic solution needed to apply (8.80) for this contact problem was derived in Chap. 2 (Sect. 2.5.9) and is described by:

$$\begin{aligned} d(a) &= a \tan \theta \arccos\left(\frac{b}{a}\right), \\ F_N^{\mathrm{el}}(a) &= 4G \tan \theta a^2 \left[\arccos\left(\frac{b}{a}\right) + \frac{b}{a}\sqrt{1 - \frac{b^2}{a^2}}\right]. \end{aligned} \tag{8.86}$$

The contact radius a as a function of time during the hardness measurement of a standard medium, therefore, results implicitly by solving the equation

$$\frac{G_0 \tan\theta a^2}{F_0}\left[\arccos\left(\frac{b}{a}\right)+\frac{b}{a}\sqrt{1-\frac{b^2}{a^2}}\right] = \frac{1}{k}\left[1-(1-k)\exp\left(-\frac{kt}{\tau}\right)\right]. \tag{8.87}$$

From the contact radius the remaining indentation depth, and thus the Shore hardness, can then be determined with the first of the two equations given in (8.86). For $b = 0$, the results for the fully conical indenter are reproduced.

8.3.2.3 Generalized Shore Test with an Indenter in the Form of a Power-Law in Contact with a Standard Body

For an indenter with the profile

$$f(r) = cr^n, \quad n \in \mathbb{R}^+, \tag{8.88}$$

where n is a positive real number and c a constant coefficient, the relationship between the normal force and the indentation depth in the elastic case has been shown in Chap. 2 of this book (Sect. 2.5.8):

$$F_N^{\mathrm{el}} \sim d^{\frac{n+1}{n}}. \tag{8.89}$$

With (8.80) and the creep function (8.81) of a standard medium, we get the following indentation depth as a function of time:

$$\frac{d(t)}{d(t=0)} = \left\{\frac{1}{k}\left[1-(1-k)\exp\left(-\frac{kt}{\tau}\right)\right]\right\}^{\frac{n}{n+1}}. \tag{8.90}$$

Then the instantaneous indentation depth $d(t = 0)$ results from the elastic solution with the glass module G_0:

$$d(t=0) = \left\{\frac{F_0[\kappa(n)c]^n}{8G}\frac{n+1}{n}\right\}^{\frac{n}{n+1}}, \tag{8.91}$$

with the stretch factor, which depends on the exponent of the profile

$$\kappa(n) := \sqrt{\pi}\frac{\Gamma(n/2+1)}{\Gamma\left[(n+1)/2\right]}, \tag{8.92}$$

where we have made use of the gamma function

$$\Gamma(z) := \int_0^\infty t^{z-1}\exp(-t)\mathrm{d}t. \tag{8.93}$$

For $n = 1$ the results of the conical indenter demonstrated here can be reproduced.

8.3.3 Non-Monotonic Indentation: Contact Radius with a Single Maximum

In this section, we consider loading histories consisting of an indentation and rebound phase. Let the contact radius have its maximum value at the time of reversal t_m and let the time $t = 0$ be zero. The latter condition presents a restriction to the generality, but it is surely the most common case. A possible graph of the contact radius is depicted in Fig. 8.13.

The first step is to solve the contact problem for the indentation phase $t \leq t_m$. We have at our disposal the results of the previous sections and, specifically, we have (8.54). Additionally, there exists an explicit function $t_1(t')$ with $t_1 < t_m$ and $t' > t_m$ so that (see the Fig. 8.13)

$$a(t_1) = a(t'). \tag{8.94}$$

For the indentation depth $d(t)$ and the normal force $F_N(t)$ during the rebound phase $t > t_m$, we obtain the expressions for arbitrary rotationally symmetric indenters determined by Ting (1966):

$$d(t) = d^{\mathrm{el}}(a(t)) - \int\limits_{t_m}^{t} \Phi(t-t')\frac{\mathrm{d}}{\mathrm{d}t'}\left[\int\limits_{t_1(t')}^{t'} G(t'-t'')\frac{\mathrm{d}}{\mathrm{d}t''}d^{\mathrm{el}}(a(t''))\mathrm{d}t''\right]\mathrm{d}t',$$

$$F_N(t) = \int\limits_{0}^{t_1(t)} G(t-t')\frac{\mathrm{d}}{\mathrm{d}t'}F_N^{\mathrm{el}}(a(t'))\mathrm{d}t', \tag{8.95}$$

with the normalized creep function Φ, the normalized relaxation function G, and the elastic solutions indicated by the superscript "el", which is discussed in Chap. 2.

In the case of the parabolic indenter, the results from (8.95) were already known to Hunter (1960). Furthermore, Ting (1968) and Graham (1967) found the solutions for arbitrary counts of maxima and minima of the contact radius. Regrettably, the analytical treatment of these increasingly interlinked differentiations and integrations is not practically viable.

Of course one can conceive various loading protocols where the contact radius has a single maximum and the contact problem thus decomposes into an indentation and rebound phase. Two technically relevant cases are discussed in the literature:

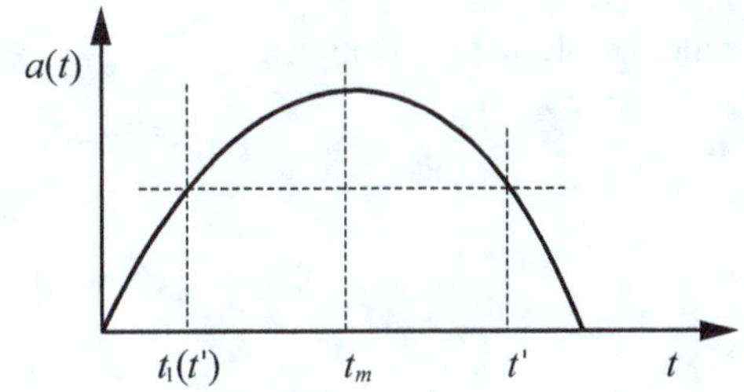

Fig. 8.13 Schematic graph of the contact radius

rebound-indentation testing in which the first phase is displacement-controlled and the force is zero in the second phase, and the linear centric impact problem, which of equal importance for various testing procedures.

8.3.3.1 Material Tests Using the Rebound-Indentation Procedure

In rebound-indentation testing, an indenter is pressed under displacement-control into a viscoelastic half-space, e.g., with a constant velocity v_0:

$$d^I(t) = v_0 t. \tag{8.96}$$

The superscript "*I*" indicates the indentation phase $t \le t_m$. The indenter is subsequently released and the half-space relaxes. By measuring the indentation depth during relaxation, it is possible to determine the time-dependent material functions of the elastomer. The protocols of the indentations are displayed in Figs. 8.14 and 8.15.

Argatov and Popov (2016) were able to prove that the indentation depth during the rebound phase fulfills the relationship

$$d^{II}(t) = \int_0^{t_m} \mathrm{K}(t - t', t_m - t') \frac{\mathrm{d}}{\mathrm{d}t'} d^I(t')\mathrm{d}t' = v_0 \int_0^{t_m} K(t - t', t_m - t')\mathrm{d}t', \tag{8.97}$$

independent of the actual form of the indenter Equation (8.97). for the flat cylindrical punch was already derived by Argatov and Mishuris (2011) and for the parabolic indenter by Argatov (2012). The function $K(t, t')$, which was previously utilized by Greenwood (2010) for an alternative presentation of the solutions by Ting (1966) and which presents the stress response via a unit deformation applied at $t = 0$ and

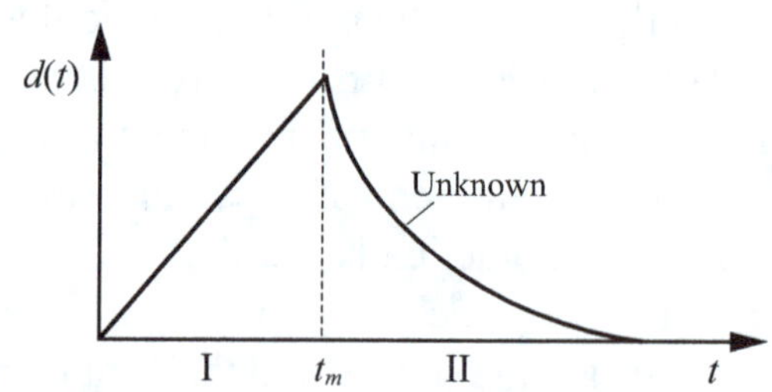

Fig. 8.14 Protocol of the indentation depth during rebound-indentation testing

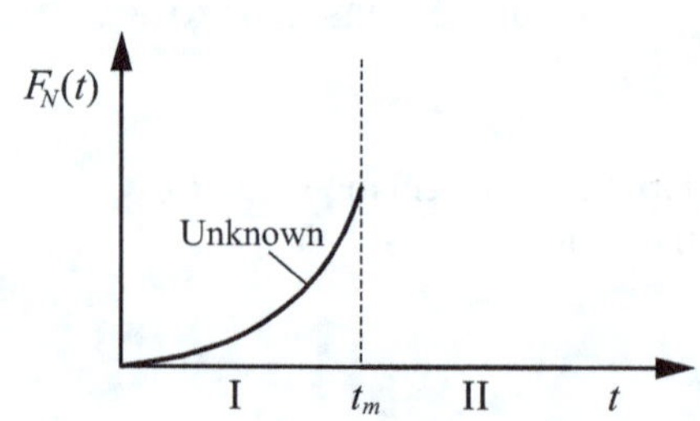

Fig. 8.15 Protocol of the normal force during rebound-indentation testing. The rebound phase occurs under zero force

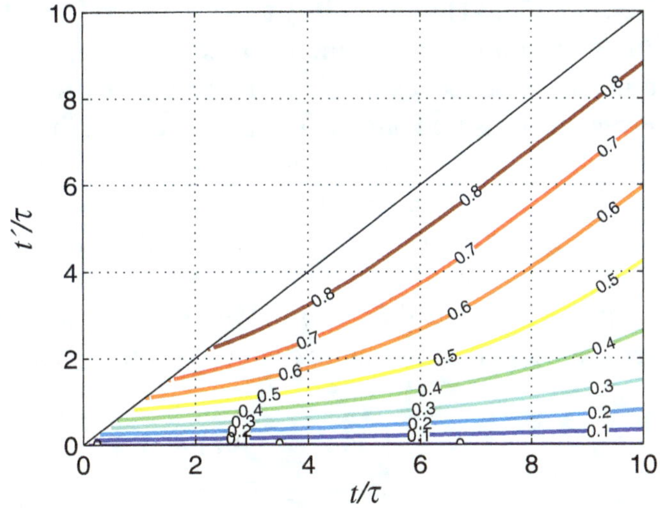

Fig. 8.16 The function $K(t, t')$ from (8.99) for the three-element standard solid with $k = 0.1$

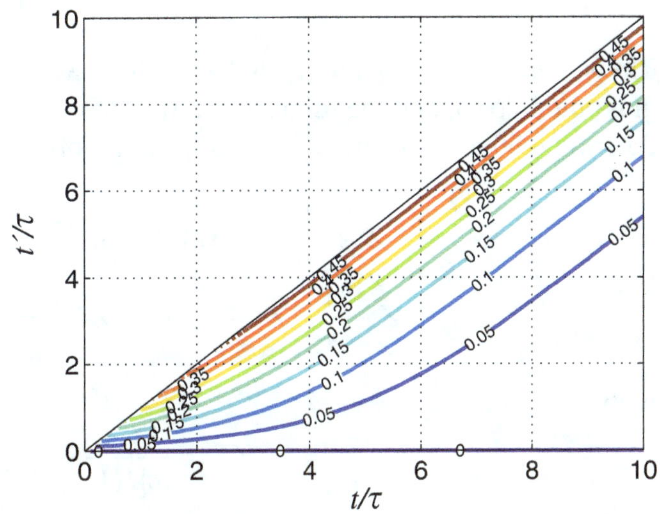

Fig. 8.17 The function $K(t, t')$ for the three-element standard solid with $k = 0.5$

removed at $t = t'$, is connected to functions Φ and Ψ by the relationship:

$$K(t, t') = 1 - \Psi(t) + \int_{t'}^{t} \Psi(t'') \frac{\partial}{\partial t''} \Phi(t - t'') \mathrm{d}t'', \quad t > t'. \tag{8.98}$$

For the three-element standard solid with $k = G_\infty / G_0$ and the characteristic relaxation time τ, the expression is:

$$K(t, t') = (1 - k) \left(1 - \exp\left(-\frac{t'}{\tau} \right) \right) \exp\left(-\frac{k(t - t')}{\tau} \right), \quad t > t'. \tag{8.99}$$

A normalized representation of this function for two different values of k is given in Figs. 8.16 and 8.17.

8.3.3.2 The Hertzian Impact Problem for a Viscoelastic Half-Space

Another simple test method for determining the viscoelastic properties of elastomers is rebound testing, which is usually carried out with spherical impact bodies. From the measured number of impacts during the experiment, conclusions about the dynamic material behavior of the rubber can be drawn.

Hunter (1960) looked at the Hertzian impact problem, i.e., the straight, centric, frictionless impact of a rigid sphere, with a viscoelastic half-space when the peak time is small compared to the smallest characteristic relaxation time τ of the elastomer. In this case, the creep and relaxation functions of the elastomer in normalized representation can be approximated by the expressions of the Maxwell body:

$$\Phi(t) = 1 + \frac{t}{\tau},$$
$$G(t) = \exp\left(-\frac{t}{\tau}\right). \tag{8.100}$$

The sphere has the radius R and the initial velocity v_0. The half-space has the glass module G_0 and the (constant in time) Poisson number ν. In this case, Hunter determined the following equations of motion for the variable $Z(t) = a^2(t)/R$:

$$\ddot{Z} + \frac{1}{\tau}(\dot{Z} - v_0) + CZ^{3/2} = 0, \quad t \le t_m,$$
$$\ddot{Z} - \frac{1}{\tau}(3\dot{Z} + v_0) + CZ^{3/2} = 0, \quad t > t_m, \tag{8.101}$$

with

$$C = \frac{8G_0 R^{1/2}}{3M(1-\nu)}. \tag{8.102}$$

During the penetration phase, Z corresponds with the indentation depth (which is the position of the sphere) and the equation of motion is exact. The presented equation of motion during the retraction phase, on the other hand, is an approximation that is accurate only for large values of τ compared to the impact time. Hunter then determined the following approximate solutions for the coefficient of restitution e and the impact duration T:

$$e = 1 - \frac{4}{9}\frac{T_0}{\tau},$$
$$T = T_0\left(1 - 0.037\frac{T_0}{\tau}\right). \tag{8.103}$$

Here, T_0 denotes the duration of impact for the purely elastic case, according to Hertz (1882). The viscoelastic impact problem was also examined in detail by Argatov (2013) using an asymptotic approximation.

8.4 Normal Contact of Compressible Viscoelastic Media

Most elastomers can, to a good approximation, be considered incompressible. This was one of the general assumptions in the previous sections of this chapter, with only the shear modulus (complex modulus, relaxation function, creep function, etc.) being the subject of rheological study. Yet there is no theoretical justification requiring this assumption of incompressibility. Indeed, elastomers generally exhibit a deformation response to hydrostatic compression, and a relaxation function and creep function associated with this volume deformation.

In the following sections, we will outline some possibilities on how to take the compressibility of the viscoelastic medium into account for contact problems. At several stages we will refer back to various concepts introduced earlier in this chapter.

8.4.1 The Compressible Viscoelastic Material Law

Equation (8.47) in Sect. 8.1.5 introduced the general material law of a linear viscoelastic medium for the traceless components (belonging to the shear components) of the stress and deformation tensor:

$$P_S[s_{ij}(t)] = Q_S[e_{ij}(t)]. \tag{8.104}$$

Here, s_{ij} denotes the traceless component of the stress tensor σ_{ij} and e_{ij} denotes the traceless component of the deformation tensor ε_{ij}. P and Q are linear operators of the form

$$\begin{aligned} P_S &:= \sum_{k=0}^{\infty} p_{S,k} \frac{\mathrm{d}^k}{\mathrm{d}t^k}, \\ Q_S &:= \sum_{k=0}^{\infty} q_{S,k} \frac{\mathrm{d}^k}{\mathrm{d}t^k}, \end{aligned} \tag{8.105}$$

with an added index "S" to emphasize that we are dealing with the shear component of the deformation. For the volume components, i.e., the trace components, an analogous material law for the most general case can be formulated:

$$P_V[\sigma_{ii}(t)] = Q_V[\varepsilon_{ii}(t)] \tag{8.106}$$

with analogous operators:

$$\begin{aligned} P_V &:= \sum_{k=0}^{\infty} p_{V,k} \frac{\mathrm{d}^k}{\mathrm{d}t^k}, \\ Q_V &:= \sum_{k=0}^{\infty} q_{V,k} \frac{\mathrm{d}^k}{\mathrm{d}t^k}. \end{aligned} \tag{8.107}$$

For example, let us consider a compressible Kelvin medium for which the stress state has only a purely linear-elastic and a purely linear viscous component. In this case, the material law that we already split into its shear and volume components is:

$$\begin{aligned}\sigma_{ll} &= 3K\varepsilon_{ll} + 3\xi\dot{\varepsilon}_{ll},\\ s_{ij} &= 2Ge_{ij} + 2\eta\dot{e}_{ij},\end{aligned} \tag{8.108}$$

which means the previously introduced differential operators have the form:

$$\begin{aligned}P_S &= P_V = 1,\\ Q_V &= 3K + 3\xi\frac{\partial}{\partial t},\\ Q_S &= 2G + 2\eta\frac{\partial}{\partial t}.\end{aligned} \tag{8.109}$$

Here, G and K are the shear and bulk modulus, while η and ξ are the dynamic shear and volume viscosity of the medium, respectively.

8.4.2 Is MDR Mapping of the Compressible Normal Contact Problem Possible?

The solution process of the normal contact problem begins, once again, with the fundamental solution for the displacement w of a half-space under the effect of a normal point force F_N, which acts on the origin starting at time $t = 0$ and is then held constant. In the elastic case, the fundamental solution to the normal displacement of the surface is the well-known solution by Boussinesq given in (2.2),

$$w^{el}(r,t) = \frac{F_N H(t)}{\pi E^* r} = \frac{F_N H(t)}{4\pi G r}\frac{3K+4G}{3K+G}, \tag{8.110}$$

with the effective elasticity modulus E^*. According to the principle of Radok (1957), solving the viscoelastic problem involves the Laplace transform of the elastic solution (owed to the analogous forms of the elastic and viscoelastic field equations) with the accompanying substitution of the elastic moduli through the operators P and Q. The following equation is obtained for the Laplace transform (all transformed quantities are indicated by a "hat") of the desired fundamental solution:

$$\hat{w}(r,s) = \frac{F_N}{2\pi r s}\frac{\hat{Q}_V + 2\hat{Q}_S}{\hat{Q}_S[\hat{Q}_V + \hat{Q}_S/2]}. \tag{8.111}$$

However, the transforms of the operators are polynomials in s,

$$\begin{aligned}\hat{Q}_S &= \sum_{k=0}^{\infty} q_{S,k}s^k,\\ \hat{Q}_V &= \sum_{k=0}^{\infty} q_{V,k}s^k.\end{aligned} \tag{8.112}$$

In other words, (8.111) can be written in the general form:

$$\hat{w}(r,s) = \frac{F_N}{4\pi r}\hat{\Phi}(s), \tag{8.113}$$

with

$$\hat{\Phi}(s) := \frac{2}{s}\frac{\sum_{k=0}^{\infty} q_{V,k}s^k + 2\sum_{k=0}^{\infty} q_{S,k}s^k}{\sum_{k=0}^{\infty} q_{S,k}s^k\left(\sum_{k=0}^{\infty} q_{V,k}s^k + \frac{1}{2}\sum_{k=0}^{\infty} q_{S,k}s^k\right)}, \tag{8.114}$$

which, after an inverse transform into the time domain, leads to the desired solution

$$w(r,t) = \frac{F_N}{4\pi r}\Phi(t). \tag{8.115}$$

Yet this is the fundamental solution for an incompressible medium with the (dimensional, shear) creep function $\Phi(t)$. This means that the normal contact problem for a compressible viscoelastic material can always be reduced to an equivalent problem of an incompressible medium, assuming that the material law of the compressible body in the form (8.104) and (8.106) is known and the inverse Laplace transform of the expression (8.114) exists.

In particular, the relationship (8.51) between the creep function and the time-dependent shear modulus $G(t)$,

$$s^2\hat{\Phi}(s)\hat{G}(s) = 1, \tag{8.116}$$

allows finding the time-dependent shear modulus by using the inverse Laplace transform of the expression

$$\hat{G}(s) = \frac{\sum_{k=0}^{\infty} q_{S,k}s^k\left(\sum_{k=0}^{\infty} q_{V,k}s^k + \frac{1}{2}\sum_{k=0}^{\infty} q_{S,k}s^k\right)}{2s\left(\sum_{k=0}^{\infty} q_{V,k}s^k + 2\sum_{k=0}^{\infty} q_{S,k}s^k\right)}. \tag{8.117}$$

Thus all previously derived MDR relations for the normal contact from previous sections can also be used in the case of compressible viscoelastic media.

8.4.3 Normal Contact of a Compressible Kelvin Element

To illustrate the relationships derived in the previous section, we will provide a brief demonstration of the transformation of the compressible normal contact to the corresponding incompressible problem for an incompressible Kelvin element with the general material law (8.108).

Inserting the operators (8.109) into (8.114) leads to the following expression for the Laplace transform of the shear creep function of the incompressible medium in question:

$$\hat{\Phi}(s) = \frac{1}{s}\frac{3K + 4G + (3\xi + 4\eta)s}{(G + \eta s)\left[3K + G + (3\xi + \eta)s\right]}. \tag{8.118}$$

The inverse transform into the time domain yields the creep function

$$\Phi(t) = \frac{1}{G}\left[1 - \exp\left(-\frac{G}{\eta}t\right)\right] + \frac{3}{3K+G}\left[1 - \exp\left(-\frac{3K+G}{3\xi+\eta}t\right)\right] \tag{8.119}$$

and the time-dependent modulus (to avoid any confusion with the shear modulus of the original compressible medium, we add the index "ink")

$$\begin{aligned} G_{\text{ink}}(t) = \eta\frac{3\xi+\eta}{3\xi+4\eta}\delta(t) + G\frac{3K+G}{3K+4G} \\ + \frac{27(G\xi - K\eta)^2}{(3K+4G)(3\xi+4\eta)^2}\exp\left(-\frac{3K+4G}{3\xi+4\eta}t\right). \end{aligned} \tag{8.120}$$

Here, $\delta(t)$ denotes the Dirac function. In the limiting case of an incompressible medium, we obtain the known solution

$$\lim_{K\to\infty} G_{\text{ink}}(t) = G + \eta\delta(t). \tag{8.121}$$

The limiting case of rapid relaxation

$$\frac{3\xi+4\eta}{3K+4G} \ll t \tag{8.122}$$

also results in an incompressible Kelvin body with

$$G_{\text{ink}}(t) \approx G\frac{3K+G}{3K+4G} + \left(\eta\frac{3\xi+\eta}{3\xi+4\eta} + \frac{27(G\xi-K\eta)^2}{(3K+4G)^2(3\xi+4\eta)}\right)\delta(t). \tag{8.123}$$

It should be noted that the creep function (8.119) and the time-dependent modulus (8.120) can be interpreted in different ways in terms of rheological models (and consequently also for the purposes of the MDR). For example, the modulus (8.120) can be produced by a parallel arrangement of a Kelvin element and a Maxwell element (see Fig. 8.18) with the material parameters of the rheological models set to (ν is the Poisson's ratio of the medium):

$$\begin{aligned} G_1 = G\frac{3K+G}{3K+4G} = \frac{G}{2(1-\nu)}, \quad G_2 = \frac{27(G\xi-K\eta)^2}{(3K+4G)\,(3\xi+4\eta)^2}, \\ \eta_1 = \eta\frac{3\xi+\eta}{3\xi+4\eta}, \quad \eta_2 = \frac{27(G\xi-K\eta)^2}{(3K+4G)^2(3\xi+4\eta)}, \\ \eta_1+\eta_2 = \frac{4\eta(1-\nu+\nu^2)+3\xi(1-2\nu)^2}{12(1-\nu)^2}. \end{aligned} \tag{8.124}$$

However, other rheological models are also possible. For instance, the form of the creep function (8.119) suggests the possibility of a serial arrangement of two

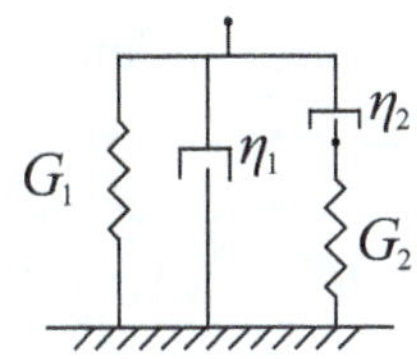

Fig. 8.18 Rheological element to model a compressible Kelvin body within the framework of the MDR

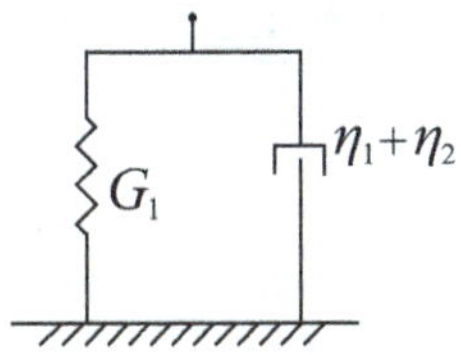

Fig. 8.19 Element for a compressible Kelvin body in the limiting case of rapid relaxation

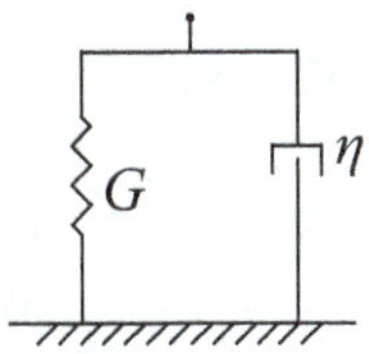

Fig. 8.20 Rheological element for an incompressible Kelvin body

suitably chosen Kelvin elements. The rheological models for the limiting case of fast relaxation and for the incompressible body are shown in Figs. 8.19 and 8.20.

Once the compressible normal contact problem has been reduced to the equivalent incompressible problem and the corresponding rheological model has been identified, all previously derived MDR rules pertaining to the solutions of incompressible problems can be applied. As an example, Brilliantov et al. (2015) derived the dissipating force during the impact of two spheres in the limiting case given by (8.122). using developments from perturbation theory. Based on the aforementioned derivations and the relationships of the incompressible MDR model, it is immediately obvious that, for an arbitrary axially symmetric indenter, this force is given by:

$$F_{\text{dis}} = 8(\eta_1 + \eta_2)a\dot{d} = \frac{2a\dot{d}}{(1-\nu)^2}\left[\frac{4}{3}\eta(1-\nu+\nu^2) + \xi(1-2\nu)^2\right], \qquad (8.125)$$

which (in the case of impacting spheres in the Hertzian approximation) naturally corresponds to the result of Brilliantov et al. In (8.125) a, as always, refers to the contact radius and d refers to the indentation depth.

8.5 Fretting Wear of Elastomers

Contacts of rigid bodies with polymers or elastomers occur in many applications. If they are moving relative to one another, then the surfaces in such contacts are subjected to wear. Counterintuitively, the endangered partner is not necessarily the softer polymer piece but may well be the metal surface (Higham et al. 1978). Whether there is wear to the more rigid or to the softer contact partner depends on many load and material parameters. In the following, we explore the idea that the "rigid" indenter wears while the elastic deformation is solely due to the elasticity of the polymer.

Let us consider an axially symmetric rigid profile which is brought into contact with a polymer (average indentation depth d_0) and is subjected to vibrations of the amplitude $\Delta u^{(0)}$ in the tangential direction and vibrations of the amplitude $\Delta w^{(0)}$ in the normal direction:

$$\begin{aligned} d(t) &= d_0 + \Delta w^{(0)} \cos(\omega t), \\ u^{(0)}(t) &= \Delta u^{(0)} \cos(\omega t + \varphi). \end{aligned} \tag{8.126}$$

Under the assumption of complete stick in the entire contact area, the tangential stresses at the edge of the contact would exhibit a singularity as in the elastic case. This means that the no-slip condition cannot be met in the vicinity of the edge of the contact. Thus, even small amplitudes lead to the formation of a ring-shaped slip zone at the edge of the contact which results in wear. The inner zone remains sticking. The existence of this stick zone was demonstrated by Barber et al. (2011) for arbitrary two-dimensional topographies (not necessarily with a singular, connected contact area) under very general assumptions and for three-dimensional topographies under the assumptions of the theory of Cattaneo and Mindlin.

In this book, we will restrict our consideration to axially symmetric contacts. Furthermore, we assume that the friction in the contact obeys the local form of the Coulomb law of friction: the surfaces remain sticking if the tangential stress τ is lower than the normal pressure p multiplied by the coefficient of friction μ, and that the tangential stress remains constant once slip sets in:

$$\begin{aligned} |\tau| &< \mu p, \quad \text{stick}, \\ |\tau| &= \mu p, \quad \text{slip}. \end{aligned} \tag{8.127}$$

We refer to the entirety of all points that fulfill the condition $|\tau| < \mu p$ at every point in time as the "permanent stick zone". Due to wear, the stress in the slip zone will decrease over time and be re-distributed to the stick zone. We can assume that the original stick zone remains sticking over the course of the wear process, while in the area of the original slip zone the slip condition is fulfilled even in the worn state. The progressing wear leads to a continual pressure decrease in the slip zone. This process finally ends when the pressure vanishes completely. In this limit state, the wear rate in the slip zone approaches zero: the system approaches a state in which no further wear occurs. The precise kinetics of this process depend on the form of the wear law. The shape of the profile in its final state, however, does not depend on the details of the wear law and it can be determined in a general form.

We will not presume any particular local wear law. Instead, we will just assume that the following very general conditions are met: (a) wear occurs only where there is a finite relative displacement of the surfaces (the existence of tangential stresses alone is not sufficient for wear); (b) wear occurs only in zones of non-zero pressure. Combined with the assumed Coulomb law of friction, these assumptions alone uniquely determine the limit shape of the worn profile. Indeed, from the

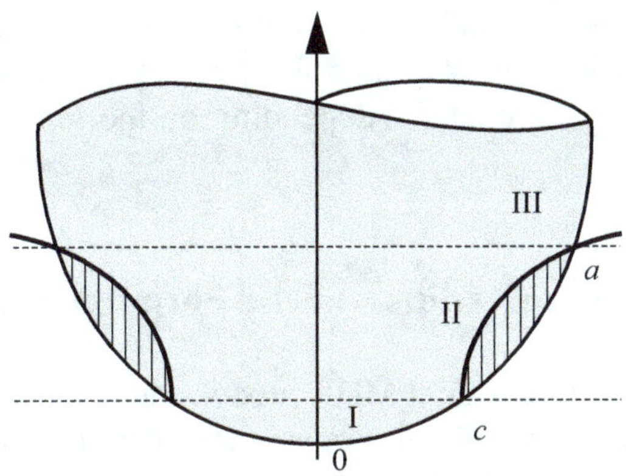

Fig. 8.21 Three zones of the worn profile in the limit state. I is the zone of permanent stick; here the original profile remains unchanged. II is the zone in which the pressure vanishes, even though at maximum indentation the surfaces are in a state of established contact with no pressure. III is the zone in which no contact is established (and consequently no change of the profile occurs)

assumption (a) it follows that the shape of the indenter in the permanent stick zone must match the original (unworn) shape (zone I in Fig. 8.21). Under assumption (b), the pressure outside the permanent stick zone vanishes in the limit state. Yet this unpressurized state should result *due to the wear*. This means that the surfaces in zone II at the maximum indentation are "just barely touching"; i.e., contact has been established with zero pressure. This in turn means that the worn indenter at the time of maximum indentation must coincide exactly with the shape of the *free* surface, which would have resulted from the indentation from the original indenter shape within the permanent stick radius (zone II in Fig. 8.21).

According to the half-space hypothesis, small oscillations parallel to the contact surface, independent of their particular oscillation mode (linear oscillations in a single direction, superposition of oscillations in two direction, or torsional oscillations), only result in a tangential displacement of the surfaces that are just barely in contact and therefore cause no wear. Since the shape of a surface that is indented by a non-varying contact area does not depend on the rheology of the medium and matches the one for an elastic body, the shape of the worn indenter in its final state is given by (6.18) which can be found in Chap. 6:

$$f_\infty(r) = \begin{cases} f_0(r), & r \le c,\ r > a, \\ \dfrac{2}{\pi}\left[\displaystyle\int_0^c \frac{g_0(x)\mathrm{d}x}{\sqrt{r^2-x^2}} + d_{\max}\arccos\left(\frac{c}{r}\right)\right], & c < r \le a. \end{cases} \tag{8.128}$$

with

$$d_{\max} = d_0 + \Delta w^{(0)} \tag{8.129}$$

in which $g_0(x)$ is the MDR transformed profile:

$$g_0(x) = |x| \int_0^{|x|} \frac{f_0'(r)\mathrm{d}r}{\sqrt{x^2-r^2}}. \tag{8.130}$$

The solution is uniquely defined by the maximum indentation depth and the radius c of the permanent stick zone. We, therefore, concern ourselves with the calculation of this determining parameter c depending on the loading procedure and the material parameters.

8.5.1 Determining the Radius c of the Permanent Stick Zone

In elements of the one-dimensional MDR model, the normal and tangential forces (in the case of an incompressible elastomer) are given by (8.35) and (8.37). The displacements of the "springs" at location x are calculated to:

$$\begin{aligned} w_{1D}(x,t) &= d_0 - g(x) + \Delta w^{(0)} \cos \omega t, \\ u_{1D}(x,t) &= \Delta u^{(0)} \cos(\omega t + \varphi_0). \end{aligned} \tag{8.131}$$

The normal and tangential forces are given by equations

$$\begin{aligned} \Delta F_N = {} & 4\Delta x G'(0)(d_0 - g(x)) \\ & + 4\Delta x \Delta w^{(0)} \left(G'(\omega)\cos\omega t - G''(\omega)\sin\omega t\right) \end{aligned} \tag{8.132}$$

and

$$\Delta F_x = \frac{8}{3}\Delta x \Delta u^{(0)} \left(G'(\omega)\cos(\omega t + \varphi_0) - G''(\omega)\sin(\omega t + \varphi_0)\right), \tag{8.133}$$

where $G(\omega) = G'(\omega) + iG''(\omega)$ is the complex shear modulus; i is the imaginary unit, $G'(\omega)$ is the real component of the complex shear modulus (storage modulus), and $G''(\omega)$ is the imaginary component of the complex modulus (loss modulus). The no-slip condition reads $|\Delta F_x| \leq \mu \Delta F_N$ or

$$\begin{aligned} & \frac{2}{3}\Delta u^{(0)} \left|G'(\omega)\cos(\omega t + \varphi_0) - G''(\omega)\sin(\omega t + \varphi_0)\right| \\ & \leq \mu\left[G'(0)(d_0 - g(c)) + \Delta w^{(0)} \left(G'(\omega)\cos\omega t - G''(\omega)\sin\omega t\right)\right]. \end{aligned} \tag{8.134}$$

It follows that:

$$\begin{aligned} g(c) \leq d_0 - \frac{1}{G'(0)} & \left[\frac{1}{\mu}\frac{2}{3}\Delta u^{(0)} \left|G'(\omega)\cos(\omega t + \varphi_0) - G''(\omega)\sin(\omega t + \varphi_0)\right|\right. \\ & \left. - \Delta w^{(0)} \left(G'(\omega)\cos\omega t - G''(\omega)\sin\omega t\right)\right]. \end{aligned} \tag{8.135}$$

The radius c of the permanent stick zone is the largest radius, for which there is no slip at any point in time or, alternatively, the smallest of the values where slip is just avoided:

$$g(c) = \min_t \left\{ d_0 - \frac{1}{G'(0)} \left[\frac{1}{\mu}\frac{2}{3}\Delta u^{(0)} \left| G'(\omega)\cos(\omega t + \varphi_0) - G''(\omega)\sin(\omega t + \varphi_0) \right| \right.\right.$$
$$\left.\left. - \Delta w^{(0)} \left(G'(\omega)\cos\omega t - G''(\omega)\sin\omega t \right) \right] \right\}. \quad (8.136)$$

This equation can also be rewritten in the more compact form, as follows:

$$g(c) = \min_t \left\{ d_0 - \frac{|G(\omega)|}{G'(0)} \left[\frac{1}{\mu}\frac{2}{3}\Delta u^{(0)} \left|\cos(\Omega + \varphi_0)\right| - \Delta w^{(0)} \cos\Omega \right] \right\}, \quad (8.137)$$

with $\Omega = \omega t + \varphi_1$, where φ_1 is defined as the phase angle of the complex shear modulus with:

$$\tan\varphi_1 = \frac{G''(\omega)}{G'(\omega)}. \quad (8.138)$$

Let us first examine the case in which the phase shift φ_0 between the tangential and normal oscillation is not fixed (e.g., slow "phase creep"). This case would simply require inserting the maximum absolute values of $|\cos(\Omega + \varphi_1)|$ and $\cos\Omega$ into (8.137):

$$g(c) = d_0 - \frac{|G(\omega)|}{G'(0)} \left[\frac{1}{\mu}\frac{2}{3}|\Delta u^{(0)}| + \Delta w^{(0)} \right], \text{ (non-fixed phase shift).} \quad (8.139)$$

This equation is equally valid for the case of two different incommensurable oscillation frequencies in the normal and the tangential direction.

If the frequencies are equal and the phase shift is fixed by some phase synchronization mechanism, the situation is more complicated. An analysis of (8.137) shows that the smallest stick radius is achieved at $\varphi_0 = 0$. In this case, the radius can also be determined from (8.139). The maximum stick radius is achieved at $\varphi_0 = \pi/2$ and is given by the following equation:

$$g(c) = d_0 - \frac{|G(\omega)|}{G'(0)} \sqrt{\left(\frac{1}{\mu}\frac{2}{3} \left|\Delta u^{(0)}\right| \right)^2 + (\Delta w^{(0)})^2}. \quad (8.140)$$

The general case of an arbitrary phase shift comprises several special cases, which are presented in research by Mao et al. (2016).

8.5.2 Fretting Wear of a Parabolic Profile on a Kelvin Body

As a concrete example for the application of (8.140), we consider a parabolic profile $f_0(r) = r^2/(2R)$ in contact with a Kelvin body, which can be represented as a parallel arrangement of an elastic body (shear modulus G) and a viscous body η. Accordingly, the elements of the one-dimensional foundation consist of the following elements:

- In the normal direction, they consist of a spring with the normal stiffness Δk_z arranged in parallel with a damper with the damping coefficient $\Delta\alpha_z$.
- In the horizontal direction, the elements have a tangential stiffness Δk_x and the damping coefficient $\Delta\alpha_x$.

Said values of stiffness and damping coefficients are determined according to (8.38) and (8.39) by the following rules:

$$\Delta k_z = 4G\Delta x, \quad \Delta\alpha_z = 4\eta\Delta x, \quad \Delta k_x = \frac{8}{3}G\Delta x, \quad \Delta\alpha_x = \frac{8}{3}\eta\Delta x. \tag{8.141}$$

The MDR profile is, in accordance with (8.130), equal to $g(x) = x^2/R$, and the complex shear modulus of the medium is equal to $\hat{G}(\omega) = G + i\omega\eta$. For the rheological quantities $|G(\omega)|$ and $G'(0)$ occurring in (8.140) we obtain:

$$|G(\omega)| = \sqrt{G^2 + \eta^2\omega^2} \quad \text{and} \quad G'(0) = G. \tag{8.142}$$

In the case of $\varphi_0 = \pi/2$, (8.140) takes the form of

$$\frac{c^2}{R} = d_0 - \sqrt{1 + (\omega\tau)^2}\sqrt{\left(\frac{1}{\mu}\frac{2}{3}|\Delta u^{(0)}|\right)^2 + (\Delta w^{(0)})^2}, \tag{8.143}$$

with the introduction of the relaxation time $\tau = \eta/G$.

References

Argatov, I.I.: An analytical solution of the rebound indentation problem for an isotropic linear viscoelastic layer loaded with a spherical punch. Acta Mech. **223**(7), 1441–1453 (2012)

Argatov, I.I.: Mathematical modeling of linear viscoelastic impact: application to drop impact testing of articular cartilage. Tribol. Int. **63**, 213–225 (2013)

Argatov, I.I., Mishuris, G.: An analytical solution for a linear viscoelastic layer loaded with a cylindrical punch: evaluation of the rebound indentation test with application for assessing viability of articular cartilage. Mech. Res. Commun. **38**(8), 565–568 (2011)

Argatov, I.I., Popov, V.L.: Rebound indentation problem for a viscoelastic half-space and axi-symmetric indenter—solution by the method of dimensionality reduction. Z. Angew. Math. Mech. **96**(8), 956–967 (2016)

Barber, J.R., Davies, M., Hills, D.A.: Frictional elastic contact with periodic loading. Int. J. Solids Struct. **48**(13), 2041–2047 (2011)

Brilliantov, N.V., Pimenova, A.V., Goldobin, D.S.: A dissipative force between colliding vis-coelastic bodies: rigorous approach. Europhys. Lett. **109**, 14005 (2015)

Graham, G.A.C.: The contact problem in the linear theory of viscoelasticity. Int. J. Eng. Sci. **3**(1), 27–46 (1965)

Graham, G.A.C.: The contact problem in the linear theory of viscoelasticity when the time-dependent contact area has any number of maxima and minima. Int. J. Eng. Sci. **5**(6), 495–514 (1967)

Greenwood, J.A.: Contact between an axi-symmetric indenter and a viscoelastic half-space. Int. J. Mech. Sci. **52**(6), 829–835 (2010)

Hertz, H.: Über die Berührung fester elastischer Körper. J. Reine Angew. Math. **92**, 156–171 (1882)

Higham, P.A., Stott, F.H., Bethune, B.: Mechanisms of wear of the metal surface during fretting corrosion of steel on polymers. Corros. Sci. **18**(1), 3–13 (1978)

Hunter, S.C.: The Hertz problem for a rigid spherical indenter and a viscoelastic half-space. J. Mech. Phys. Solids **8**(4), 219–234 (1960)

Lee, E.H.: Stress analysis in viscoelastic bodies. Q. Appl. Math. **13**(2), 183–190 (1955)

Lee, E.H., Radok, J.R.M.: The contact problem for viscoelastic bodies. J. Appl. Mech. **27**(3), 438–444 (1960)

Mao, X., Liu, W., Ni, Y., Popov, V.L.: Limiting shape of profile due to dual-mode fretting wear in contact with an elastomer. Pro Inst. Mech. Eng. C J. Mech. Eng. Sci. **230**(9), 1417–1423 (2016)

Popov, V.L., Heß, M.: Method of dimensionality reduction in contact mechanics and friction. Springer, Heidelberg (2015). ISBN 978-3-642-53875-9

Radok, J.R.M.: Viscoelastic stress analysis. Q. Appl. Math. **15**(2), 198–202 (1957)

Ting, T.C.T.: The contact stresses between a rigid indenter and a viscoelastic half-space. J. Appl. Mech. **33**(4), 845–854 (1966)

Ting, T.C.T.: Contact problems in the linear theory of viscoelasticity. J. Appl. Mech. **35**(2), 248–254 (1968)

Vandamme, M., Ulm, F.-J.: Viscoelastic solutions for conical indentation. Int. J. Solids Struct. **43**(10), 3142–3165 (2006)

Contact Problems of Functionally Graded Materials

9

Modern technological developments require innovative materials to meet ever increasing performance demands. An example of this is functionally graded materials (FGMs) whose material composition or micro-structure varies continually within the volume in a pre-defined way. In this manner, the material properties can be set (possibly independently of each other) to the optimal values. A controlled gradient of the elasticity modulus has been proven to lead to a greater resistance to contact and friction damage (Suresh 2001). For instance, Hertzian cone cracks are suppressed due to the reduction of the maximum tensile stresses in the surface (Jitcharoen et al. 1998) and the wear resistance is increased (Suresh et al. 1999). In mechanical engineering, components such as cutting tools, gears, and parts of roller bearings or turbine blades are made from FGMs. This is just a small sample of the constantly expanding range of applications of FGMs (Miyamoto et al. 1999). In the biomedical arena, for endoprosthetics in particular, the use of FGMs in artificial knee and hip joints is intended to improve biocompatibility and reduce wear, extending the service life of the endoprosthesis and, thereby, increasing quality of life (Sola et al. 2016). A high degree of biocompatibility is also essential for dental implants (Mehrali et al. 2013).

Many insects and animals such as geckos possess highly effective attachment devices, allowing them to stick to and move on surfaces of widely varying ranges of roughness and topographies. This has inspired a great amount of research into the adhesion of biological structures with the aim of manufacturing artificial surfaces with similar adhesive properties (Boesel et al. 2010; Gorb et al. 2007). In fact, many biological structures exhibit functional material gradients which serve to optimize the adhesive properties (Peisker et al. 2013; Liu et al. 2017). Furthermore, the adhesive properties of FGMs are of importance to the fields of innovative nano-electromechanical and micro-electromechanical systems (NEMS, MEMS).

Although the term "functionally graded material" was only coined in 1986 (Miyamoto et al. 1999), analytical and experimental research of their contact mechanical behavior had already been conducted much earlier. The origin lies in the field of geomechanics where the influence of the elasticity modulus, which increases with the depth of the soil foundation, on the stresses and displacements was

V.L. Popov, M. Heß, E. Willert, *Handbook of Contact Mechanics*,
https://doi.org/10.1007/978-3-662-58709-6_9

of particular interest (Fröhlich 1934; Holl 1940). Over the years, different functions for the varying elasticity modulus were in use. On this issue, the works of Selvadurai (2007) and Aleynikov (2011) offer a good overview and we can safely skip over providing a complete list of references. In nearly all cases, the calculations are exceedingly complex and only allow numerical solutions. The majority of the publications deal with an exponential increase or a power-law dependent increase of the elasticity modulus. For an exponential change of the elasticity modulus, only approximate solutions exist (so far) (Giannakopoulos and Suresh 1997). However, contact problems of elastically inhomogeneous materials that obey the law

$$E(z) = E_0 \left(\frac{z}{c_0}\right)^k \quad \text{with } -1 < k < 1 \tag{9.1}$$

can be solved exactly in an analytical manner. The Poisson's ratio is assumed to be constant. Due to the fact that the elasticity modulus of the foundation always increases with the depth, law (9.1) was initially restricted to positive exponents. The first solutions go back to Holl (1940) and Rostovtsev (1961), who imposed the additional restriction $\nu = 1/(2+k)$. Complete solutions of frictionless normal contacts were provided by Booker et al. (1985) and Giannakopoulos and Suresh (1997). A special case of (9.1) is the famous Gibson medium: the linear-inhomogeneous, incompressible half-space ($k \to 1, \nu \to 1/2$). Gibson (1967) managed to prove that such a medium behaves like a Winkler foundation (see also the work of Awojobi and Gibson 1973). If the solutions of the contact without adhesion are known, the work required to transfer that understanding to the adhesive normal contact between graded materials is not too great. Still, for an elastic inhomogeneity according to (9.1), the problem was only solved in the previous decade and remains the subject of current research (Chen et al. 2009; Jin et al. 2013, 2016; Willert 2018). However, all of these works assume a positive value of the exponent, which restricts their theoretical application to the class of graded materials where the elasticity modulus grows with increasing depth. Yet, as already mentioned in this chapter, plenty of practical applications exist (e.g., cutting tools, dental implants) which require a hard surface along with a softer core. A careful examination of the literature revealed that the theory is equally valid for negative exponents $-1 < k < 0$ (Rostovtsev 1964; Fabrikant and Sankar 1984). Although law (9.1) now permits the representation of positive and negative material gradients, and barring the special case of $k = 0$, the law remains physically unrealistic for homogeneous materials due to the vanishing or infinitely large elasticity modulus at the surface and at infinite depth, respectively. Nevertheless, FEM calculations by Lee et al. (2009) confirmed that it yields qualitatively correct results for functional gradients, if the modulus is described piecewise by a power-law.

Analytical solutions of tangential contacts between materials exhibiting an elastic inhomogeneity according to (9.1) were only recently developed by Heß (2016b) and Heß and Popov (2016). Up until then, merely the plane tangential contact problem between a rigid, infinitely long cylinder and the elastically inhomogeneous half-space was considered completely solved (Giannakopoulos and Pallot 2000). In very recent times, the MDR-based theory of Heß and Popov enabled the fast calculation of even complicated impact problems of FGMs (Willert and Popov 2017a,b).

Following the order in which the applications of FGMs are presented in this chapter, we will now discuss frictionless normal contacts (Sect. 9.1), adhesive normal contacts (Sect. 9.2), and tangential contacts with partial slip (Sect. 9.3). We restrict the consideration to the elastic inhomogeneity given by (9.1). Furthermore, the parameter studies of the majority of the graphical solutions were conducted for positive exponents $0 \leq k < 1$ only. Yet it must be expressly noted that all provided solutions are also valid for negative exponents $-1 < k < 0$. Additionally, our examination is usually limited to the contact of a rigid indenter and a functionally graded material. However, the theory is equally applicable to the contact between two elastically inhomogeneous bodies of the same exponent k and of the same characteristic depths c_0. In this case, the Poisson's ratios ν_i and the elastic parameters E_{0i} are allowed to differ. It should also be noted that the special case $k = 0$ yields many of the solutions for the contact problems of elastically homogeneous materials previously examined in Chaps. 2, 3, and 4.

9.1 Frictionless Normal Contact Without Adhesion

The frictionless normal contact between a rigid indenter of the shape $f(r)$ and a functionally graded material is shown in Fig. 9.1. Here, we differentiated between an elasticity modulus which drops with increasing depth (left) and one which rises with increasing depth (right). Law (9.1) has been demonstrated using a graphical representation.

9.1.1 Basis for Calculation of the MDR

For the solution of frictionless normal contacts without adhesion under consideration of the elastic inhomogeneity given by (9.1), we make use of the mapping rules and calculation formulas of the MDR developed by Heß (2016a). Accordingly, the

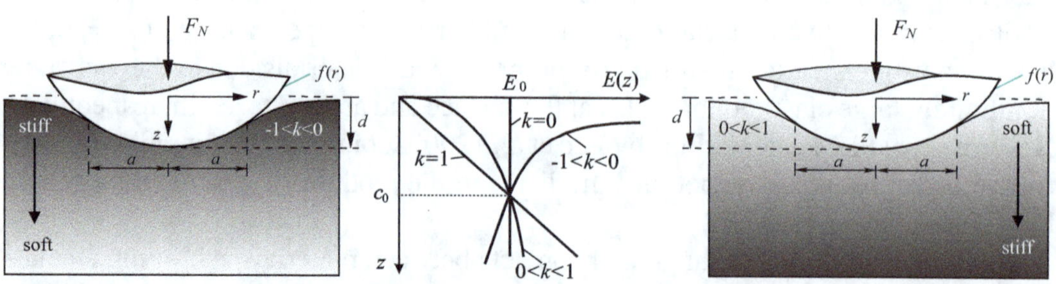

Fig. 9.1 Indentation of a rigid indenter with the profile $f(r)$ in an elastically inhomogeneous half-space, whose elasticity modulus drops (left) or rises (right) with increasing depth according to the power-law (9.1), depending on whether the exponent k is negative or positive

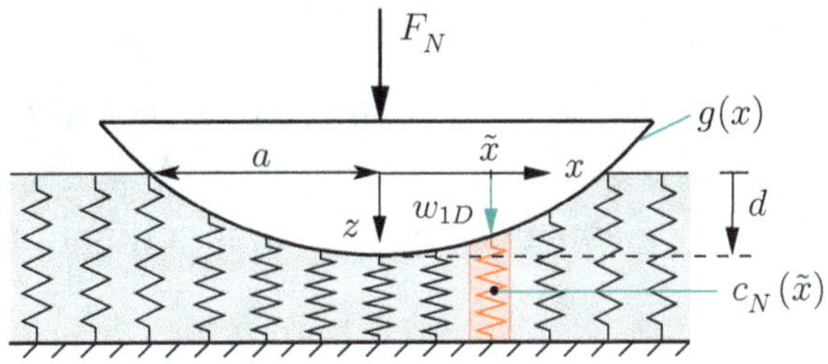

Fig. 9.2 The equivalent substitute model for the normal contact between two elastically inhomogeneous half-spaces whose elasticity moduli satisfy condition (9.1) while assuming that both media have equal exponents k

contact problem shown in Fig. 9.1 is equivalent to the indentation of a rigid, planar profile of the shape $g(x)$ in a one-dimensional Winkler foundation with respect to the relationships between normal force, contact radius, and indentation depth. This is valid for Winkler foundations whose stiffness depends on the coordinate x, which is expressed by a laterally varying foundation modulus $c_N(x)$ (spring stiffness divided by spring distance Δx). The equivalent substitute model is demonstrated in Fig. 9.2.

The planar profile $g(x)$, which is sometimes called the equivalent, one-dimensional substitute profile, is calculated according to:

$$g(x) = |x|^{1-k} \int_0^{|x|} \frac{f'(r)}{(x^2 - r^2)^{\frac{1-k}{2}}} \mathrm{d}r. \tag{9.2}$$

The spring stiffness is given by equation:

$$\Delta k_z(x) = c_N(x) \cdot \Delta x. \tag{9.3}$$

Here, $c_N(x)$ refers to the foundation modulus (stiffness per unit of distance) which, in this case, depends on the distance to the contact mid-point

$$c_N(x) = \left(\frac{1-\nu_1^2}{h_N(k,\nu_1)E_{01}} + \frac{1-\nu_2^2}{h_N(k,\nu_2)E_{02}} \right)^{-1} \left(\frac{|x|}{c_0} \right)^k. \tag{9.4}$$

The coefficient h_N is dependent on the Poisson's ratio ν and the exponent k of the inhomogeneity in the following way:

$$h_N(k,\nu) = \frac{2(1+k)\cos\left(\frac{k\pi}{2}\right)\Gamma\left(1+\frac{k}{2}\right)}{\sqrt{\pi}C(k,\nu)\beta(k,\nu)\sin\left(\frac{\beta(k,\nu)\pi}{2}\right)\Gamma\left(\frac{1+k}{2}\right)}, \tag{9.5}$$

with

$$C(k,\nu) = \frac{2^{1+k}\Gamma\left(\frac{3+k+\beta(k,\nu)}{2}\right)\Gamma\left(\frac{3+k-\beta(k,\nu)}{2}\right)}{\pi\Gamma(2+k)} \tag{9.6}$$

and

$$\beta(k, \nu) = \sqrt{(1+k)\left(1 - \frac{k\nu}{1-\nu}\right)}. \tag{9.7}$$

Thereby making use of the Gamma function:

$$\Gamma(z) := \int_0^\infty t^{z-1} \exp(-t) \mathrm{d}t. \tag{9.8}$$

Due to the mutual independence of the springs, the vertical displacement of the springs at location x is given by the obvious equation

$$w_{1D}(x) = d - g(x). \tag{9.9}$$

The indentation depth d is determined from the condition of a vanishing displacement at the edge of the contact:

$$w_{1D}(a) = 0 \quad \Rightarrow \quad d = g(a). \tag{9.10}$$

The normal force F_N is the sum of all spring forces in the contact:

$$F_N(a) = \int_{-a}^{a} c_N(x) w_{1D}(x) \mathrm{d}x. \tag{9.11}$$

Additionally, the normal displacement of the Winkler foundation $w_{1D}(x) = d - g(x)$ uniquely defines the pressure distribution and the normal surface displacement:

$$p(r; a) = -\frac{c_N(c_0)}{\pi c_0^k} \int_r^\infty \frac{w'_{1D}(x)}{(x^2 - r^2)^{\frac{1-k}{2}}} \mathrm{d}x,$$
$$w(r; a) = \begin{cases} \dfrac{2\cos\left(\frac{k\pi}{2}\right)}{\pi} \displaystyle\int_0^r \frac{x^k w_{1D}(x)}{(r^2 - x^2)^{\frac{1+k}{2}}} \mathrm{d}x & \text{for } r \leq a, \\ \dfrac{2\cos\left(\frac{k\pi}{2}\right)}{\pi} \displaystyle\int_0^a \frac{x^k w_{1D}(x)}{(r^2 - x^2)^{\frac{1+k}{2}}} \mathrm{d}x & \text{for } r > a. \end{cases} \tag{9.12}$$

These equations also solve sets of problems related to situations where the stresses at the half-space surface are known and the displacements are the desired quantities. This requires the preliminary calculation of the displacement of the Winkler foundation using the stresses according to

$$w_{1D}(x,a) = \frac{2c_0^k \cos\left(\frac{k\pi}{2}\right)}{c_N(c_0)} \int_x^a \frac{rp(r)}{(r^2 - x^2)^{\frac{1+k}{2}}} \mathrm{d}r, \tag{9.13}$$

which are then inserted into (9.12).

Here it should be noted that contact problems between FGMs with *arbitrary* elastic inhomogeneities can always be represented by the MDR-based model from Fig. 9.2. The difficulty lies in finding the correct calculation formula for the one-dimensional profile and determining the foundation modulus depending on the coordinate x (Argatov et al. 2018). Taking advantage of the expanded model from Fig. 9.20, these statements even extend to adhesive contact problems.

9.1.2 The Cylindrical Flat Punch

The complete solution of the contact problem between a rigid, cylindrical flat punch and an inhomogeneous half-space (shown in Fig. 9.3) is attributed to Booker et al. (1985). The simple geometry of the contact allows for an extremely simple derivation of the solution using the MDR. Since the profile function is always measured from the indenter tip $f(r) = 0$, it follows from (9.2) that $g(x) = 0$. The surface displacement of the 1D Winkler foundation is then

$$w_{1D}(x) = d\left[\mathrm{H}(x+a) - \mathrm{H}(x-a)\right], \tag{9.14}$$

where $\mathrm{H}(\cdot)$ represents the Heaviside function. For the contact of a flat punch, the indentation depth is independent of the (fixed) contact radius. Thus the evaluation

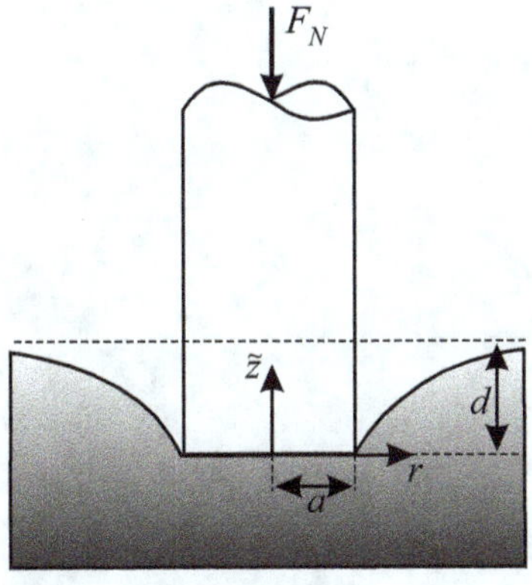

Fig. 9.3 Normal indentation of the elastically inhomogeneous half-space by a flat cylindrical punch

of (9.10) is not necessary and the solution of the contact problem merely requires applying (9.11) and (9.12). Under consideration of the derivative

$$w'_{1D}(x) = d\left[\delta(x+a) - \delta(x-a)\right], \tag{9.15}$$

with the Delta distribution $\delta(\cdot)$ and its filter property, it follows that:

$$\begin{aligned}
F_N(d) &= \frac{2h_N(k,\nu)E_0 d a^{1+k}}{(1-\nu^2)(1+k)c_0^k},\\
p(r;d) &= -\frac{h_N(k,\nu)E_0}{\pi(1-\nu^2)c_0^k}\int\limits_r^\infty \frac{d\left[\delta(x+a)-\delta(x-a)\right]}{(x^2-r^2)^{\frac{1-k}{2}}}\mathrm{d}x\\
&= \frac{h_N(k,\nu)E_0 d}{\pi(1-\nu^2)c_0^k(a^2-r^2)^{\frac{1-k}{2}}}\\
w(r;d) &= \frac{2\cos\left(\frac{k\pi}{2}\right)}{\pi}\int\limits_0^a \frac{x^k d}{(r^2-x^2)^{\frac{1+k}{2}}}\mathrm{d}x\\
&= \frac{\cos\left(\frac{k\pi}{2}\right)d}{\pi}\mathrm{B}\left(\frac{a^2}{r^2};\frac{1+k}{2},\frac{1-k}{2}\right),
\end{aligned} \tag{9.16}$$

where $\mathrm{B}(z;x,y)$ represents the incomplete Beta function according to:

$$\mathrm{B}(z;x,y) := \int\limits_0^z t^{x-1}(1-t)^{y-1}\mathrm{d}t \quad \forall x,y \in \mathbb{R}^+. \tag{9.17}$$

The pressure distribution normalized to the average pressure in the contact area is shown in Fig. 9.4. It is clear to see that a rising exponent of the elastic inhomo-

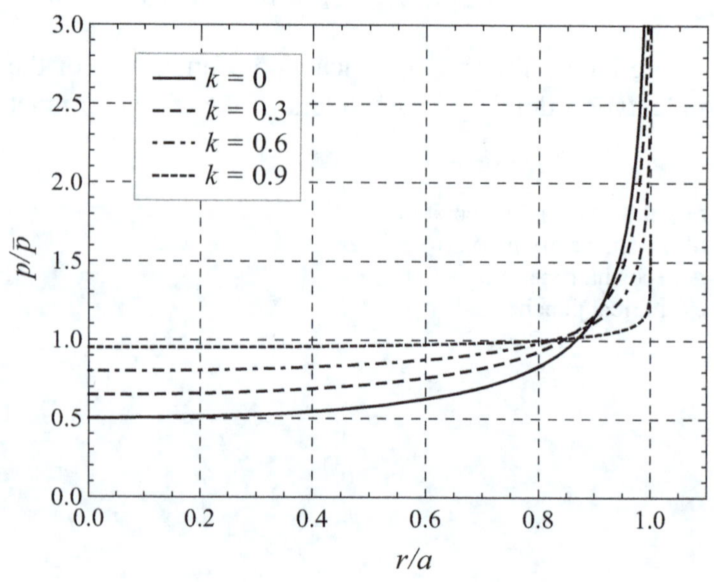

Fig. 9.4 Pressure distribution for the indentation by a flat cylindrical punch for different exponents of the elastic inhomogeneity k, normalized to the average pressure

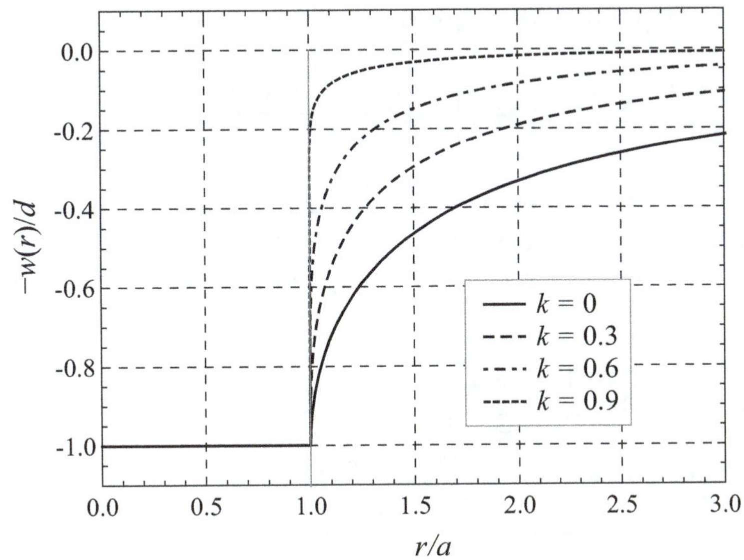

Fig. 9.5 Normal surface displacement of the inhomogeneous half-space for the contact of a flat punch and different exponents of the elastic inhomogeneity k, normalized to the indentation depth

geneity reduces the singularity at the contact edge. Figure 9.5 displays the normal surface displacements for different k at equal indentation depth d. It shows that the displacement of the half-space surface outside the contact area drops with increasing k.

9.1.3 The Cone

Most solutions and the insights derived therefrom for the conical contact as shown in Fig. 9.6 can be found in the work of Giannakopoulos and Suresh (1997). The shape of the rigid conical indenter is given by:

$$f(r) = r \tan\theta. \tag{9.18}$$

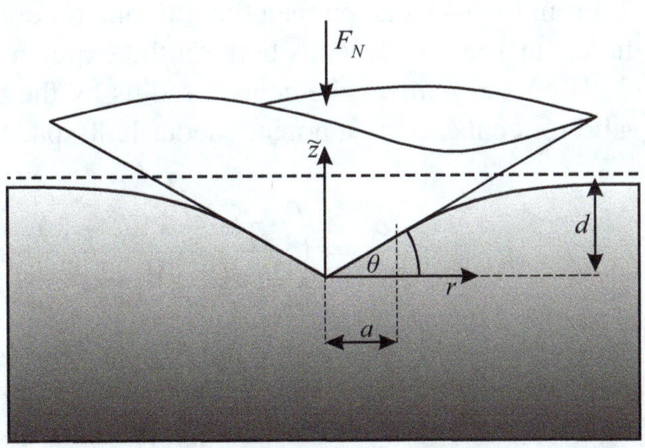

Fig. 9.6 Normal indentation of the elastically inhomogeneous half-space by a conical indenter

Applying the mapping rule (9.2) leads to the equivalent plane profile

$$g(x) = \frac{1}{2}|x|\mathrm{B}\left(\frac{1}{2}, \frac{1+k}{2}\right)\tan\theta, \tag{9.19}$$

where $\mathrm{B}(x, y)$ denotes the complete beta function, which is related to the incomplete beta function (9.17), according to $\mathrm{B}(x, y) := \mathrm{B}(1; x, y)$. Substituting (9.19) into (9.10)–(9.12), gives the solution of the contact problem, after a short calculation:

$$\begin{aligned}
d(a) &= \frac{1}{2}a\tan\theta\mathrm{B}\left(\frac{1}{2}, \frac{1+k}{2}\right), \\
F_N(a) &= \frac{h_N(k,\nu)\tan\theta\mathrm{B}\left(\frac{1}{2}, \frac{1+k}{2}\right)E_0 a^{k+2}}{(1-\nu^2)c_0^k(k+1)(k+2)}, \\
p(r;a) &= \frac{h_N(k,\nu)\tan\theta\mathrm{B}\left(\frac{1}{2}, \frac{1+k}{2}\right)E_0}{4\pi(1-\nu^2)c_0^k}r^k \\
&\quad\cdot\left[\mathrm{B}\left(\frac{-k}{2}, \frac{1+k}{2}\right) - \mathrm{B}\left(\frac{r^2}{a^2}; \frac{-k}{2}, \frac{1+k}{2}\right)\right], \\
w(r;a) &= \frac{\tan\theta\mathrm{B}\left(\frac{1}{2}, \frac{1+k}{2}\right)\cos\left(\frac{k\pi}{2}\right)a}{2\pi} \\
&\quad\cdot\left[\mathrm{B}\left(\frac{a^2}{r^2}; \frac{1+k}{2}, \frac{1-k}{2}\right) - \frac{r}{a}\mathrm{B}\left(\frac{a^2}{r^2}; \frac{3+k}{2}, \frac{1-k}{2}\right)\right] \quad \text{for } r > a.
\end{aligned} \tag{9.20}$$

Since the beta function in the pressure distribution partially contains negative arguments, we expand the definition of (9.17) by using (for negative arguments) the representation via the hypergeometric series

$$\mathrm{B}(z; x, y) = \frac{z^x}{x}{}_2\mathrm{F}_1(x, 1-y; 1+x; z). \tag{9.21}$$

The definition of the hypergeometric series is given in Chap. 11 by (11.93). To obtain the limiting case of the homogeneous half-space, $k \to 0$ should be set.

From (9.20) it can be seen that (among other things) with the same normal force the contact radius depends both on the exponent k and on the characteristic depth c_0. If one normalizes the contact radius by the contact radius a_h, which would be valid for contact with a homogeneous half-space, then it is

$$\frac{a}{a_h} = \left(\frac{\pi(1+k)(2+k)}{2h_N(k,\nu)\mathrm{B}\left(\frac{1}{2}, \frac{1+k}{2}\right)}\right)^{\frac{1}{2+k}}\left(\frac{c_0}{a_h}\right)^{\frac{k}{2+k}}. \tag{9.22}$$

Depending on the choice of k and c_0, a larger or smaller contact radius can result in comparison to the homogeneous half-space. This can be seen from the abscissa in Figs. 9.7 and 9.8 which illustrate the pressure distribution for various characteristic

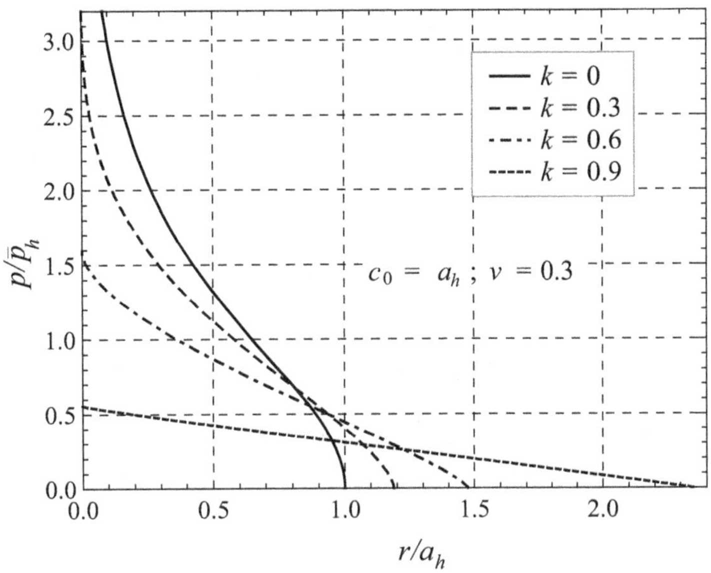

Fig. 9.7 Stress distribution in a contact with a cone for different exponents of elastic inhomogeneity k, normalized to the mean pressure $\overline{p}_h$, which results from contact with the homogeneous half-space. The characteristic depth is $c_0 = a_h$

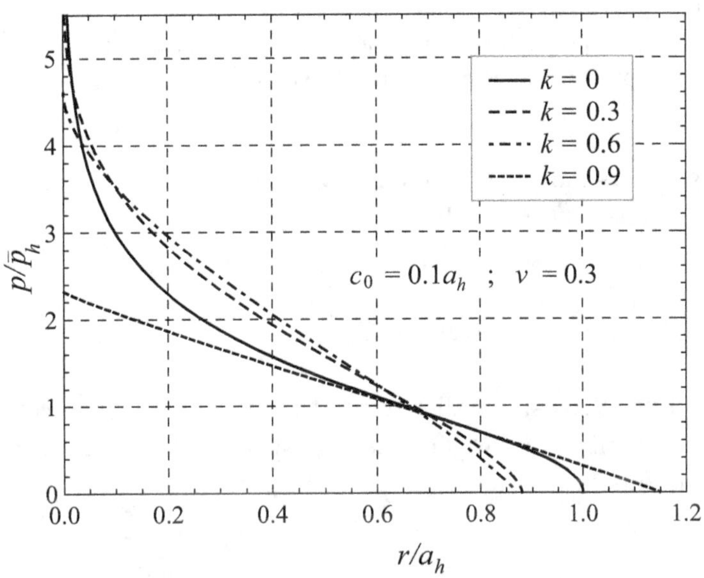

Fig. 9.8 Stress distribution in a contact with a cone for different exponents of elastic inhomogeneity k, normalized to the mean pressure $\overline{p}_h$, which results from contact with the homogeneous half-space. The characteristic depth is $c_0 = 0.1a_h$

depths. Thereby the pressure was normalized to the average pressure $\overline{p}_h$ in conical contact with a homogeneous half-space, which is known to be independent of the contact radius:

$$\overline{p}_h := \frac{F_N}{\pi a_h^2} = \frac{E_0 \tan\theta}{2(1-\nu^2)}. \tag{9.23}$$

The influence of the elastic inhomogeneity is reflected, above all, in the maximum pressure in the center of the contact area. Because the pressure singularity in the homogeneous case is suppressed the pressure maximum takes a *finite* value.

In Fig. 9.9, the surface normal displacements of the inhomogeneous half-space are graphically compared for different exponents k at the same contact radii. They were normalized to the indentation depth d_h, which results in contact with a homo-

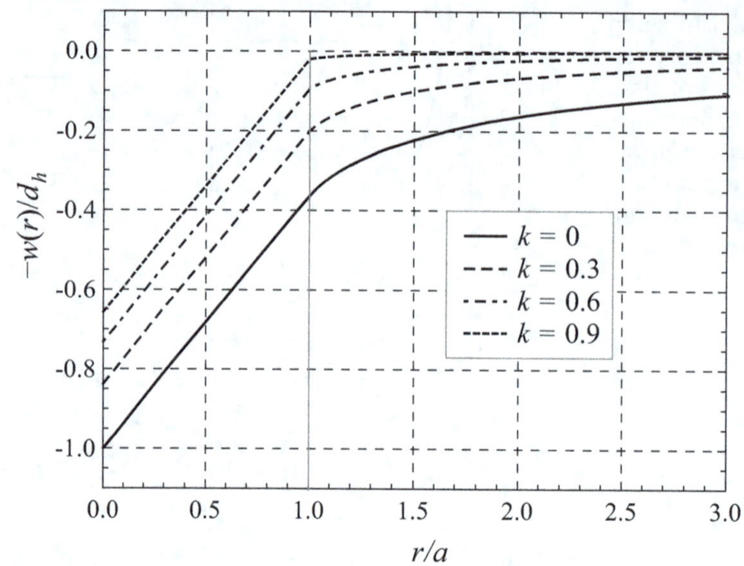

Fig. 9.9 Normalized normal displacements of the surface of the inhomogeneous half-space for a contact with a cone for different exponents of elastic inhomogeneity k; d_h is the indentation depth, corresponding to the homogeneous half-space

geneous half-space. By increasing k the indentation depth and the displacements decrease successively.

9.1.4 The Paraboloid

Figure 9.10 shows the normal contact between a rigid parabolic indenter and an elastically inhomogeneous half-space. The shape of the rigid indenter is given by the function

$$f(r) = \frac{r^2}{2R} \tag{9.24}$$

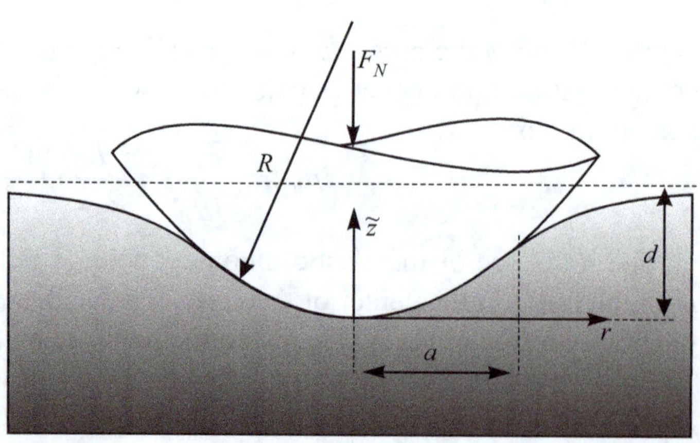

Fig. 9.10 Normal indentation of the elastically inhomogeneous half-space by a parabolic indenter

and can be understood as a (parabolic) approximation for a spherical contact with the radius of curvature R. The equivalent planar profile follows from (9.2):

$$g(x) = \frac{x^2}{(k+1)R}. \tag{9.25}$$

Considering (9.25), the evaluation of (9.10)–(9.12) provides the solution to the contact problem:

$$\begin{aligned}
d(a) &= \frac{a^2}{(k+1)R},\\
F_N(a) &= \frac{4h_N(k,\nu)E_0 a^{k+3}}{(1-\nu^2)c_0^k(k+1)^2(k+3)R},\\
p(r;a) &= \frac{2h_N(k,\nu)E_0 a^{k+1}}{\pi(1-\nu^2)c_0^k(k+1)^2 R}\left[1-\left(\frac{r}{a}\right)^2\right]^{\frac{1+k}{2}},\\
w(r;a) &= \frac{a^2\cos\left(\frac{k\pi}{2}\right)}{(k+1)\pi R}\left[\mathrm{B}\left(\frac{a^2}{r^2};\frac{1+k}{2},\frac{1-k}{2}\right)\right.\\
&\quad\left.-\frac{r^2}{a^2}\mathrm{B}\left(\frac{a^2}{r^2};\frac{3+k}{2},\frac{1-k}{2}\right)\right] \quad \text{for } r>a,
\end{aligned} \tag{9.26}$$

where $\mathrm{B}(z;x,y)$ is the incomplete beta function from (9.17).

Figure 9.11 shows the pressure distribution in the parabolic contact normalized to the maximum pressure in the Hertzian contact. At the same normal force, a Poisson's number of $\nu = 0.3$ and a characteristic depth c_0 (which was chosen equal to

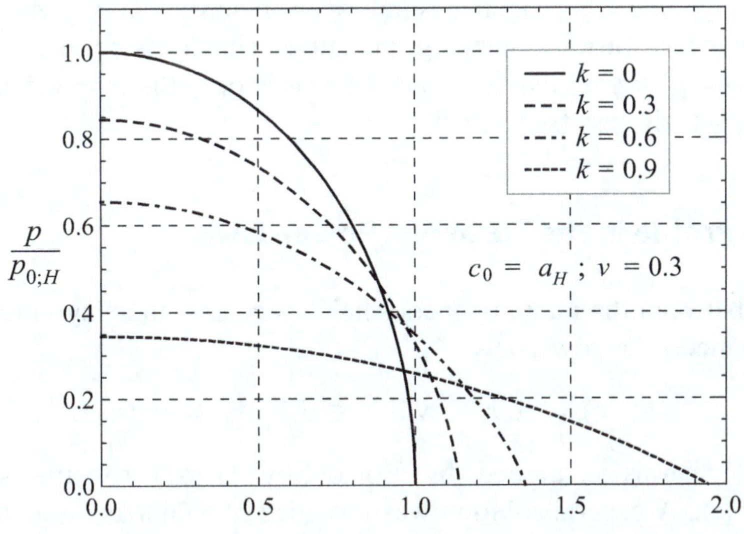

Fig. 9.11 Pressure distribution in parabolic contact for different exponents of elastic inhomogeneity k, normalized to the maximum pressure of Hertzian contact $p_{0,H}$

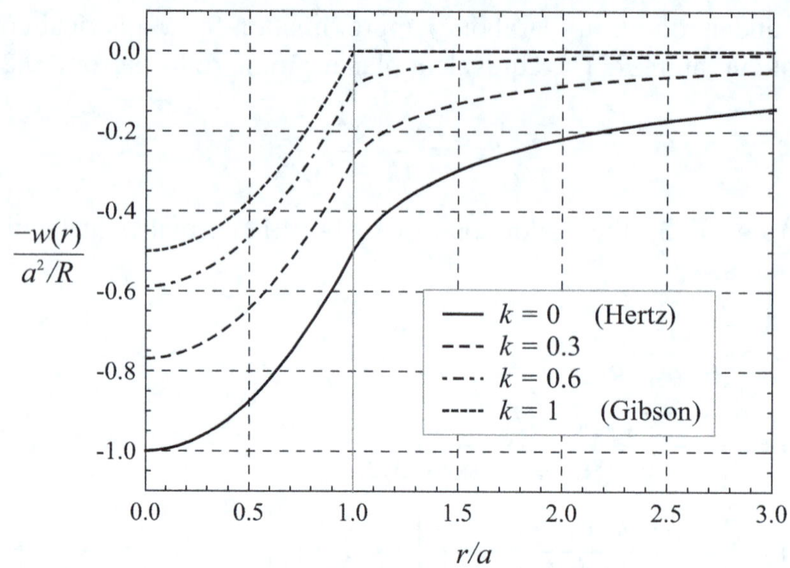

Fig. 9.12 Normalized surface normal displacements of the inhomogeneous half-space in a parabolic contact and different exponents of elastic inhomogeneity k

the contact radius a_H in Hertzian contact), the maximum pressure decreases with increasing exponent k of the elastic inhomogeneity with simultaneous enlargement of the contact area. It should be noted that, depending on the choice of the characteristic depth and the Poisson's number, an opposite effect can also occur. In Fig. 9.12, the normalized normal displacements of the surface of the inhomogeneous half-space are visualized for different exponents k at the same contact radii. By increasing k the indentation depth and the displacements decrease successively.

In the limiting case $k = 1$ the displacements outside the contact area disappear completely—the inhomogeneous half-space behaves like a (two-dimensional) Winkler foundation. Strictly speaking, the latter behavior is coupled to the *linear-inhomogeneous incompressible* half-space because only the $\nu = 0.5$ contact has a non-zero contact stiffness (see Sect. 9.1.8).

9.1.5 The Profile in the Form of a Power-Law

The contact between the inhomogeneous half-space and an axisymmetric indenter whose shape meets the power-law

$$f(r) = A_n r^n \quad \text{with } n \in \mathbb{R}^+, A_n = \text{const} \tag{9.27}$$

(see Fig. 9.13) was examined by Rostovtsev (1961) for the special case $\nu = 1/(2+k)$. A general solution was provided by Giannakopoulos and Suresh (1997). Their derivation by means of the MDR requires the calculation of the equivalent plane profile according to (9.2):

$$g(x) = \kappa(n,k) A_n |x|^n = \kappa(n,k) f(|x|). \tag{9.28}$$

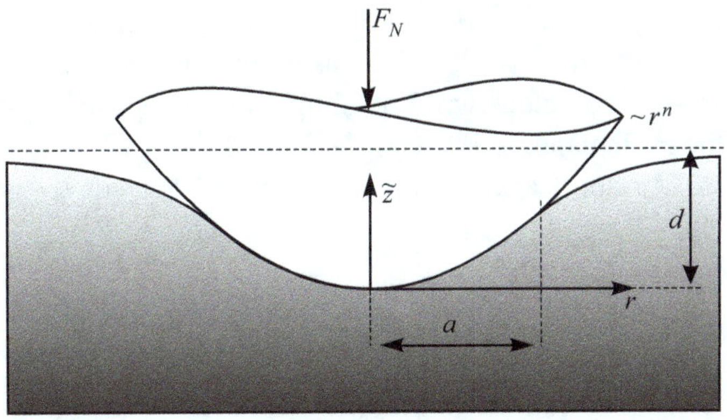

Fig. 9.13 Normal indentation of the elastically inhomogeneous half-space by an indenter whose profile is described by a power function

Equation (9.28) clearly shows that the equivalent profile results from a simple stretch from the original profile. The stretch factor is dependent on the exponent of the power function and the exponent of the elastic inhomogeneity,

$$\kappa(n,k) = \int_0^1 \frac{n\varsigma^{n-1}}{(1-\varsigma^2)^{\frac{1-k}{2}}} \mathrm{d}\varsigma = \frac{n}{2}\int_0^1 t^{\frac{n}{2}-1}(1-t)^{\frac{k+1}{2}-1}\mathrm{d}t$$
$$=: \frac{n}{2}\mathrm{B}\left(\frac{n}{2}, \frac{k+1}{2}\right), \tag{9.29}$$

where $\mathrm{B}(x, y)$ denotes the complete beta function, which follows as a special case $\mathrm{B}(x, y) := \mathrm{B}(1; x, y)$ from the definition of the incomplete beta function (9.17).

The stretch factor increases as the exponent of the power profile increases (see Fig. 9.14). In homogeneous cases, the known values $\kappa(1, 0) = \pi/2$ for the conical

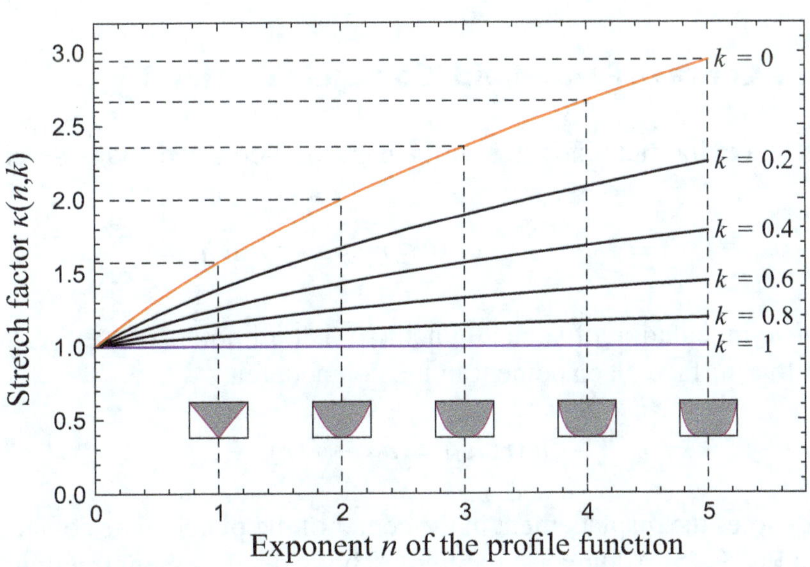

Fig. 9.14 Dependence of the stretch factor κ on the exponent n of the power-law profile function for different exponents k of the elastic inhomogeneity (from Heß 2016a)

and $\kappa(2,0) = 2$ for the parabolic indenter are reproduced. As the exponent of elastic inhomogeneity increases, the stretch factor decreases. The limiting case, $k \to 1$ coincides with the simple cross-section of the three-dimensional profile in the x–z-plane.

The application of rules (9.10)–(9.12) leads to the solution of the contact problem:

$$\begin{aligned}
d(a) &= \kappa(n,k)A_n a^n, \\
F_N(a) &= \frac{2h_N(k,\nu)\kappa(n,k)nA_nE_0a^{n+k+1}}{(1-\nu^2)c_0^k(k+1)(n+k+1)}, \\
p(r;a) &= \frac{h_N(k,\nu)\kappa(n,k)nA_nE_0}{2\pi(1-\nu^2)c_0^k}r^{n+k-1} \\
&\quad \cdot \left[\mathrm{B}\left(\frac{1-k-n}{2},\frac{1+k}{2}\right) - \mathrm{B}\left(\frac{r^2}{a^2};\frac{1-k-n}{2},\frac{1+k}{2}\right)\right], \\
w(r;a) &= \frac{\cos\left(\frac{k\pi}{2}\right)\kappa(n,k)A_na^n}{\pi} \\
&\quad \cdot \left[\mathrm{B}\left(\frac{a^2}{r^2};\frac{1+k}{2},\frac{1-k}{2}\right)\right. \\
&\quad \left. -\left(\frac{r}{a}\right)^n \mathrm{B}\left(\frac{a^2}{r^2};\frac{1+k+n}{2},\frac{1-k}{2}\right)\right] \quad \text{for } r > a.
\end{aligned} \tag{9.30}$$

As already discussed in the context of the investigation of the conical contact (Sect. 9.1.3), the expanded definition of the beta function according to (9.21) should be used.

9.1.6 The Concave Paraboloid (Complete Contact)

The shape of a cylindrical indenter with a parabolic-concave end is described by:

$$f(r) = -h_0\frac{r^2}{a^2} \quad \text{for } 0 \le r \le a \quad \text{with } h_0 = \frac{a^2}{2R}. \tag{9.31}$$

The *complete* indentation of such a punch (so that all points of the face are in contact) leads to a surface displacement in the contact area

$$w(r;d_0) = d_0 - f(r), \tag{9.32}$$

where d_0 denotes the displacement in the center of the punch. The contact geometry is shown in Fig. 9.15. Adding a suitable rigid body displacement fraction and taking

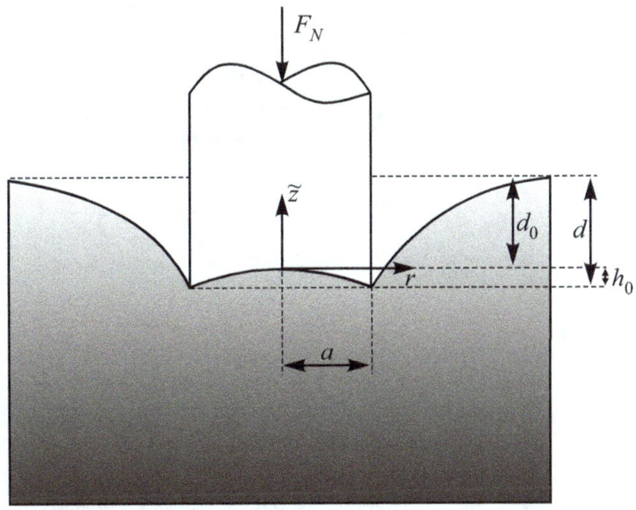

Fig. 9.15 Normal indentation of the elastically inhomogeneous half-space by a parabolic concave indenter

into account (9.31), (9.32) can be written as follows:

$$w(r; d_0) = \underbrace{d_0 + \frac{a^2}{(k+1)R}}_{=w_1} - \underbrace{\left(\frac{a^2}{(k+1)\,R} - \frac{r^2}{2R}\right)}_{=w_2}. \tag{9.33}$$

According to (9.33), the displacement can be represented as the difference between the rigid body displacement w_1 and the displacement w_2 resulting from a parabolic contact. For both separate fractions the solutions have already been developed (see (9.16) and (9.26)). Due to the validity of the superposition principle, the overall solution is the difference of the partial solutions

$$\begin{aligned} F_N(d_0) &= F_1 - F_2 = \frac{2h_N(k,\nu)E_0 a^{k+1}}{(1-\nu^2)(1+k)c_0^k}\left(d_0 + \frac{2h_0}{k+3}\right), \\ p(r; d_0) &= p_1 - p_2 \\ &= \frac{h_N(k,\nu)E_0}{\pi(1-\nu^2)c_0^k(a^2-r^2)^{\frac{1-k}{2}}}\left[d_0 + \frac{2h_0}{(k+1)^2}\left(k - 1 + 2\frac{r^2}{a^2}\right)\right], \\ w(r; d_0) &= w_1 - w_2 \\ &= \frac{\cos\left(\frac{k\pi}{2}\right)}{\pi}\left[d_0 \mathrm{B}\left(\frac{a^2}{r^2}; \frac{1+k}{2}, \frac{1-k}{2}\right)\right. \\ &\qquad \left. + \frac{2h_0}{k+1}\frac{r^2}{a^2}\mathrm{B}\left(\frac{a^2}{r^2}; \frac{3+k}{2}, \frac{1-k}{2}\right)\right]. \end{aligned} \tag{9.34}$$

With the exeption of the displacements, solutions (9.34) were derived by Jin et al. (2013). It should be emphasized that these solutions are only valid under the condition of *complete* contact; for this, the requirement $p(r = 0) > 0$ must be fulfilled,

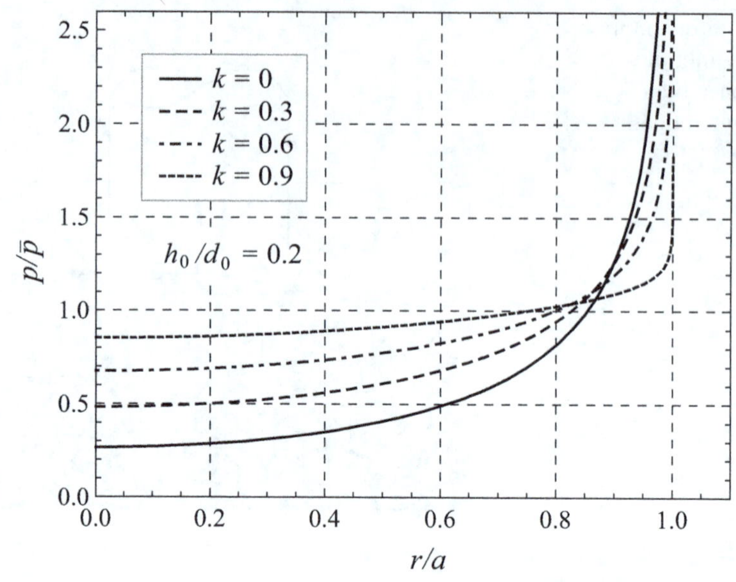

Fig. 9.16 Pressure distribution when indented by a cylindrical punch with parabolic-concave tip for different exponents of elastic inhomogeneity k and $h_0/d_0 = 0.2$, normalized to the mean pressure $\overline{p}$ in the contact area

which suggests:

$$d_0 > \frac{2(1-k)}{(1+k)^2} h_0 \quad \text{or} \quad F_N > \frac{16 h_N(k,\nu) E_0 a^{1+k} h_0}{(1-\nu^2) c_0^k (1+k)^3 (3+k)}. \tag{9.35}$$

Figure 9.16 shows the pressure distribution in the contact area, normalized to the mean pressure $\overline{p}$ for $h_0 = 0.2 d_0$. As the k increases the graph progressively approaches the shape of the indenter. For the same specification, the surface displacements are illustrated in Fig. 9.17.

Finally, to solve contact problems with concave profiles, the MDR rules (9.2), (9.11), and (9.12) are still valid. However, care must be taken to ensure that the center displacement d_0 is used instead of the indentation depth. In addition, complete contact in the replacement model does not necessarily result in full contact with the original problem so the requirement $p(r) > 0$ always has to be checked. For the contact problem investigated here the following is valid:

$$g(x) = -\frac{2h_0}{k+1}\frac{x^2}{a^2} \quad \Rightarrow \quad w_{1D}(x) := d_0 - g(x) = d_0 + \frac{2h_0}{k+1}\frac{x^2}{a^2}. \tag{9.36}$$

An evaluation of (9.11) and (9.12), and taking into account (9.36), gives solutions (9.34) exactly.

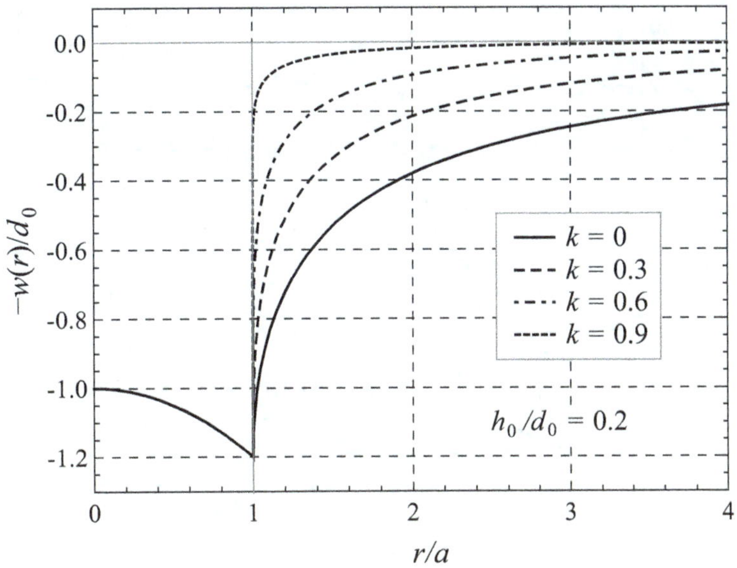

Fig. 9.17 Surface normal displacements normalized by the center displacement d_0 on impression with a cylindrical indenter with parabolic-concave tip for different exponents of elastic inhomogeneity k and $h_0/d_0 = 0.2$

9.1.7 The Profile That Generates Constant Pressure

In contrast to the previous contact problems, where the indenter shape was known *a priori*, the issue now lies in determining the surface displacement caused by a known constant pressure distribution:

$$p(r) = p_0 \quad \text{for} \quad 0 \leq r \leq a. \tag{9.37}$$

This requires first determining the 1D displacement of the foundation according to (9.13):

$$\begin{aligned} w_{1D}(x) &= \frac{2c_0^k(1-\nu^2)\cos\left(\frac{k\pi}{2}\right)}{h_N(k,\nu)E_0}\int_x^a \frac{rp_0}{(r^2-x^2)^{\frac{1+k}{2}}}\mathrm{d}r \\ &= \frac{2c_0^k(1-\nu^2)\cos\left(\frac{k\pi}{2}\right)p_0(a^2-x^2)^{\frac{1-k}{2}}}{h_N(k,\nu)(1-k)E_0}. \end{aligned} \tag{9.38}$$

The displacement at location $r = 0$ in the original must coincide with the displacement at location $x = 0$ in the substitute model, as follows:

$$w_c := w(r=0) \equiv w_{1D}(x=0) = \frac{2c_0^k\left(1-\nu^2\right)\cos\left(\frac{k\pi}{2}\right)}{h_N(k,\nu)(1-k)E_0\pi}\frac{F_N}{a^{1+k}}, \tag{9.39}$$

taking into account that $F_N = p_0\pi a^2$. When the stresses are specified instead of the indenter shape, the equivalent quantity to the indentation depth is the center displacement. Inserting (9.38) in (9.12) yields, after a short calculation, the following

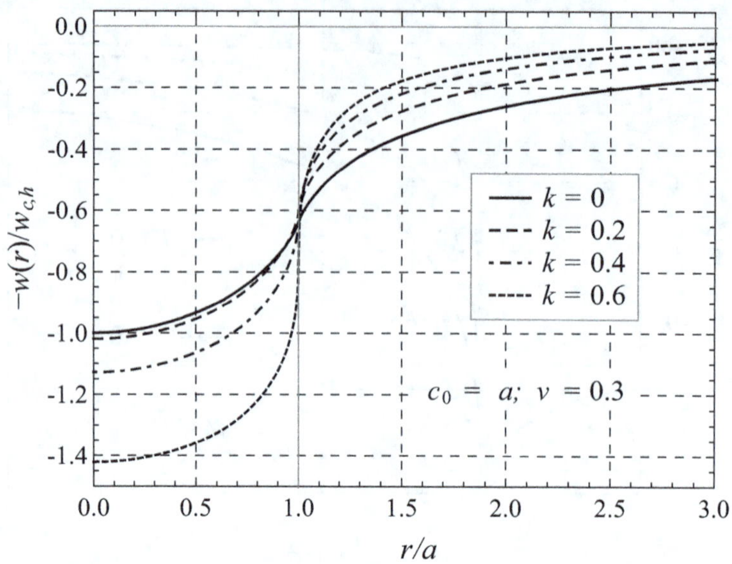

Fig. 9.18 Normalized normal surface displacement of the inhomogeneous half-space for a constant pressure distribution and for different exponents of the elastic inhomogeneity k and fixed c_0

surface displacements of the inhomogeneous half-space:

$$w(r;a) = \begin{cases} w_c\,{}_2\mathrm{F}_1\left(\dfrac{k-1}{2}, \dfrac{1+k}{2}; 1; \dfrac{r^2}{a^2}\right) & \text{for } r \le a, \\ w_c\dfrac{1-k}{2}\left(\dfrac{a}{r}\right)^{1+k} {}_2\mathrm{F}_1\left(\dfrac{1+k}{2}, \dfrac{1+k}{2}; 2; \dfrac{a^2}{r^2}\right) & \text{for } r > a, \end{cases} \tag{9.40}$$

which coincide with ones calculated by Booker et al. (1985). The displacements normalized to the center displacement of the homogeneous half-space,

$$w_{c,h} = \frac{2(1-\nu^2)p_0 a}{E_0}, \tag{9.41}$$

are displayed in Figs. 9.18 and 9.19. Fig. 9.18 studies the influence of the elastic inhomogeneity exponent at a fixed characteristic depth c_0. For the chosen characteristic depth, an increasing exponent k causes a rise of the displacements within the load zone and a drop outside this zone. We consciously avoided particularly large exponents k since it results in unbounded displacements within the load zone (see the discussion in Sect. 9.1.8). Figure 9.19, on the other hand, demonstrates decreasing displacements for a reduction of the characteristic depth, which is fixed at $k = 0.2$. It should be noted that these graphs are offered as examples and do not present all characteristics. However, a complete analysis of the fundamental displacement behavior depending on k, c_0, and p_0 can already be gained from the power-law of the elastic inhomogeneity from (9.1).

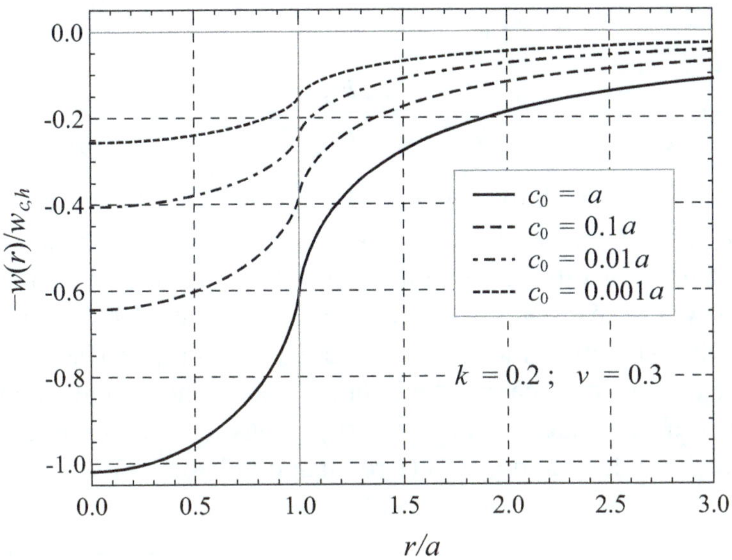

Fig. 9.19 Normalized normal surface displacement of the inhomogeneous half-space for a constant pressure distribution and for different characteristic depths c_0, and fixed exponent k of the elastic inhomogeneity

9.1.8 Notes on the Linear-Inhomogeneous Half-Space—the Gibson Medium

In a linear-inhomogeneous half-space, the elasticity modulus increases proportionally with the depth z. This dependency is covered by the general power-law (9.1); it merely requires the setting of $k = 1$. Accounting for this condition in (9.2)–(9.12) exposes some interesting characteristics, independent of the geometry of the contact. For instance, it follows that the equivalent planar profile coincides with the cross-section of the real profile in the x-z-plane,

$$g(x) = f(|x|). \tag{9.42}$$

Hence the rule for the calculation of the indentation depth (9.10) already returns an unusual result of

$$d = f(a) \quad \Rightarrow \quad w(r) = f(a) - f(r) \quad \text{for } 0 \leq r \leq a. \tag{9.43}$$

According to (9.43), the displacement of the half-space surface at the contact edge is zero like in the 1D model. The calculation formula for the displacements of (9.12) even reveals that the half-space surface exterior of the contact area remains in its original, undeformed state. Such behavior is typical for a (two-dimensional) Winkler foundation, whose surface points are displaced in proportion to the normal stresses acting locally at these points. However, an evaluation of the rule for the calculation of the pressure distribution from (9.12) reveals that this proportionality is valid only for the special case of a linear-homogeneous, incompressible half-space since the factor $h(1, \nu)$ takes on a non-zero value only for $\nu = 0.5$. With

$h(1, 0.5) = \pi/2$ it follows that:

$$p(x, y) = \frac{2E_0}{3c_0} w(x, y). \tag{9.44}$$

The latter finding goes back to Gibson (1967), which is why the linear-inhomogeneous *incompressible* half-space is also named the *Gibson medium*. A normal load on a linear-inhomogeneous, *compressible* half-space results in unbounded surface displacements within the load zone (Awojobi and Gibson 1973). This is indicated by (9.11). The foundation modulus vanishes according to (9.4). Therefore the effects of an external force cannot be balanced out. The indeterminateness of the displacements is a consequence of the vanishing elasticity modulus at the half-space surface in a medium defined by (9.1) (Brown and Gibson 1972).

9.2 Frictionless Normal Contact with JKR Adhesion

9.2.1 Basis for Calculation of the MDR and General Solution

The framework provided by the JKR theory (see Sects. 3.2 and 3.3 in Chap. 3 of this book) permits a particularly easy solution to contact problems with adhesion since the basic idea relies on a simple superposition of the corresponding non-adhesive contact and a rigid-body translation. The latter does depend on the particular contact radius but not on the shape of the indenter. Application of this approach to contact problems related to the elastically inhomogeneous half-space gives the indentation depth as a function of the contact radius

$$d(a) = d_{\text{n.a.}}(a) - \Delta\ell(a) \quad \text{with} \quad \Delta\ell(a) := \sqrt{\frac{2\pi\Delta\gamma c_0^k a^{1-k}}{E^* h_N(k, \nu)}}. \tag{9.45}$$

Here, $d_{\text{n.a.}}$ refers to the indentation depth of the contact without adhesion (which would lead to the same contact radius as the one of the adhesive contact). The unusual notation of the superimposed rigid-body translation $\Delta\ell(a)$ stems from the substitute model of the MDR displayed in Fig. 9.20.

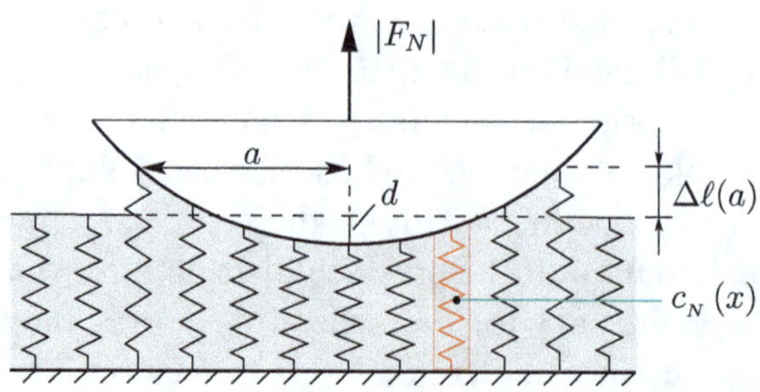

Fig. 9.20 Equivalent substitute model for the adhesive normal contact between two elastically inhomogeneous half-spaces, whose elasticity moduli satisfy condition (9.1), assuming equal exponents k of both media

For the complete solution of a contact problem with adhesion via the MDR, all rules and calculation formulas from Sect. 9.1.1 remain valid with the exception of (9.10). It must be replaced by the condition

$$w_{1D}(a) = -\Delta\ell(a) = -\sqrt{\frac{2\pi\Delta\gamma c_0^k a^{1-k}}{E^* h_N(k,\nu)}} \tag{9.46}$$

with the displacement of the spring elements given by the same equation as in the non-adhesive case:

$$w_{1D}(x) = d - g(x). \tag{9.47}$$

This means that (under the condition that all springs in the contact area remain in contact) the critical detachment state of the edge springs is achieved when the elongation of the springs at the edge reaches the pre-defined value $\Delta\ell(a)$ (see Fig. 9.20).

For convex profiles, the solution approach via the MDR additionally offers a simple way to calculate the critical contact radii (after achieving this, no equilibrium state of the contact as a whole exists). The critical contact radii satisfies the following equation:

$$\frac{\Delta\ell(a_c)}{a_c} = \tilde{C}(k)\left.\frac{\partial g(a)}{\partial a}\right|_{a_c} \quad \text{with } \tilde{C}(k) := \begin{cases} \dfrac{2}{3+k} & \text{for } F_N = \text{const}, \\ \dfrac{2}{1-k} & \text{for } d = \text{const}. \end{cases} \tag{9.48}$$

The definition of the coefficient $\tilde{C}(k)$ varies depending on whether the experiment is force-controlled or displacement-controlled (Heß 2016a).

Taking (9.47) into consideration, (9.11) and (9.12) yield the general solution of the contact with adhesion:

$$\begin{aligned} F_N(a) &= F_{\text{n.a.}}(a) - \frac{2h_N(k,\nu)E^*\Delta\ell(a)a^{1+k}}{(1+k)c_0^k}, \\ p(r;a) &= p_{\text{n.a.}}(r;a) - \frac{h_N(k,\nu)E^*\Delta\ell(a)}{\pi c_0^k (a^2-r^2)^{\frac{1-k}{2}}}, \\ w(r;a) &= \begin{cases} d_{\text{n.a.}}(a) - f(r) - \Delta\ell(a) & \text{for } r \le a, \\ w_{\text{n.a.}}(r;a) - \dfrac{\cos\left(\frac{k\pi}{2}\right)\Delta\ell(a)}{\pi}\mathrm{B}\left(\dfrac{a^2}{r^2}; \dfrac{1+k}{2}, \dfrac{1-k}{2}\right) & \text{for } r > a. \end{cases} \end{aligned} \tag{9.49}$$

The quantities with the indices "n.a." indicate the solutions of the non-adhesive contact. The additional terms result from the rigid-body translation and correspond to the solutions of the flat punch contact (if, instead of the indentation depth d, the (negative) rigid-body translation $-\Delta\ell(a)$ is used).

9.2.2 The Cylindrical Flat Punch

When applying a tensile force to a cylindrical flat punch of radius a, which is adhering to the surface of the inhomogeneous half-space, *all* points in the contact area are subjected to the same displacement $w(r) = d < 0$. Therefore, except for the negative value of the indentation depth d, the boundary conditions match those of the indentation of a cylindrical flat punch in an inhomogeneous half-space. Therefore, we can adopt the solutions from Sect. 9.1.2:

$$\begin{aligned} F_N(d) &= \frac{2h_N(k,\nu)E_0 d a^{1+k}}{(1-\nu^2)(1+k)c_0^k}, \\ p(r;d) &= \frac{h_N(k,\nu)E_0 d}{\pi(1-\nu^2)c_0^k(a^2-r^2)^{\frac{1-k}{2}}}, \\ w(r;d) &= \frac{\cos\left(\frac{k\pi}{2}\right)d}{\pi}\mathrm{B}\left(\frac{a^2}{r^2};\frac{1+k}{2},\frac{1-k}{2}\right). \end{aligned} \tag{9.50}$$

Since the contact radius cannot (stably) decrease in a tensile test, the adhesive contact radius of the flat punch becomes unstable once the "indentation depth" reaches the critical value

$$d_c(a) = -\Delta\ell(a) = -\sqrt{\frac{2\pi\Delta\gamma c_0^k a^{1-k}}{E^* h_N(k,\nu)}}. \tag{9.51}$$

In this case, the flat punch will instantly completely detach from the half-space surface. The corresponding normal force, the absolute value of which is the maximum pull-off force, is:

$$F_c(a) = -\int_{-a}^{a} c_N(x)\Delta\ell(a)\mathrm{d}x = -\sqrt{\frac{8\pi\Delta\gamma h_N(k,\nu)E^* a^{3+k}}{(k+1)^2 c_0{}^k}}. \tag{9.52}$$

A comparison of (9.51) and (9.52) with the equilibrium conditions (9.49) for curved profiles reveals that the solution of the contact without adhesion is calculated by summing the aforementioned terms. This is a pre-requisite step for the solution of the adhesive contact.

9.2.3 The Paraboloid

The solution of the contact problem between a parabolic, rigid indenter and the elastically inhomogeneous half-space, displayed in Fig. 9.21, is gained from the

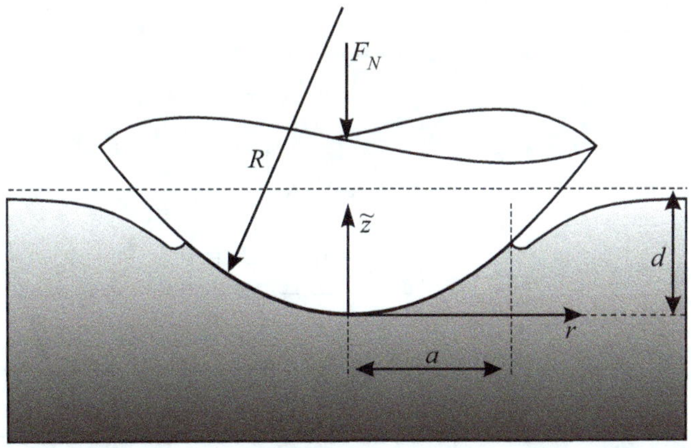

Fig. 9.21 Adhesive normal contact between a rigid parabolic indenter and an elastically inhomogeneous half-space

universal superposition (9.49) by incorporating the solution of the non-adhesive contact problem (9.26):

$$
\begin{aligned}
d(a) &= \frac{a^2}{(k+1)R} - \sqrt{\frac{2\pi\Delta\gamma c_0^k a^{1-k}}{E^* h_N(k,\nu)}},\\
F_N(a) &= \frac{4h_N(k,\nu)E^*a^{k+3}}{c_0^k(k+1)^2(k+3)R} - \sqrt{\frac{8\pi\Delta\gamma h_N(k,\nu)E^*a^{3+k}}{(k+1)^2{c_0}^k}},\\
p(r;a) &= \frac{2h_N(k,\nu)E^*a^{k+1}}{\pi c_0^k(k+1)^2R}\left[1-\left(\frac{r}{a}\right)^2\right]^{\frac{1+k}{2}}\\
&\quad - \sqrt{\frac{2h_N(k,\nu)\Delta\gamma E^*a^{k-1}}{\pi {c_0}^k}}\left[1-\left(\frac{r}{a}\right)^2\right]^{\frac{k-1}{2}},\\
w(r;a) &= \frac{a^2\cos\left(\frac{k\pi}{2}\right)}{(k+1)\pi R}\left[\left(1-(1+k)R\sqrt{\frac{2\pi\Delta\gamma c_0^k a^{-3-k}}{E^*h_N(k,\nu)}}\right)\right.\\
&\quad \left.\cdot \mathrm{B}\left(\frac{a^2}{r^2};\frac{1+k}{2},\frac{1-k}{2}\right) - \frac{r^2}{a^2}\mathrm{B}\left(\frac{a^2}{r^2};\frac{3+k}{2},\frac{1-k}{2}\right)\right].
\end{aligned}
\tag{9.53}
$$

The calculation of the *critical* contact radii from the condition of global stability (9.48) merely requires a knowledge of the slope of the equivalent substitute profile at the contact edge. From (9.25) it follows that

$$
g'(a) = \frac{2a}{(1+k)R}, \tag{9.54}
$$

and thus, with (9.48), a short calculation yields:

$$
a_c = \left(\frac{\pi(1+k)^2R^2\Delta\gamma {c_0}^k}{2\tilde{C}(k)^2h_N(k,\nu)E^*}\right)^{\frac{1}{3+k}}. \tag{9.55}
$$

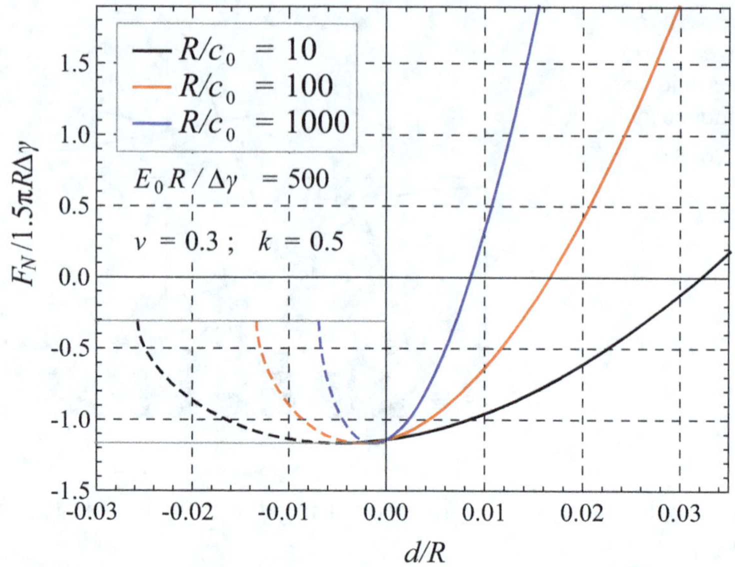

Fig. 9.22 Normal force as a function of the indentation depth in normalized representation for $k = 0.5$ at different characteristic depths c_0

Inserting the critical contact radii (9.55) into the first two equations of (9.53) returns the following critical indentation depths and normal forces:

$$
\begin{aligned}
d_c &= \frac{1-2\tilde{C}(k)}{(1+k)R}\left(\frac{\pi(1+k)^2R^2\Delta\gamma c_0^k}{2\tilde{C}(k)^2 h_N(k,\nu)E^*}\right)^{\frac{2}{3+k}}, \\
F_c &= \left(\frac{1-\tilde{C}(k)(3+k)}{\tilde{C}(k)^2(3+k)}\right)2\pi\Delta\gamma R \\
&= \begin{cases} -\dfrac{k+3}{2}\pi\Delta\gamma R & \text{(force-control)} \\ -\dfrac{(1-k)(5+3k)}{2(3+k)}\pi\Delta\gamma R & \text{(displacement-control).} \end{cases}
\end{aligned}
\tag{9.56}
$$

Solutions (9.53)–(9.56) were derived by Chen et al. (2009), and are focused on force-controlled experiments. They noted that, according to (9.56), the maximum pull-off force (absolute value of the critical normal force under force-control) is independent of the elasticity parameters and characteristic depth, as for homogeneous cases. This is made clear in Fig. 9.22, which provides a graph of the normal force in relation to the maximum pull-off force in the classic JKR theory as a function of the normalized indentation depth. Equilibrium states corresponding to the dotted parts of the curves can only be realized for displacement-control. Additionally, Fig. 9.23 shows that the maximum pull-off force rises by increasing k, while the absolute value of the pull-off force decreases under displacement-controlled test conditions.

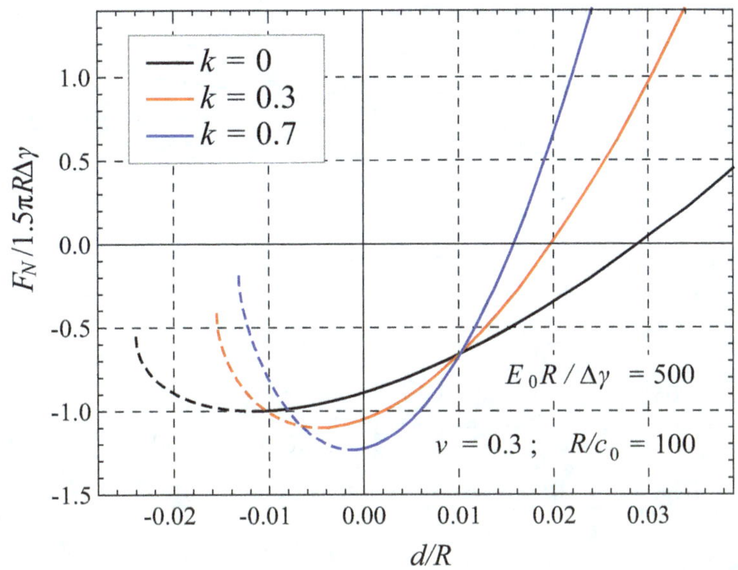

Fig. 9.23 Normal force as a function of the indentation depth in normalized representation for different values of k and fixed characteristic depth c_0

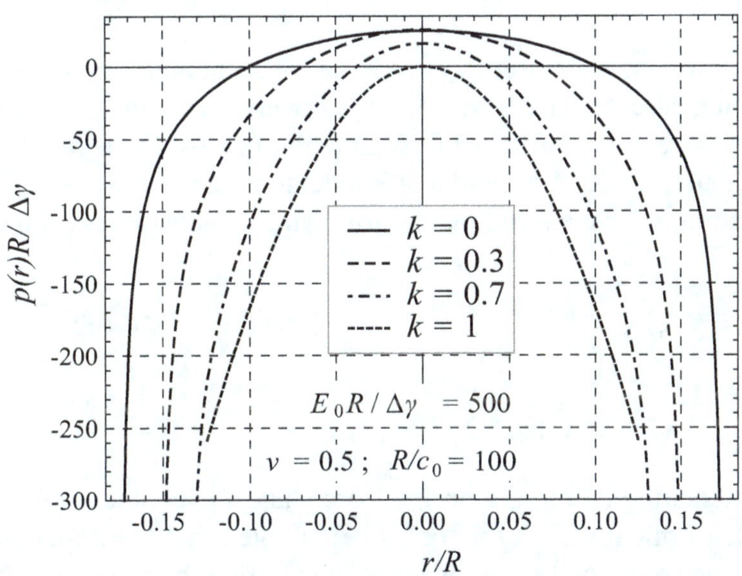

Fig. 9.24 Normalized pressure distribution in the contact area for a varying exponent k and an incompressible material at a fixed characteristic depth c_0

The pressure distribution in the critical state, i.e., under the effect of the maximum pull-off force, is demonstrated for an incompressible material in Fig. 9.24. For the chosen parameters, the critical radius drops by increasing k. Figure 9.25 demonstrates that this property is not universally valid but instead dependent on the characteristic depth. Here we can also see that, for a fixed k, a decrease in the characteristic depth also leads to a decrease of the critical contact radius. Furthermore,

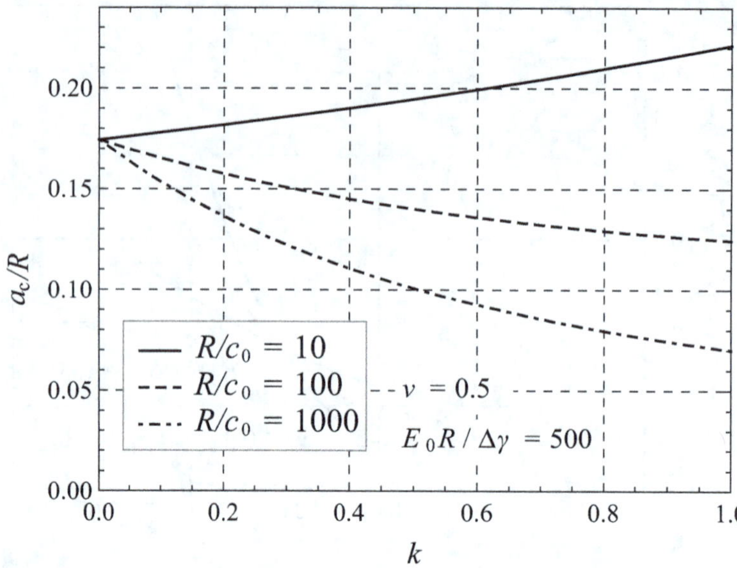

Fig. 9.25 Critical contact radius (*fixed-load*) as a function of k for select characteristic depths c_0 and incompressible material

the maximum pressure is extremely sensitive to parameter variations and must be determined on a case-by-case basis. For the given parameters, the maximum pressure is reached at $k \approx 0.19$. In the Gibson medium ($k \to 1$) the maximum pressure and, consequently, the (critical) indentation depth is zero.

As a last point, we have provided the following dimensionless equations:

$$\tilde{d}(\tilde{a}) = \frac{1}{1 - 2\tilde{C}(k)} \left(\tilde{a}^2 - 2\tilde{C}(k)\tilde{a}^{\frac{1-k}{2}} \right) \quad \text{with } \tilde{d} := \frac{d}{d_c} \text{ and } \tilde{a} := \frac{a}{a_c},$$
$$\tilde{F}_N(\tilde{a}) = \frac{1}{1 - (3+k)\tilde{C}(k)} \left[\tilde{a}^{3+k} - (3+k)\tilde{C}(k)\tilde{a}^{\frac{3+k}{2}} \right] \quad \text{with } \tilde{F}_N := \frac{F_N}{F_c}, \tag{9.57}$$

where the indentation depth, the normal force, and the contact radius are given in relation to their critical value. Jin et al. (2016) noted that the dimensionless forms of (9.57) depend solely on the exponent of the elastic inhomogeneity k.

9.2.4 The Profile in the Form of a Power-Law

Taking into consideration solutions (9.30) for the contact without adhesion, the universal superposition (9.49) provides the solution of the adhesive contact. For the adhesive normal contact of a rigid indenter with a profile in the shape of a power-law and an elastically inhomogeneous half-space, displayed schematically in Fig. 9.26, we obtain the following relationships between the indentation depth, contact radius,

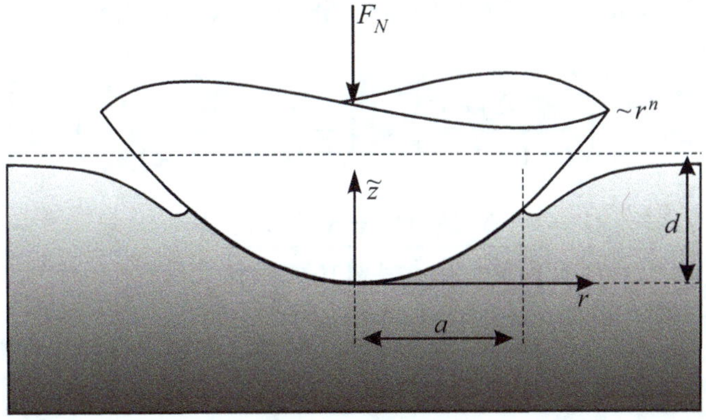

Fig. 9.26 Adhesive normal contact between a rigid indenter with a profile in the form of a power-law and an elastically inhomogeneous half-space

and normal force:

$$d(a) = \kappa(n,k)A_n a^n - \sqrt{\frac{2\pi\Delta\gamma c_0^k a^{1-k}}{E^* h_N(k,\nu)}},$$

$$F_N(a) = \frac{2h_N(k,\nu)\kappa(n,k)nA_nE^*a^{n+k+1}}{(k+1)(n+k+1)c_0^k} - \sqrt{\frac{8\pi h_N(k,\nu)\Delta\gamma E^* a^{3+k}}{(k+1)^2 c_0{}^k}}. \quad (9.58)$$

The pressure distribution in the contact area and the normal surface displacement outside of the contact area are given by:

$$\begin{aligned} p(r;a) = {} & \frac{h_N(k,\nu)\kappa(n,k)nA_nE^*}{2\pi c_0^k} r^{n+k-1} \\ & \cdot \left[\mathrm{B}\left(\frac{1-k-n}{2}, \frac{1+k}{2}\right) - \mathrm{B}\left(\frac{r^2}{a^2}; \frac{1-k-n}{2}, \frac{1+k}{2}\right)\right] \\ & - \sqrt{\frac{2h_N(k,\nu)\Delta\gamma E^* a^{k-1}}{\pi c_0{}^k}} \left[1 - \left(\frac{r}{a}\right)^2\right]^{\frac{k-1}{2}}, \\ w(r;a) = {} & \frac{\cos\left(\frac{k\pi}{2}\right)\kappa(n,k)A_n a^n}{\pi} \\ & \cdot \left[\mathrm{B}\left(\frac{a^2}{r^2}; \frac{1+k}{2}, \frac{1-k}{2}\right) - \left(\frac{r}{a}\right)^n \mathrm{B}\left(\frac{a^2}{r^2}; \frac{1+k+n}{2}, \frac{1-k}{2}\right)\right] \\ & - \frac{\cos\left(\frac{k\pi}{2}\right)}{\pi} \sqrt{\frac{2\pi\Delta\gamma c_0^k a^{1-k}}{E^* h_N(k,\nu)}} \mathrm{B}\left(\frac{a^2}{r^2}; \frac{1+k}{2}, \frac{1-k}{2}\right). \end{aligned} \quad (9.59)$$

The critical contact radii are determined by evaluating condition (9.48), which requires the slope of the equivalent profile at the contact edge as its input. Differentiation of (9.28) yields $g'(a) = n\kappa(n,k)A_n a^{n-1}$, resulting in the following critical radii:

$$a_c = \left(\frac{2\pi\Delta\gamma c_0{}^k}{h_N(k,\nu)E^*\tilde{C}(k)^2\kappa(n,k)^2 n^2 A_n{}^2}\right)^{\frac{1}{2n+k-1}}. \quad (9.60)$$

Substituting (9.60) into (9.58) leads to critical indentation depths and forces,

$$d_c = \left[\left(\frac{2\pi\Delta\gamma c_0{}^k}{h_N(k,\nu)E^*}\right)^n\left(\frac{1}{\tilde{C}(k)n\kappa(n,k)A_n}\right)^{1-k}\right]^{\frac{1}{2n+k-1}}\left(\frac{1-n\tilde{C}(k)}{n\tilde{C}(k)}\right),$$

$$F_c = 2\left(\frac{1-\tilde{C}(k)(n+k+1)}{\tilde{C}(k)(n+k+1)(k+1)}\right)$$
$$\cdot\left[(2\pi\Delta\gamma)^{n+k+1}\left(\frac{c_0{}^k}{h_N(k,\nu)E^*}\right)^{2-n}\left(\frac{1}{\tilde{C}(k)n\kappa(n,k)A_n}\right)^{k+3}\right]^{\frac{1}{2n+k-1}}. \quad (9.61)$$

Jin et al. (2013) presented a detailed examination of the maximum pull-off force depending on the shape exponents n. For scaling reasons, the indenter shape was specified according to slightly modified function $f(r) = r^n/(nR^{n-1})$ instead of (9.27). Normalizing their result to the maximum pull-off force of the classical JKR problem, they derived the following relationship:

$$\frac{F_c}{1.5\pi\Delta\gamma R} = \frac{2(1-k-2n)}{3\pi(n+k+1)(k+1)}$$
$$\cdot\left[(2\pi)^{n+k+1}\left(\frac{1-\nu^2}{h_N(k,\nu)}\right)^{2-n}\left(\frac{3+k}{2\kappa(n,k)}\right)^{k+3}\right.$$
$$\left.\cdot\left(\frac{R}{c_0}\right)^{k(n-2)}\left(\frac{E_0R}{\Delta\gamma}\right)^{n-2}\right]^{\frac{1}{2n+k-1}}, \quad (9.62)$$

which is visualized for selected parameters and shape exponents in Fig. 9.27. It is clear to see that the maximum pull-off force changes linearly with k only for the parabolic contact ($n = 2$) and otherwise exhibits a non-linear dependency.

Putting the indentation depth and normal force in relation to their critical values results in the following dimensionless representations, which depend solely on the exponents k and n (see Heß 2016a):

$$\tilde{d}(\tilde{a}) = \frac{1}{1-\tilde{C}(k)n}\left(\tilde{a}^n - n\tilde{C}(k)\tilde{a}^{\frac{1-k}{2}}\right) \quad \text{with } \tilde{d} := \frac{d}{d_c} \text{ and } \tilde{a} := \frac{a}{a_c},$$
$$\tilde{F}_N(\tilde{a}) = \frac{1}{1-\tilde{C}(k)(n+k+1)}\left[\tilde{a}^{n+k+1} - \tilde{C}(k)(n+k+1)\tilde{a}^{\frac{k+3}{2}}\right]$$
$$\text{with } \tilde{F}_N := \frac{F_N}{F_c}. \quad (9.63)$$

9.2.5 The Concave Paraboloid (Complete Contact)

In the following, we will assume that a rigid, cylindrical, concave punch is initially pressed into the inhomogeneous half-space to the point of complete contact. During

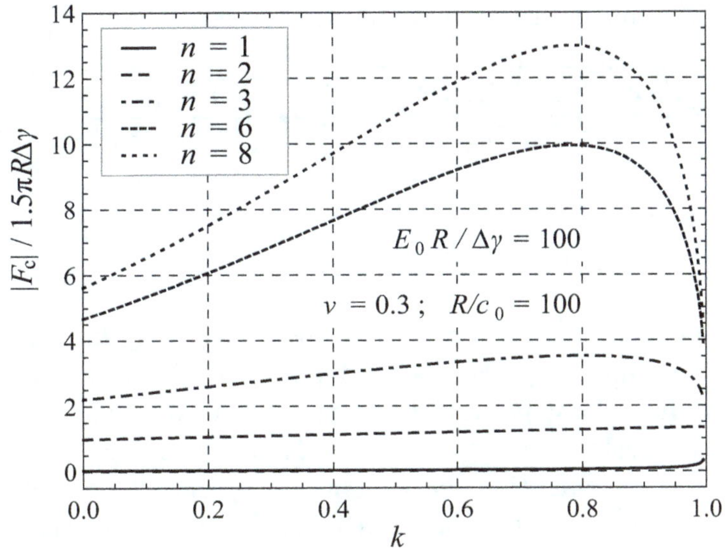

Fig. 9.27 Maximum pull-off force as a function of the exponent of the elastic inhomogeneity k for different exponents of the power-law profile according to Jin et al. (2013)

the subsequent unloading, two competing mechanisms lead to the separation of the surfaces:

1. For slightly concave punch profiles, the separation of the punch is caused (as in convex profiles) by the stress singularity at the contact edge.
2. For highly concave punch profiles, the separation occurs in the center of the contact area because of too high tensile stresses, whose absolute values exceed the theoretical tensile strength (of the adhesion) σ_{th}.

The two states, whose accompanying quantities we denote by the indices "P" (Periphery) and "C" (Center), were examined in greater detail by Jin et al. (2013). The critical state, at which the separation at the edge occurs, satisfies the equilibrium condition (9.45). The displacement of the Winkler foundation is taken from the solution of the corresponding contact without adhesion, (9.36). By replacing d by d_{P} we get:

$$w_{1D}(x) = d_{\text{P}} - h_0 + \frac{2h_0}{k+1}\frac{x^2}{a^2}. \tag{9.64}$$

The evaluation of condition (9.45) yields the critical indentation depth d_{P} at which the edge begins to separate:

$$d_{\text{P}} = -\sqrt{\frac{2\pi\Delta\gamma c_0^k a^{1-k}}{h_N(k,\nu)E^*}} - h_0\frac{1-k}{1+k}. \tag{9.65}$$

The remaining terms of the complete solution of the contact problem are obtained by simply inserting (9.65) into (9.34), taking into account that $d_0 = d_{\text{P}} - h_0$. Here, we restrict the presentation to the normal force and the pressure distribution:

$$F_{\mathrm{P}}(a) = -\sqrt{\frac{8\pi h_N(k,\nu)\Delta\gamma E^* a^{3+k}}{(1+k)^2 c_0^k}} - \frac{8 h_N(k,\nu) E^* h_0 a^{1+k}}{(k+1)^2(k+3) c_0^k},$$

$$p_{\mathrm{P}}(r;a) = \frac{h_N(k,\nu)E^*}{\pi c_0^k (a^2-r^2)^{\frac{1-k}{2}}}\left[-\sqrt{\frac{2\pi\Delta\gamma c_0^k a^{1-k}}{h_N(k,\nu)E^*}} - \frac{4h_0}{(k+1)^2}\left(1-\frac{r^2}{a^2}\right)\right]. \quad (9.66)$$

The first term of the critical force in (9.66) returns exactly the adhesive force of the flat punch according to (9.52). The second term results from the concave shape. Since both terms have the same sign, the maximum pull-off force of weakly concave profiles is always greater than the one of the flat punch contact.

For strongly concave profiles, the separation begins in the center of the contact area upon reaching the theoretical adhesive tensile strength. The critical indentation depth is gained from the condition $p(r = 0) = -\sigma_{\mathrm{th}}$, where once again, in the pressure distribution of the contact without adhesion (9.34), d is replaced by d_{C}. Therefore, we obtain:

$$d_{\mathrm{C}} = -\frac{\pi c_0^k a^{1-k}}{h_N(k,\nu)E^*}\sigma_{\mathrm{th}} + \frac{3+k^2}{(1+k)^2}h_0, \quad (9.67)$$

from which the normal force and the pressure distribution in the critical state follow as:

$$F_{\mathrm{C}}(a) = -\frac{2\pi a^2}{(1+k)}\sigma_{\mathrm{th}} + \frac{16 h_N(k,\nu) h_0 E^* a^{k+1}}{(k+1)^3(k+3)c_0^k},$$

$$p_{\mathrm{C}}(a) = \frac{1}{(a^2-r^2)^{\frac{1-k}{2}}}\left[-a^{1-k}\sigma_{\mathrm{th}} + \frac{4h_N(k,\nu)h_0E^*}{\pi(k+1)^2c_0^k}\frac{r^2}{a^2}\right]. \quad (9.68)$$

For $k = 0$, these solutions coincide with those of Waters et al. (2011).

9.2.6 The Indenter Which Generates a Constant Adhesive Tensile Stress

The optimal profile of a rigid, concave, cylindrical indenter is defined by the condition that, in the critical state, the constant tensile stresses in the contact area are (in absolute terms) equal to the theoretical adhesive tensile strength (see Gao and Yao 2004):

$$p_{\mathrm{opt}}(r) = -\sigma_{\mathrm{th}} \quad \text{for } r \le a, \quad (9.69)$$

The maximum pull-off force is then

$$F_{\mathrm{opt}}(a) = -\sigma_{\mathrm{th}}\pi a^2. \quad (9.70)$$

Determining the corresponding optimal shape is rendered particularly easy when accounting for the results from Sect. 9.1.7. Equations (9.39) and (9.40) determine the necessary surface displacement to generate a constant pressure in the contact area. Replacing p_0 with $-\sigma_{\text{th}}$ or, alternatively, F_N by F_{opt} according to (9.70) leads to the surface displacement of the considered problem:

$$w_{\text{opt}}(r;a) = \begin{cases} w_{c,\text{opt}}\ {}_2\text{F}_1\left(\frac{k-1}{2}, \frac{1+k}{2}; 1; \frac{r^2}{a^2}\right) & \text{for } r \leq a, \\ w_{c,\text{opt}}\frac{1-k}{2}\left(\frac{a}{r}\right)^{1+k}\ {}_2\text{F}_1\left(\frac{1+k}{2}, \frac{1+k}{2}; 2; \frac{a^2}{r^2}\right) & \text{for } r > a, \end{cases} \tag{9.71}$$

with

$$w_{c,\text{opt}} = -\frac{2c_0^k \cos\left(\frac{k\pi}{2}\right) a^{1-k}}{h_N(k,\nu)(1-k)E^*}\sigma_{\text{th}}. \tag{9.72}$$

The optimal indenter shape f_{opt} can basically be extracted from (9.71) and (9.72). Accounting for $f_{\text{opt}}(r=0)=0$ returns:

$$f_{\text{opt}}(r) = w_{c,\text{opt}}\left[1 - {}_2\text{F}_1\left(\frac{k-1}{2}, \frac{1+k}{2}; 1; \frac{r^2}{a^2}\right)\right]. \tag{9.73}$$

For the purposes of a graphical representation of the influence of the elastic inhomogeneity k and the normalized characteristic depth c_0/a, it seems more appropriate to use the function shifted by the boundary value $f_{\text{opt}}(a)$

$$\tilde{f}_{\text{opt}}(r) := f_{\text{opt}}(r) - f_{\text{opt}}(a) \tag{9.74}$$

In Figs. 9.28 and 9.29 the optimal profile functions are plotted for different values of k and c_0/a, according to the publication by Jin et al. (2013). For a fixed exponent

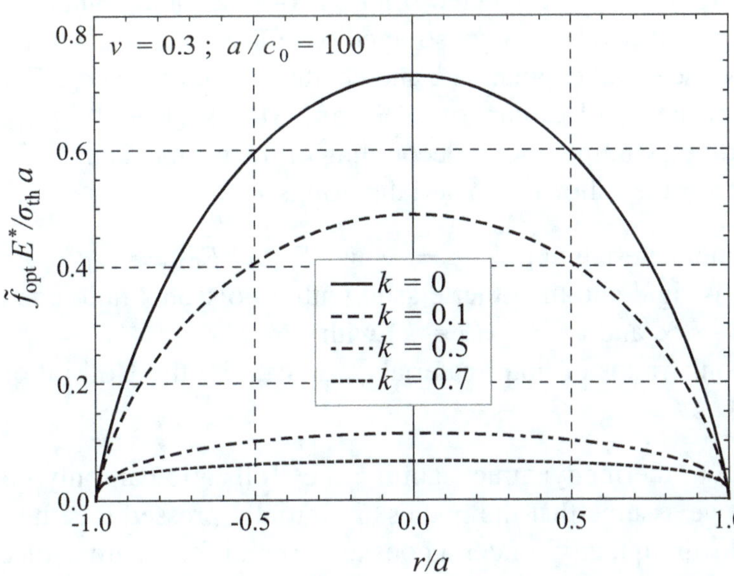

Fig. 9.28 Optimal concave profile for a fixed characteristic depth c_0 and different set parameters k

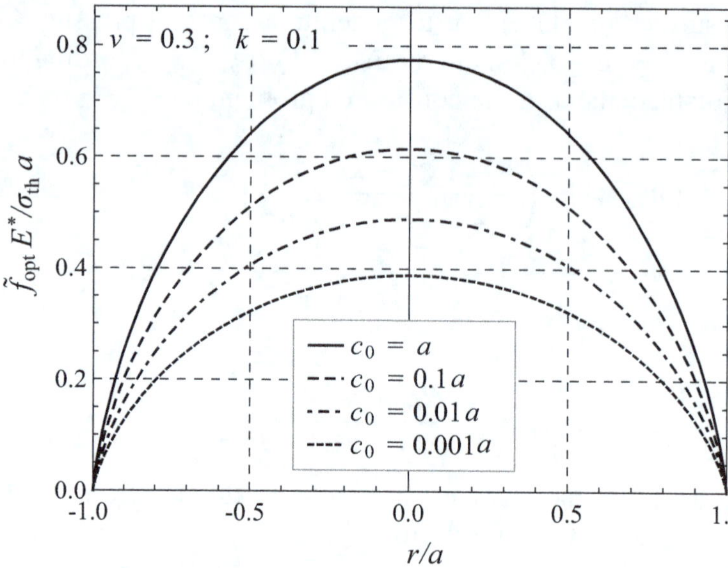

Fig. 9.29 Optimal concave profile for a fixed exponent of the elastic inhomogeneity $k = 0.1$ and different characteristic depths c_0

k, a decrease of the characteristic depth c_0 causes a drop in the height of the concave section.

9.3 Tangential Contact

9.3.1 Basis of Calculation and Restricting Assumptions

In the following, we present selected solutions of tangential contact problems with partial slip between two bodies constructed of FGMs, according to (9.1). For both bodies we assume equal exponents k and characteristic depths c_0. Furthermore, as in the classical theory of Cattaneo (1938) and Mindlin (1949) (see Chap. 4), we operate under the assumption of a decoupling of the normal and tangential contact, which is valid for the following material pairings:

1. Equal elastic materials: $\nu_1 = \nu_2 =: \nu$ and $E_{01} = E_{02} =: E_0$
2. One body is rigid and the other elastic with a Poisson's ratio equal to the Holl ratio: $E_{0i} \to \infty$ and $\nu_j = 1/(2+k)$ with $i \neq j$
3. The Poisson's ratios of both materials are given by the Holl ratio: $\nu_1 = \nu_2 = 1/(2+k)$

Note that, due to thermodynamic stability, the Holl ratio can only be fulfilled for positive k. If we assume that the bodies are initially pressed together by a normal force F_N and subsequently (under a constant normal force) are subjected to a tangential force, then the contact area will be composed of an inner stick zone and an outer slip zone (see Fig. 9.30).

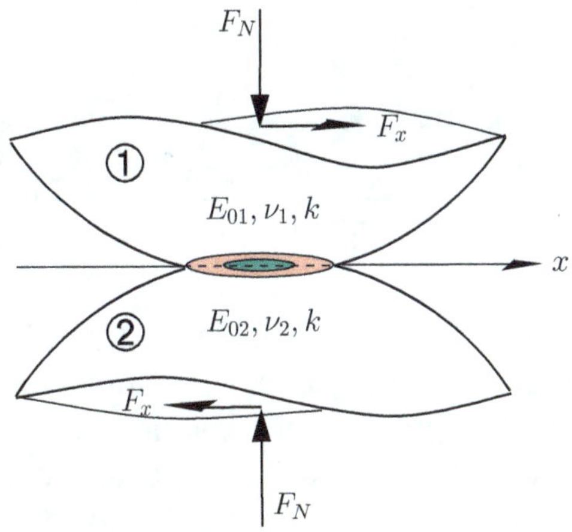

Fig. 9.30 Tangential contact with partial slip between two elastically inhomogeneous bodies; the contact area is composed of an inner stick zone and an outer slip zone

The boundary conditions are determined by the rigid-body translation of the points of the stick zone

$$u(x, y) = \text{const.} \quad \text{and} \quad v(x, y) = 0 \quad \text{for } 0 < r < c \tag{9.75}$$

and by the Coulomb law of friction

$$\begin{aligned} \left|\vec{\tau}(r)\right| &< \mu p(r) \quad \text{for } 0 \le r < c, \\ \left|\vec{\tau}(r)\right| &= \mu p(r) \quad \text{for } c \le r \le a. \end{aligned} \tag{9.76}$$

In the slip zone, the tangential stresses must be additionally directed opposite to the relative tangential displacement of the surface points. With increasing tangential force, the stick zone shrinks until, finally, gross slip sets in. For the calculation of this problem using the generalized Ciavarella–Jäger theorem is recommended, according to which the tangential contact can be represented as the superposition of two normal contacts. For the tangential stress, the tangential force, and the relative tangential displacement $u^{(0)}$ between two remote material points—within the contacting bodies and far from the contact interface—the following is valid (see Heß 2016b):

$$\begin{aligned} \tau(r) &= \mu[p(r, a) - p(r, c)], \\ F_x &= \mu[F_N(a) - F_N(c)], \\ u^{(0)} &= \mu\alpha[d(a) - d(c)]. \end{aligned} \tag{9.77}$$

We have defined $\tau(r) := -\tau_{zx}(r)$ and assume that a tangential force in the x-direction only leads to tangential stresses in the same x-direction. α represents the ratio of normal stiffness to tangential stiffness:

$$\alpha(k, \nu_i, E_{0i}) := \frac{\dfrac{1}{h_T(k, \nu_1)E_{01}} + \dfrac{1}{h_T(k, \nu_2)E_{02}}}{\dfrac{1-\nu_1^2}{h_N(k, \nu_1)E_{01}} + \dfrac{1-\nu_2^2}{h_N(k, \nu_2)E_{02}}}, \tag{9.78}$$

where

$$h_T(k,\nu) = \frac{2\beta(k,\nu)\cos\left(\frac{k\pi}{2}\right)\Gamma\left(1+\frac{k}{2}\right)}{(1-\nu^2)\sqrt{\pi}C(k,\nu)\sin\left(\frac{\beta(k,\nu)\pi}{2}\right)\Gamma\left(\frac{3+k}{2}\right)+\beta(k,\nu)(1+\nu)\Gamma\left(1+\frac{k}{2}\right)}, \quad (9.79)$$

and the additional functional relationships are defined as an addendum to (9.5). With (9.77) we are able to solve the tangential contact between arbitrarily shaped bodies, provided the solutions of the corresponding normal contact problem are known. It is noted that tangential contacts can be solved in an equally simple fashion using the MDR (Heß and Popov 2016).

9.3.2 Tangential Contact Between Spheres (Parabolic Approximation)

Applying (9.77) to the (parabolic) contact of two equally elastically inhomogeneous spheres of radius R leads to the following results:

$$\tau(r;a,c) = \mu\frac{2h_N(k,\nu)E_0}{\pi(1-\nu^2)c_0^k(k+1)^2R}\begin{cases}(a^2-r^2)^{\frac{1+k}{2}}-(c^2-r^2)^{\frac{1+k}{2}} & \text{for } 0\le r\le c,\\ (a^2-r^2)^{\frac{1+k}{2}} & \text{for } c<r\le a,\end{cases}$$

$$F_x(c,a) = \mu F_N(a)\left[1-\left(\frac{c}{a}\right)^{k+3}\right],$$

$$u^{(0)}(c,a) = \mu\frac{h_N(k,\nu)}{(1-\nu^2)h_T(k,\nu)}d(a)\left[1-\left(\frac{c}{a}\right)^2\right], \quad (9.80)$$

where $F_N(a)$ and $d(a)$ are taken from the solutions of the normal contact (9.26).

Figure 9.31 shows the values of the tangential stresses along the x-axis under the assumption that the normal force is kept constant for all selected exponents k of the elastic inhomogeneity. The quantities are normalized to the maximum pressure $p_{0,H}$ and the contact radius a_H of the Hertzian contact problem. It is clearly visible that, compared to the respective Hertzian contact radius (contact of homogeneous materials), the contact radii are smaller for negative exponents and greater for positive exponents. In contrast, the tangential stress is greater for negative exponents and vice-versa. However, this behavior of the stresses is only observed in the slip zone and not necessarily true for the entire contact area. The curves presented in Fig. 9.32 make it clear that gross slip sets in at a small tangential displacement $u^{(0)}$ for negative exponents, and at a substantially greater displacement for positive exponents. For ease of comparison, the quantities are normalized to the tangential displacement $u^{(0)}_{c,\text{hom}}$, which marks the onset of gross slip of elastically *homogeneous* materials.

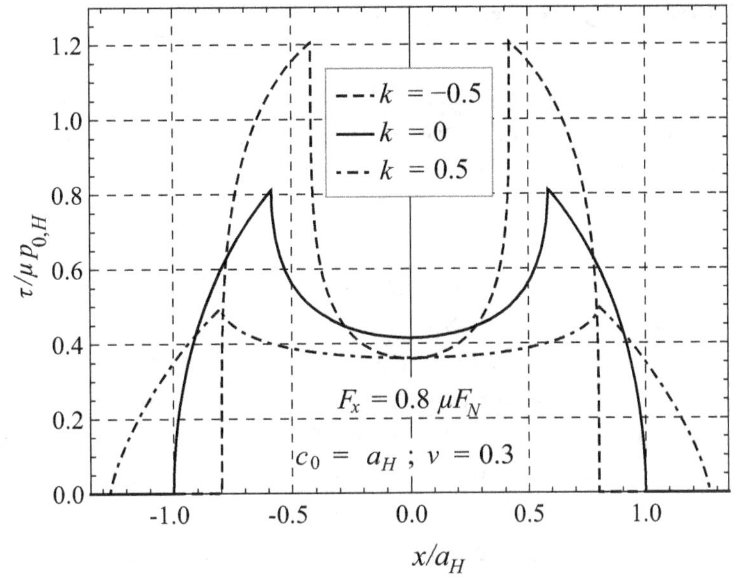

Fig. 9.31 Normalized tangential stresses along the *x*-axis for different exponents of the elastic inhomogeneity for set values of $F_x = 0.8\mu F_N$, $c_0 = a_H$ and $\nu = 0.3$

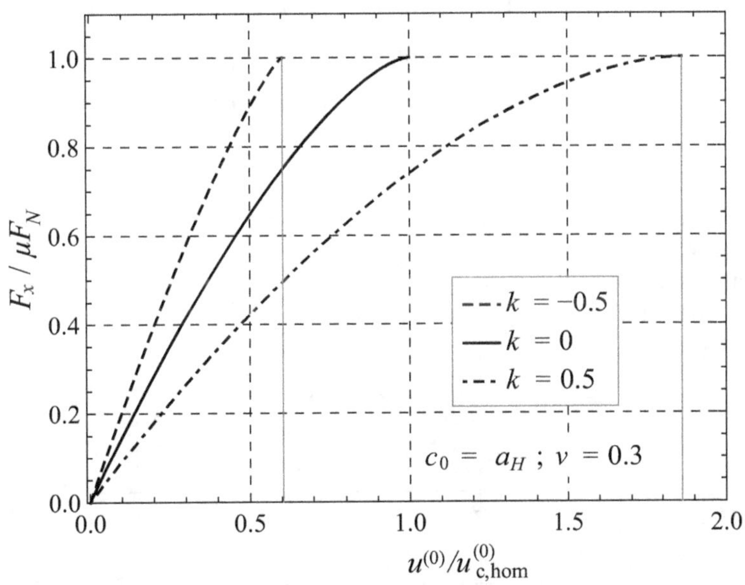

Fig. 9.32 Normalized tangential force as a function of the normalized tangential displacement for different exponents of the elastic inhomogeneity for set values of $c_0 = a_H$, $\nu = 0.3$

9.3.3 Oscillating Tangential Contact of Spheres

Two bodies with parabolically curved surfaces (curvature radii R) of the same elastically inhomogeneous material according to (9.1) are initially pressed together by a normal force, and subsequently (under a constant normal force) are subjected to the oscillating tangential force given in Fig. 9.33. Accordingly, the tangential force varies within the bounds $\pm 0.9\mu F_N$ to avoid gross slip at every point in time.

The solutions of the contact problem for the initial rise to Point A on the load curve were covered in Sect. 9.3.2. Slip will occur, beginning at the edge of the contact, during a subsequent successive reduction of the tangential force. It is directed opposite to the initial slip and propagates inwards during the unloading process. In the new ring-shaped slip zone $b < r \leq a$, frictional stresses develop, which are again directed opposite to the initial frictional stresses. Therefore, a reduction in the tangential force leads to a *change* of the frictional stresses in the ring-shaped slip zone by $-2\mu p(r)$. Yet since no slip occurs in the remaining contact area, all surface points in that domain are subjected to the same *change* in their tangential displacement. Therefore the *change* of the tangential force leads to *changes* of the tangential stresses and tangential displacements, which solve the contact problem formulated in Sect. 9.3.1. Thus, except for the factor "2" and a negative sign, the changes satisfy (9.77):

$$
\begin{aligned}
\Delta\tau(r) &= -2\mu[p(r,a) - p(r,b)], \\
\Delta F_x &= -2\mu[F_N(a) - F_N(b)], \\
\Delta u^{(0)} &= -2\mu\alpha[d(a) - d(b)].
\end{aligned}
\tag{9.81}
$$

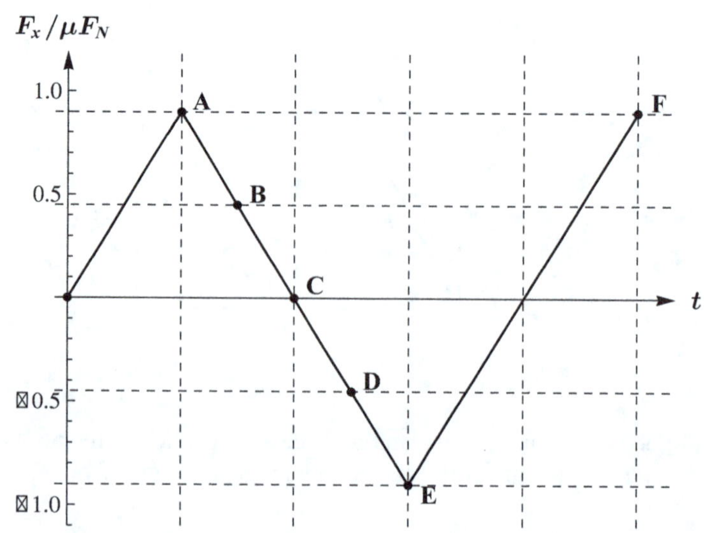

Fig. 9.33 Loading history of the tangential force: oscillation of the tangential force between the extreme values $\pm 0.9\mu F_N$

The absolute values result from the summation of the state variables in Point A and the changes (9.81). For the stresses, this leads to:

$$\tau(r) = \mu \frac{2h_N(k,\nu)E_0}{\pi(1-\nu^2)c_0^k(k+1)^2 R} \cdot \begin{cases} -(a^2-r^2)^{\frac{1+k}{2}} + 2(b^2-r^2)^{\frac{1+k}{2}} - (c^2-r^2)^{\frac{1+k}{2}} & \text{for } 0 \le r \le c, \\ -(a^2-r^2)^{\frac{1+k}{2}} + 2(b^2-r^2)^{\frac{1+k}{2}} & \text{for } c < r \le b, \\ -(a^2-r^2)^{\frac{1+k}{2}} & \text{for } b < r \le a, \end{cases} \tag{9.82}$$

and for the tangential force and the tangential displacements, this leads to:

$$F_x = F_A - 2\mu F_N(a)\left[1-\left(\frac{b}{a}\right)^{k+3}\right] \quad \text{with} \quad F_A = \mu F_N(a)\left[1-\left(\frac{c}{a}\right)^{k+3}\right],$$
$$u^{(0)} = u_A^{(0)} - 2\mu\alpha d(a)\left[1-\left(\frac{b}{a}\right)^{2}\right] \quad \text{with} \quad u_A^{(0)} = \mu\alpha d(a)\left[1-\left(\frac{c}{a}\right)^{2}\right]. \tag{9.83}$$

F_A and $u_A^{(0)}$ represent the load and displacement quantities in Point A, and c is the radius of the corresponding stick zone. For the selected exponents of the elastic inhomogeneity $k = 0.5$ and $k = -0.5$, Figs. 9.34 and 9.35 display the tangential stress distribution along the x-axis for the characteristic points A through E of the unloading curve (see Fig. 9.33). In both cases, the same normal force was applied and the representation was normalized to the radius a_H of the Hertzian contact corresponding to this normal force. For positive k, the increase of the contact

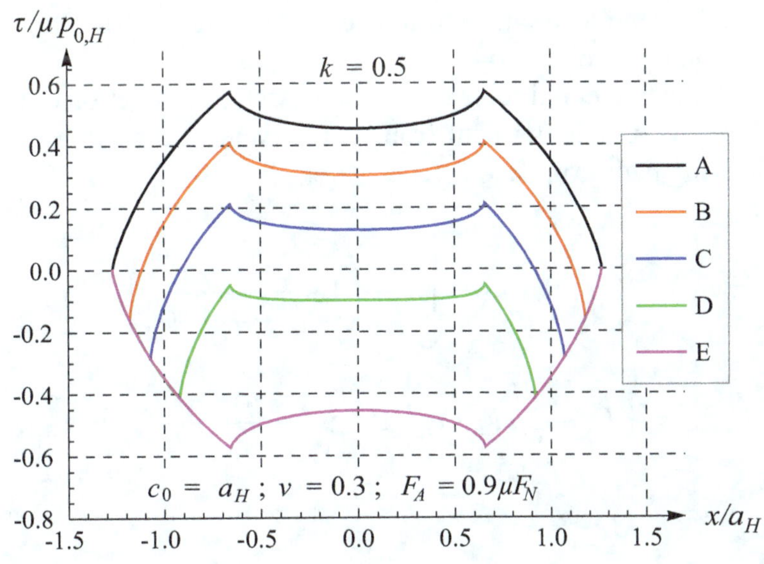

Fig. 9.34 Curve of the normalized tangential stresses along the x-axis for different points of the load curve from Fig. 9.33 and positive exponents of the elastic inhomogeneity $k = 0.5$

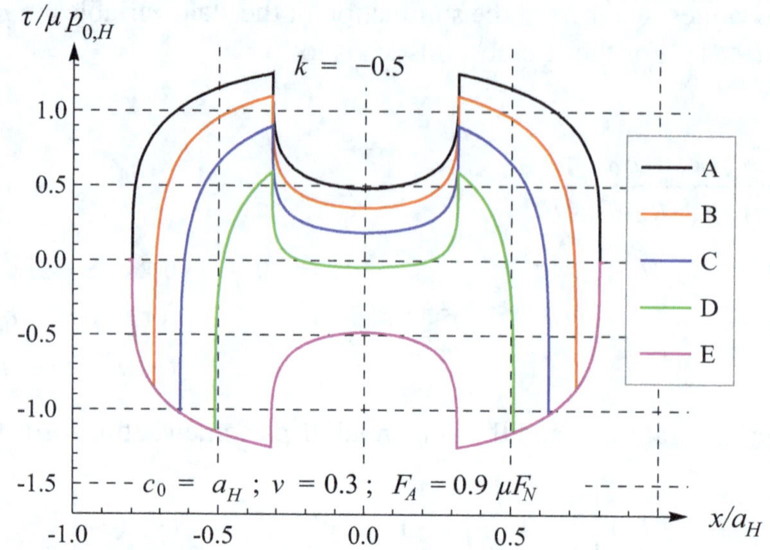

Fig. 9.35 Curve of the normalized tangential stresses along the x-axis for different points of the load curve from Fig. 9.33 and negative exponents of the elastic inhomogeneity $k = -0.5$

area is coupled with a decrease in the tangential stresses. On the other hand, negative exponents reduce the contact area, resulting in an increase of the tangential stresses compared to the homogeneous case $k = 0$. However, this effect is extremely sensitive to changes in the characteristic depth and can even manifest in the inverse behavior. In Figs. 9.34 and 9.35 it is assumed that $c_0 = a_H$. The characteristic difference in the shape of the curves for positive and negative exponents remain unaffected. The dependency of the stress distribution on the loading history is demonstrated best at the load point C. Here, non-zero tangential stresses arise, even though no external tangential force is applied.

The solutions for the load curve from E to F (see Fig. 9.33) follow in complete analogy to the solution from A to E.

Using (9.83), the tangential force can be expressed as a function of the tangential displacement for the unloading path from A to E and in the same manner for the subsequent loading path from E to F. These are:

$$\mathrm{A} \rightarrow \mathrm{E}: F_x = F_A - 2\mu F_N(a)\left[1 - \left(1 + \frac{u^{(0)} - u_A^{(0)}}{2\alpha\mu d(a)}\right)^{\frac{k+3}{2}}\right],$$

$$\mathrm{E} \rightarrow F: F_x = -F_A + 2\mu F_N(a)\left[1 - \left(1 - \frac{u^{(0)} + u_A^{(0)}}{2\alpha\mu d(a)}\right)^{\frac{k+3}{2}}\right]. \tag{9.84}$$

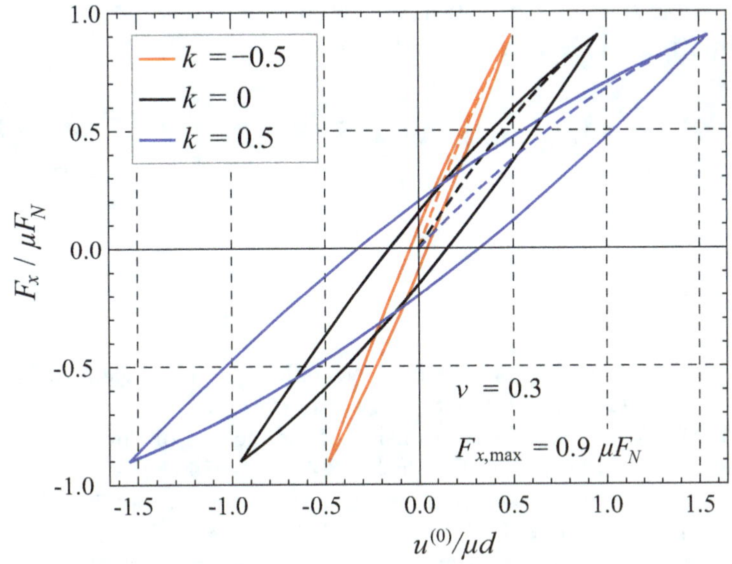

Fig. 9.36 Hysteresis loops as a result of an oscillating tangential contact for different exponents of the elastic inhomogeneity

where the tangential displacement in Point A depends on the corresponding tangential force according to:

$$u_A^{(0)} = \mu a d(a) \left[1 - \left(1 - \frac{F_A}{\mu F_N} \right)^{\frac{2}{k+3}} \right]. \tag{9.85}$$

Equations (9.84) characterize a hysteresis curve, a graphical representation of which is given in Fig. 9.36 for different values of the exponent k. In the chosen normalization, it is quite visible that, compared to the homogeneous case, the gradient of the hysteresis is greater for negative exponents and smaller for positive exponents. Additionally, an increasing k is connected with an increase of the area enclosed by the hysteresis loop and, therefore, the dissipated energy per loading cycle. An explicit calculation of the energy dissipated per cycle yields:

$$\Delta W = \frac{4(k+3)^2(k+1)\mu^2 F_N^2 c_0^k}{(k+5)h_T(k,\nu)E_0 a^{k+1}} \cdot \left[1 - \left(1 - \frac{F_A}{\mu F_N} \right)^{\frac{k+5}{k+3}} - \frac{k+5}{2(k+3)} \frac{F_A}{\mu F_N} \left(1 + \left(1 - \frac{F_A}{\mu F_N} \right)^{\frac{2}{k+3}} \right) \right]. \tag{9.86}$$

The special case of elastically homogeneous bodies ($k = 0$) leads to the result derived by Mindlin and Deresiewicz (1953):

$$\Delta W = \frac{18\mu^2 F_N^2}{5G^* a}\left[1-\left(1-\frac{F_A}{\mu F_N}\right)^{\frac{5}{3}}-\frac{5}{6}\frac{F_A}{\mu F_N}\left(1+\left(1-\frac{F_A}{\mu F_N}\right)^{\frac{2}{3}}\right)\right]. \quad (9.87)$$

References

Aleynikov, S.M.: Spatial contact problems in Geotechnics: boundary-element method. Springer, Heidelberg, pp 55–83 (2011). ISBN 978-3-540-25138-5

Argatov, I., Heß, M., Popov, V.L.: The extension of the method of dimensionality reduction to layered elastic media. Z. Angew. Math. Mech. **98**(4), 622–634 (2018)

Awojobi, A.O., Gibson, R.E.: Plane strain and axially-symmetric problems of a linearly non-homogeneous elastic half-space. Q. J. Mech. Appl. Math. **26**(3), 285–302 (1973)

Boesel, L.F., Greiner, C., Arzt, E., Del Campo, A.: Gecko-inspired surfaces: a path to strong and reversible dry adhesives. Adv. Mater. **22**(19), 2125–2137 (2010)

Booker, J.R., Balaam, N.P., Davis, E.H.: The behavior of an elastic non-homogeneous half-space. Part II—circular and strip footings. Int. J. Numer. Anal. Methods. Geomech **9**(4), 369–381 (1985)

Brown, P.T., Gibson, R.E.: Surface settlement of a deep elastic stratum whose modulus increases linearly with depth. Can. Geotech. J. **9**(4), 467–476 (1972)

Cattaneo, C.: Sul contatto di due corpi elastici: distribuzione locale degli sforzi. Rendiconti Dell'accademia Nazionale Dei Lincei **27**, 342–348, 434–436, 474–478 (1938)

Chen, S., Yan, C., Zhang, P., Gao, H.: Mechanics of adhesive contact on a power-law graded elastic half-space. J. Mech. Phys. Solids **57**(9), 1437–1448 (2009)

Fabrikant, V.I., Sankar, T.S.: On contact problems in an inhomogeneous half-space. Int. J. Solids Struct. **20**(2), 159–166 (1984)

Fröhlich, O.K.: XI Das elastische Verhalten der Böden. In: Druckverteilung im Baugrunde, pp. 86–108. Springer, Vienna (1934)

Gao, H., Yao, H.: Shape insensitive optimal adhesion of nanoscale fibrillar structures. Proc. Natl. Acad. Sci. U.S.A. **101**(21), 7851–7856 (2004)

Giannakopoulos, A.E., Pallot, P.: Two-dimensional contact analysis of elastic graded materials. J. Mech. Phys. Solids **48**(8), 1597–1631 (2000)

Giannakopoulos, A.E., Suresh, S.: Indentation of solids with gradients in elastic properties: Part II. Axi-symmetric indentors. Int. J. Solids Struct. **34**(19), 2393–2428 (1997)

Gibson, R.E.: Some results concerning displacements and stresses in a non-homogeneous elastic half-space. Geotechnique **17**(1), 58–67 (1967)

Gorb, S., Varenberg, M., Peressadko, A., Tuma, J.: Biomimetic mushroom-shaped fibrillar adhesive micro-structure. J. R. Soc. Interface **4**(13), 271–275 (2007)

Heß, M.: A simple method for solving adhesive and non-adhesive axi-symmetric contact problems of elastically graded materials. Int. J. Eng. Sci. **104**, 20–33 (2016a)

Heß, M.: Normal, tangential and adhesive contacts between power-law graded materials. Presentation at the Workshop on Tribology and Contact Mechanics in Biological and Medical Applications, TU-Berlin, 14.–17. Nov. 2016. (2016b)

Heß, M., Popov, V.L.: Method of dimensionality reduction in contact mechanics and friction: a user's handbook. II. Power-law graded materials. Facta Univ. Ser. Mech. Eng. **14**(3), 251–268 (2016)

Holl, D.L.: Stress transmission in earths. Highway Res. Board Proc. **20**, 709–721 (1940)

Jin, F., Guo, X., Zhang, W.: A unified treatment of axi-symmetric adhesive contact on a power-law graded elastic half-space. J. Appl. Mech. **80**(6), 61024 (2013)

Jin, F., Zhang, W., Wan, Q., Guo, X.: Adhesive contact of a power-law graded elastic half-space with a randomly rough rigid surface. Int. J. Solids Struct. **81**, 244–249 (2016)

Jitcharoen, J., Padture, P.N., Giannakopoulos, A.E., Suresh, S.: Hertzian-crack suppression in ceramics with elastic-modulus-graded surfaces. J. Am. Ceram. Soc. **81**(9), 2301–2308 (1998)

Lee, D., Barber, J.R., Thouless, M.D.: Indentation of an elastic half-space with material properties varying with depth. Int. J. Eng. Sci. **47**(11), 1274–1283 (2009)

Liu, Z., Meyers, M.A., Zhang, Z., Ritchie, R.O.: Functional gradients and heterogeneities in biological materials: design principles, functions, and bioinspired applications. Prog. Mater. Sci. **88**, 467–498 (2017)

Mehrali, M., Shirazi, F.S., Mehrali, M., Metselaar, H.S.C., Kadri, N.A.B., Osman, N.A.A.: Dental implants from functionally graded materials. J. Biomed. Mater. Res. A **101**(10), 3046–3057 (2013)

Mindlin, R.D.: Compliance of elastic bodies in contact. J. Appl. Mech. **16**(3), 259–268 (1949)

Mindlin, R.D., Deresiewicz, H.: Elastic spheres in contact under varying oblique forces. J. Appl. Mech. **20**, 327–344 (1953)

Miyamoto, Y., Kaysser, W.A., Rabin, B.H., Kawasaki, A., Ford, R.G.: Functionally graded materials: design, processing and applications. Kluwer Academic Publishers, Boston, Dordrecht, London (1999)

Peisker, H., Michels, J., Gorb, S.N.: Evidence for a material gradient in the adhesive tarsal setae of the ladybird beetle Coccinella septempunctata. Nat. Commun. **4**, 1661 (2013)

Rostovtsev, N.A.: An integral equation encountered in the problem of a rigid foundation bearing on non-homogeneous soil. J. Appl. Math. Mech. **25**(1), 238–246 (1961)

Rostovtsev, N.A.: On certain solutions of an integral equation of the theory of a linearly deformable foundation. J. Appl. Math. Mech. **28**(1), 127–145 (1964)

Selvadurai, A.P.S.: The analytical method in geomechanics. Appl. Mech. Rev. **60**(3), 87–106 (2007)

Sola, A., Bellucci, D., Cannillo, V.: Functionally graded materials for orthopedic applications—an update on design and manufacturing. Biotechnol. Adv. **34**(5), 504–531 (2016)

Suresh, S.: Graded materials for resistance to contact deformation and damage. Sci. Compass Rev. **292**(5526), 2447–2451 (2001)

Suresh, S., Olsson, M., Giannakopoulos, A.E., Padture, N.P., Jitcharoen, J.: Engineering the resistance to sliding-contact damage through controlled gradients in elastic properties at contact surfaces. Acta. Mater. **47**(14), 3915–3926 (1999)

Waters, J.F., Gao, H.J., Guduru, P.R.: On adhesion enhancement due to concave surface geometries. J. Adhes. **87**(3), 194–213 (2011)

Willert, E.: Dugdale-Maugis adhesive normal contact of axi-symmetric power-law graded elastic bodies. Facta Univ. Ser. Mech. Eng. **16**(1), 9–18 (2018)

Willert, E., Popov, V.L.: Adhesive tangential impact without slip of a rigid sphere and a power-law graded elastic half-space. Z. Angew. Math. Mech. **97**(7), 872–878 (2017a)

Willert, E., Popov, V.L.: The oblique impact of a rigid sphere on a power-law graded elastic half-space. Mech. Mater. **109**, 82–87 (2017b)

10 Annular Contacts

In this chapter, we will turn our attention back to contact problems with an ideally elastic, homogeneous, isotropic half-space. However, now the contact area is not compact but instead ring-shaped. The simplest example of such a problem is the contact between a flat, hollow cylindrical punch and the half-space. Even for this simplest of examples, there exists only an extremely complicated solution. Nonetheless, for this class of problems there exists a range of analytical approaches which will be documented in this chapter. These approaches are for the frictionless normal contact with and without adhesion and the non-slipping, purely torsional contact.

10.1 Frictionless Normal Contact without Adhesion

Due to its complexity, the Boussinesq problem of ring-shaped contact areas, i.e., the frictionless normal contact without adhesion between a rigid indenter and an elastic half-space with the effective elasticity modulus E^*, has rarely been considered in literature. Of most practical importance is the case of indenters in the form of hollow cylinders. Some concave rotationally symmetric indenter profiles have been studied as well. The problem here lies in the fact that, for a small normal load and the corresponding ring-shaped contact area, the inner contact radius is unknown and must be determined—as is generally the case for contact problems—as part of the solution. Barber (1974) observed that the current contact area A_c must be the one which maximized the normal force F_N. From the resulting condition

$$\left.\frac{\partial F_N}{\partial A}\right|_{A=A_c} = 0 \tag{10.1}$$

one can determine the contact area if the relationship $F_N(A)$ is known. Barber (1983b) also examined the boundary value problem of a flat-ended cylindrical punch and a central circular recess. Finally, Argatov and Nazarov (1996, 1999) and Argatov et al. (2016) performed a study of toroidal indenters.

V.L. Popov, M. Heß, E. Willert, *Handbook of Contact Mechanics*,
https://doi.org/10.1007/978-3-662-58709-6_10

Support structures are frequently constructed as hollow cylinders to save weight and material. The interior of the hollow cylinder can also be filled with a very soft matrix—similar to bones. The corresponding normal contact problem is considered in Sect. 10.1.1. Some tools feature conical (see Sect. 10.1.2) or parabolic (see Sect. 10.1.3) concave heads, e.g., for stamping purposes. The circular, central recess at the tip of a cylinder (see Sect. 10.1.4) is a classic engineering solution to ensure form-fit connections. The tori examined in Sects. 10.1.5 and 10.1.6 also see wide-spread use. As an example, direct current transformers are increasingly constructed in a toroidal shape instead of the classic E-I design.

10.1.1 The Hollow Flat Cylindrical Punch

The simplest problem considered in this chapter is the normal contact of a flat, hollow cylinder with an elastic half-space. The use of the word "simple" is meant in a relative sense, since even this problem proves to be extremely complex. However, there are a number of solutions to the problem. A schematic diagram of the problem is displayed in Fig. 10.1. Let the cylinder have the outer radius a and the inner radius b and be pressed by the normal force F_N into the half-space to the depth d.

Shibuya et al. (1974) performed a series expansion of the stresses in the contact area $b \leq r \leq a$ proceeding from the singularities at $r = b$ and $r = a$, thereby reducing the problem to an infinite system of coupled linear equations. The coefficients of this system of equations are integrals of a product of four Bessel functions, which does not make the solution particularly easy. Nonetheless, the system can of course be solved numerically and the authors provide a great number of graphical representations of the obtained numerical solutions for the stresses and displacements.

Gladwell and Gupta (1979) made use of an especially elegant superposition of appropriate, known potentials to find an approximate analytical solution of the problem. The potentials result from the solution of the Dirichlet–Neumann problem

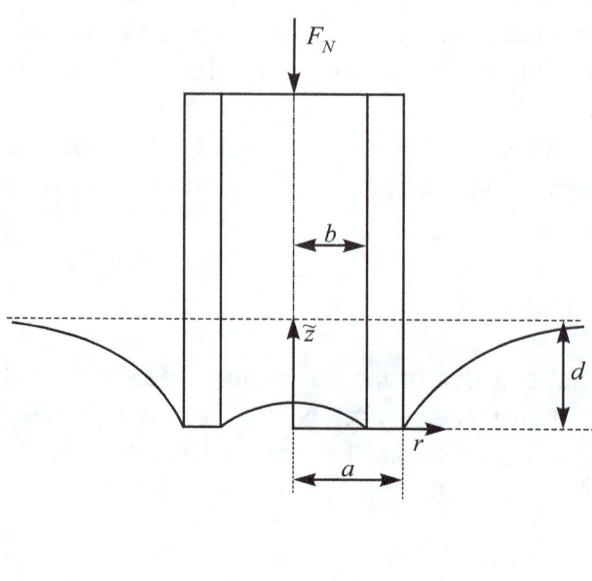

Fig. 10.1 Normal contact of a hollow flat cylinder with an elastic half-space

within and outside the circles $r = b$ and $r = a$. For the relationship between F_N and d they obtain the expression:

$$F_N = 2E^* d a \gamma, \tag{10.2}$$

with

$$\gamma = 1 - \frac{4}{3\pi^2}\left(\frac{b}{a}\right)^3 - \frac{1}{8}\left(\frac{b}{a}\right)^4 + O\left[\left(\frac{b}{a}\right)^5\right]. \tag{10.3}$$

The approximation coincides with the ratio of the radii of the exact solution, except for a fourth-order term which is erroneous. The authors also applied their method to the normal contact of a concave paraboloid (see Sect. 10.1.3) and to the Reissner–Sagoci problem for the hollow cylinder (see Sect. 10.3.1).

Using the superposition of harmonic potentials, Gubenko and Mossakovskii (1960) and Collins (1962, 1963) managed to reduce the problem to a Fredholm integral equation, which can be solved iteratively. For γ, Collins found the expression:

$$\gamma = 1 - \frac{4}{3\pi^2}\left(\frac{b}{a}\right)^3 - \frac{8}{15\pi^2}\left(\frac{b}{a}\right)^5 - \frac{16}{27\pi^4}\left(\frac{b}{a}\right)^6 + O\left[\left(\frac{b}{a}\right)^7\right]. \tag{10.4}$$

All series terms given are exactly correct.

Borodachev and Borodacheva (1966a) utilized a similar approach as Collins and determined numerical solutions for the indentation depth $d = d(F_N)$ and the stress $\sigma_{zz} = \sigma_{zz}(r, F_N)$ in a series expansion to the ratio b/a. Borodachev (1976) additionally obtained an asymptotic solution for the stresses in the vicinity of the singularities.

The complete exact solution of the problem was finally worked out by Roitman and Shishkanova (1973), albeit in the form of recursive formulas. Through series expansions and coefficient comparisons they obtained the following expression for the stresses in the contact area:

$$\sigma_{zz}(r; d) = -\frac{E^* d}{2a}\sum_{k=0}^{\infty}\sum_{p=0}^{\infty}\left(\frac{b}{a}\right)^p\left[\alpha_{pk}\left(\frac{r}{a}\right)^{2k} + \beta_{pk}\left(\frac{b}{r}\right)^{2k+3}\right], \tag{10.5}$$

with the recursively determined coefficients

$$\alpha_{pk} = \frac{(2k+1)!!}{(2k)!!}\tilde{\alpha}_{pk}, \quad \beta_{pk} = \frac{(2k+1)!!}{(2k)!!}\tilde{\beta}_{pk},$$

$$\tilde{\alpha}_{0k} = \frac{2}{\pi}\frac{1}{2k+1}, \quad \tilde{\beta}_{0k} = \frac{2}{\pi}\frac{1}{2k+1},$$

$$\tilde{\alpha}_{1k} = \tilde{\alpha}_{2k} = \tilde{\beta}_{1k} = 0, \quad \tilde{\beta}_{2k} = \frac{4}{\pi^2}\frac{1}{3(2k+5)},$$

$$\tilde{\alpha}_{pk} = \frac{2}{\pi}\sum_{q=0}^{I\left(\frac{p-3}{2}\right)}\frac{\tilde{\beta}_{p-2q-3,q}}{2k+2q+3}, \tilde{\beta}_{pk} = \frac{2}{\pi}\sum_{q=0}^{I\left(\frac{p}{2}\right)}\frac{\tilde{\alpha}_{p-2q,q}}{2k+2q+3}, p \geq 3. \tag{10.6}$$

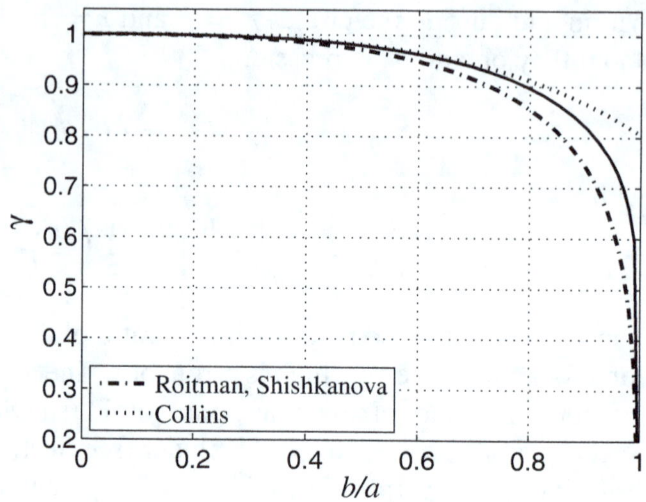

Fig. 10.2 Normalized normal force as a function of the ratio of the radii for a normal contact with a hollow cylindrical punch, according to (10.4) and (10.9). The solid line represents the approximation (10.10)

Here, $(\cdot)!!$ denotes the double faculty

$$n!! := \begin{cases} 2 \cdot 4 \cdot \ldots \cdot n, & n \text{ even}, \\ 1 \cdot 3 \cdot 5 \cdot \ldots \cdot n, & n \text{ odd} \end{cases} \tag{10.7}$$

and $I(x)$ is the largest integer which x does not exceed. For $b = 0$ it returns the familiar result of the flat cylindrical punch:

$$\sigma_{zz}(r; d; b = 0) = -\frac{E^* d}{2a} \sum_{k=0}^{\infty} \alpha_{0k} \left(\frac{r}{a}\right)^{2k} = -\frac{E^* d}{\pi \sqrt{a^2 - r^2}}. \tag{10.8}$$

As expected, the series in (10.5) converges in the open interval $b < r < a$. Term by term integration gives the following expression for γ:

$$\gamma = \frac{\pi}{2} \sum_{k=0}^{\infty} \sum_{p=0}^{\infty} \left(\frac{b}{a}\right)^p \left[\frac{\alpha_{pk}}{2k+2}\left(1 - \left(\frac{b}{a}\right)^{2k+2}\right) + \frac{\beta_{pk}}{2k+1} \frac{b^2}{a^2} \left(1 - \left(\frac{b}{a}\right)^{2k+1}\right)\right]. \tag{10.9}$$

It should be noted that the result for γ in (2.18) of the publication by Roitman and Shishkanova (1973) is incorrect. Grouping the right-hand side of the (10.9) by terms $(b/a)^n$ returns the series expansion from (10.4).

The solutions for γ and the pressure distribution from the (10.9) and (10.5) are shown in Figs. 10.2 and 10.3. The coefficients in (10.6) were evaluated only to $p, k = 150$, with the curve, then scaled to the value $\gamma(b = 0) = 1$.

A closed-form analytical solution was later published by Antipov (1989), the complexity of which would exceed the scope of this handbook. A very good approximation for $\gamma(\varepsilon)$,

$$\gamma(\varepsilon) \approx (1 - \varepsilon^m)^n, \quad m = 2.915;\ n = 0.147 \tag{10.10}$$

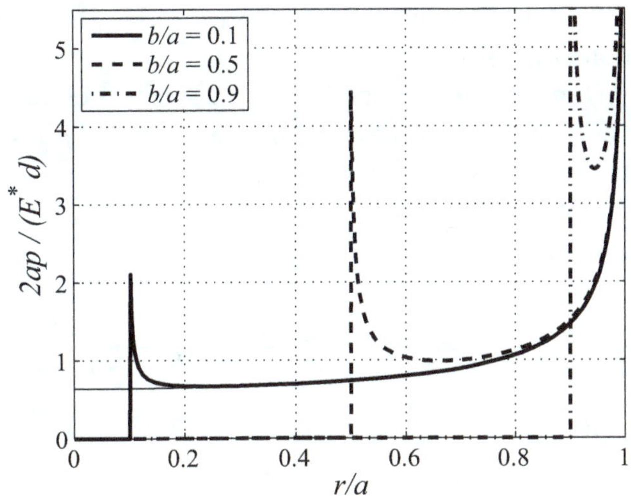

Fig. 10.3 Normalized pressure distribution for a normal contact with a hollow cylindrical punch. The thin solid line represents the solution of the solid cylinder

was given by Willert et al. (2016) on the basis of a very precise boundary element calculation. This is also displayed in Fig. 10.2.

10.1.2 The Concave Cone

The Boussinesq problem of a concave cone was examined by Barber (1976) and Shibuya (1980). A schematic diagram of the problem is shown in Fig. 10.4. Let the base cylinder have the radius a and a concave, conical tip of the depth h. A normal force F_N presses the body into the elastic half-space to the depth d, with the (*a priori* unknown) inner contact radius of b.

Shibuya utilized the same solution approach as in his 1974 publication on the annular punch, i.e., he reduced the problem via a series expansion at the stress singularity at $r = a$ to an infinite system of coupled linear equations and then

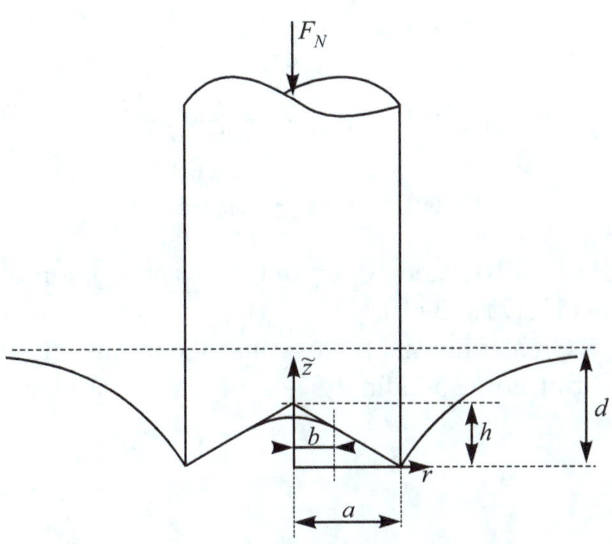

Fig. 10.4 Normal contact of a concave cone with an elastic half-space

solved these numerically. He provided a variety of graphical representations of the numerical solution.

Barber (1976) examined the contact problem in two limiting cases: firstly, for very small values of $\varepsilon = b/a$, and secondly, for the case $\varepsilon \to 1$. For the first approach he used the method by Collins (1963), and for the second approach he used the one by Grinberg and Kuritsyn (1962). For small values of ε, the relationship between F_N, d and ε is given by:

$$F_N(d,\varepsilon) = 2E^*a\left[(d-h)\gamma_1(\varepsilon) + \frac{\pi h}{4}\left(\gamma_2(\varepsilon) - \frac{8\varepsilon^3 \ln\varepsilon}{3\pi^2}\gamma_3(\varepsilon)\right)\right], \quad (10.11)$$

with

$$\begin{aligned}\gamma_1(\varepsilon) &= 1 - \frac{4\varepsilon^3}{3\pi^2} - \frac{8\varepsilon^5}{15\pi^2} - \frac{16\varepsilon^6}{27\pi^4} - \frac{92\varepsilon^7}{315\pi^2} - \frac{448\varepsilon^8}{675\pi^4} + O(\varepsilon^9),\\ \gamma_2(\varepsilon) &= 1 + \frac{8\varepsilon^3}{9\pi^2} - \frac{52\varepsilon^5}{75\pi^2} - \frac{32\varepsilon^6}{81\pi^4} - \frac{326\varepsilon^7}{735\pi^2} - \frac{496\varepsilon^8}{10125\pi^4} + O(\varepsilon^9),\\ \gamma_3(\varepsilon) &= 1 + \frac{\varepsilon^2}{5} - \frac{4\varepsilon^3}{9\pi^2} + \frac{3\varepsilon^4}{35} - \frac{92\varepsilon^5}{45\pi^2} + O(\varepsilon^6).\end{aligned} \quad (10.12)$$

The condition stemming from (10.1)

$$\frac{\partial F_N}{\partial \varepsilon} = 0 \quad (10.13)$$

leads to the desired relationship between d and ε,

$$\frac{d}{h} = 1 - \frac{\pi}{4}\frac{\varepsilon^2\gamma_4(\varepsilon) + 2\ln\varepsilon\,\gamma_5(\varepsilon)}{\gamma_6(\varepsilon)}, \quad (10.14)$$

with

$$\begin{aligned}\gamma_4(\varepsilon) &= 1 + \frac{16\varepsilon}{81\pi^2} + \frac{5\varepsilon^2}{6} - \frac{14792\varepsilon^3}{10125\pi^2} + O(\varepsilon^4),\\ \gamma_5(\varepsilon) &= 1 + \frac{\varepsilon^2}{3} - \frac{8\varepsilon^3}{9\pi^2} + \frac{\varepsilon^4}{5} - \frac{736\varepsilon^5}{135\pi^2} + O(\varepsilon^6),\\ \gamma_6(\varepsilon) &= 1 + \frac{2\varepsilon^2}{3} + \frac{8\varepsilon^3}{9\pi^2} + \frac{23\varepsilon^4}{45} + \frac{896\varepsilon^5}{675\pi^2} + O(\varepsilon^6).\end{aligned} \quad (10.15)$$

Figures 10.5 and 10.6 provide a visualization of the shorthand symbols introduced in (10.12) and (10.15).

Greater normal forces cause the radius b to increase and the expression $1 - \varepsilon$ to shrink correspondingly. For this case, Barber (1976) gave the following solution for

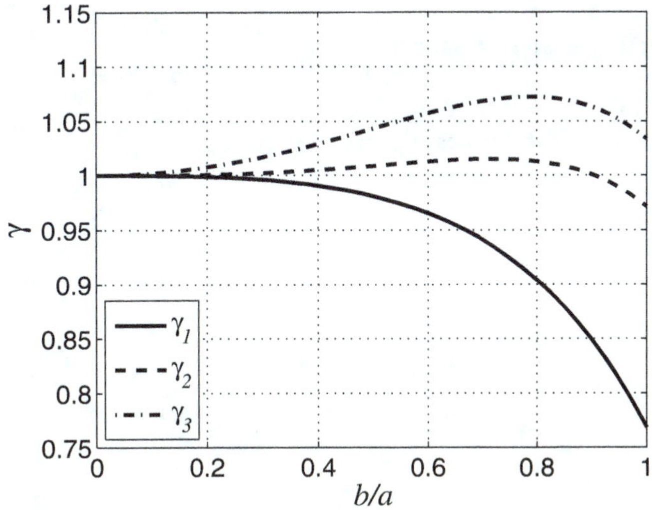

Fig. 10.5 The parameters γ_1 to γ_3 from (10.12) as functions of $\varepsilon = b/a$

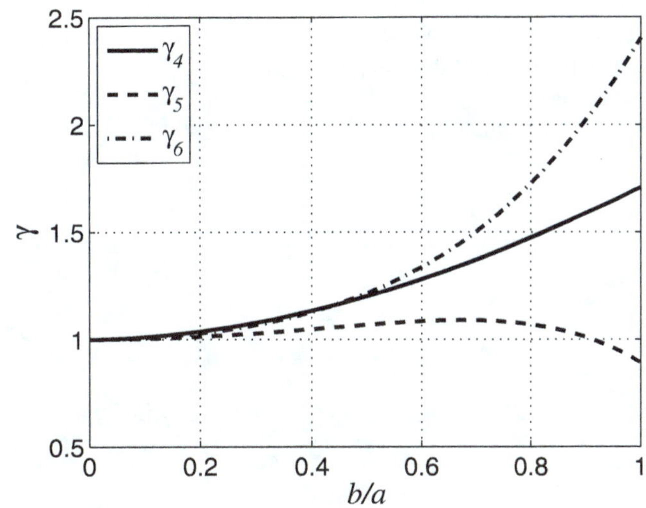

Fig. 10.6 The parameters γ_4 to γ_6 from (10.15) as functions of $\varepsilon = b/a$

the relationship between the global quantities:

$$F_N(d,\delta) = \frac{2\pi^2 E^* a h}{(1+\delta^2)} \left\{ \frac{\delta^2}{8} + \left[(1+\delta)\frac{d}{h} - \delta \right] \left[\frac{1}{2L} + \frac{\delta^2}{32}\left(2 - \frac{5}{L} + \frac{4}{L^2} \right) \right] \right\},$$

$$\frac{d}{h} = 1 - \frac{\left[\frac{1}{L} - \frac{3\delta}{2} + \frac{\delta^2}{16}\left(19 - \frac{13}{2L} + \frac{4}{L^2}\right) - \frac{\delta^3}{64}(9 - 6L) + O(\delta^4) \right]}{(1+\delta)\left[\frac{1}{L} - \frac{\delta}{2} + \frac{\delta^2}{16}\left(3 - \frac{13}{2L} + \frac{4}{L^2}\right) - \frac{\delta^3}{64}(3 - 2L) + O(\delta^4) \right]}, \tag{10.16}$$

with the expressions

$$\delta = \frac{1-\varepsilon}{1+\varepsilon},$$
$$L = \ln \frac{16}{\delta}. \tag{10.17}$$

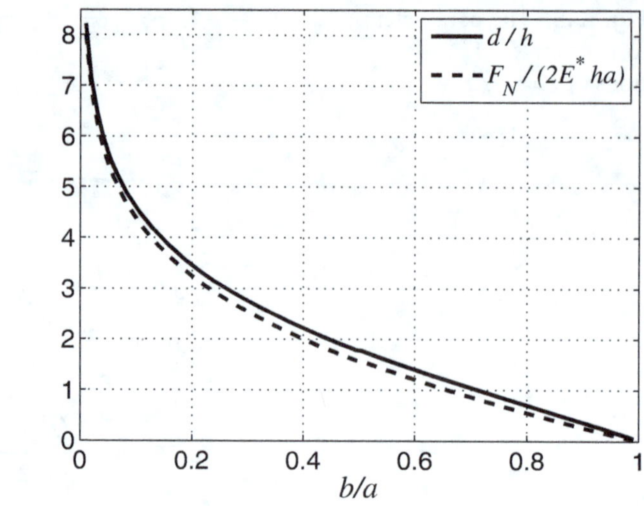

Fig. 10.7 Normalized indentation depth d/h and normal force $F_N/(2E^*ha)$ as functions of $\varepsilon = b/a$ for normal contact with a concave conical indenter

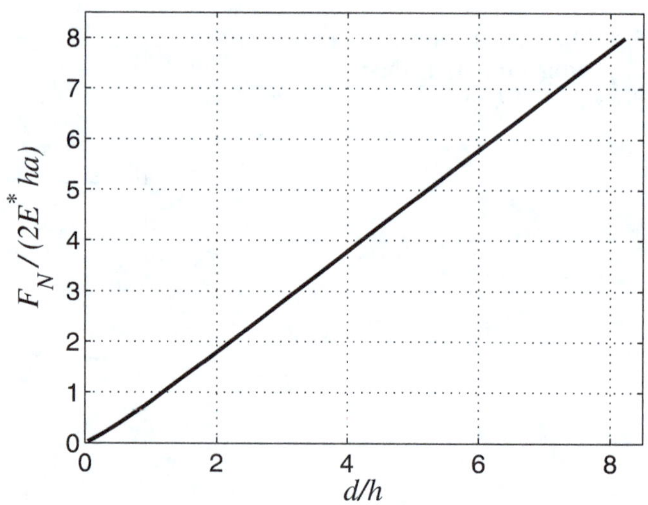

Fig. 10.8 Normalized normal force as a function of the normalized indentation depth for the normal contact with a concave conical indenter

A graphical representation of the relationships between the global contact quantities is given in Figs. 10.7 and 10.8. For $\varepsilon < 0.5$, (10.11) and (10.14) were used; for all others (10.16) was used. They achieve a near perfect match at the transition point $\varepsilon = 0.5$. It can also be seen that, quantitatively, the curves are nearly identical, thus the implicitly given relationship $F_N = F_N(d)$ can be written as:

$$F_N \approx 2E^* da, \tag{10.18}$$

which means that the relationship between the normal force and the indentation depth roughly corresponds to the one of the flat cylindrical punch! The case $b = 0$, i.e., complete contact, is impossible since it would imply an infinitely large normal force.

10.1.3 The Concave Paraboloid

The normal contact with a concave paraboloid was investigated by Barber (1976), Gladwell and Gupta (1979), and Shibuya (1980). Shibuya formulated the problem as a coupled system of infinitely many linear equations and then solved these numerically (see the beginning of the previous section). Gladwell and Gupta utilized an elegant approximation method which relied on a superposition of appropriate potentials (see Sect. 10.1.1). They compared their results to that of Barber's which was the source of the only completely analytical calculations.

A schematic diagram of the contact problem is displayed in Fig. 10.9. Let a flat punch of radius a have a parabolic depression of height h at its tip. Under small loads, the contact area will be ring-shaped with an inner radius b. As already demonstrated in Chap. 2 (see Sect. 2.5.15), fulfilment of the condition

$$d \geq d_c = 3h, \tag{10.19}$$

or alternatively

$$F_N \geq F_c = \frac{16}{3} E^* a h, \tag{10.20}$$

leads to the formation of complete contact ($b = 0$), in which case solution is much simpler. In the following section, we will present the solution of the incomplete contact. Barber (1976) examined, as in the case of the concave conical indenter, the limiting cases $\varepsilon = b/a \to 0$ (using the method of Collins 1963) and $\varepsilon \to 1$ (with the method of Grinberg and Kuritsyn 1962). Small values of ε return the following equations as the solution of the contact problem:

$$\begin{aligned} F_N(d, \varepsilon) &= 2E^* a \left[(d-h)\gamma_1(\varepsilon) + \frac{2h}{3}\gamma_2(\varepsilon) \right], \\ \frac{d}{h} &= 1 + 2\frac{\gamma_3(\varepsilon)}{\gamma_4(\varepsilon)}, \end{aligned} \tag{10.21}$$

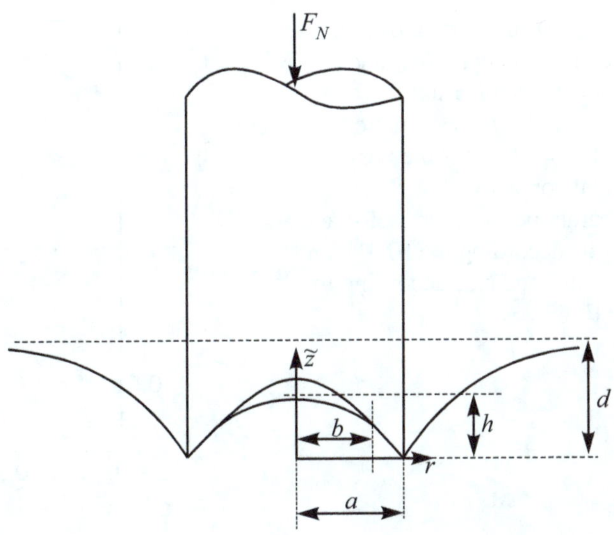

Fig. 10.9 Normal contact of a concave paraboloid with an elastic half-space

with the previously introduced shorthand symbols:

$$\begin{aligned}\gamma_1(\varepsilon) &= 1 - \frac{4\varepsilon^3}{3\pi^2} - \frac{8\varepsilon^5}{15\pi^2} - \frac{16\varepsilon^6}{27\pi^4} - \frac{92\varepsilon^7}{315\pi^2} - \frac{448\varepsilon^8}{675\pi^4} + O(\varepsilon^9),\\ \gamma_2(\varepsilon) &= 1 + \frac{4\varepsilon^3}{\pi^2} - \frac{8\varepsilon^5}{5\pi^2} - \frac{16\varepsilon^6}{9\pi^4} - \frac{4\varepsilon^7}{5\pi^2} - \frac{32\varepsilon^8}{225\pi^4} + O(\varepsilon^9),\\ \gamma_3(\varepsilon) &= 1 - \frac{2\varepsilon^2}{3} - \frac{8\varepsilon^3}{9\pi^2} - \frac{7\varepsilon^4}{15} - \frac{64\varepsilon^5}{675\pi^2} + O(\varepsilon^6),\\ \gamma_4(\varepsilon) &= 1 + \frac{2\varepsilon^2}{3} + \frac{8\varepsilon^3}{9\pi^2} + \frac{23\varepsilon^4}{45} + \frac{896\varepsilon^5}{675\pi^2} + O(\varepsilon^6)\end{aligned} \tag{10.22}$$

For values of $\varepsilon \to 1$ we obtain:

$$\begin{aligned}&F_N(d,\delta) =\\ &\frac{2\pi^2E^*ah}{(1+\delta^2)}\left\{\frac{\delta^2}{4}\left(1+\frac{1}{L}\right)\right.\\ &\qquad\left.+\left[(1+\delta)^2\left(\frac{d}{h}-1\right)+1\right]\left[\frac{1}{2L}+\frac{\delta^2}{32}\left(2-\frac{5}{L}+\frac{4}{L^2}\right)\right]\right\},\\ &\frac{d}{h} = 1 - \frac{\left[\frac{1}{L}-\frac{5\delta}{2}+\frac{\delta^2}{16}\left(51+\frac{3}{2L}+\frac{4}{L^2}\right)-\frac{\delta^3}{64}(55-10L)+O(\delta^4)\right]}{(1+\delta)^2\left[\frac{1}{L}-\frac{\delta}{2}+\frac{\delta^2}{16}\left(3-\frac{13}{2L}+\frac{4}{L^2}\right)-\frac{\delta^3}{64}(3-2L)+O(\delta^4)\right]},\end{aligned} \tag{10.23}$$

with the expressions:

$$\begin{aligned}\delta &= \frac{1-\varepsilon}{1+\varepsilon},\\ L &= \ln\frac{16}{\delta}.\end{aligned} \tag{10.24}$$

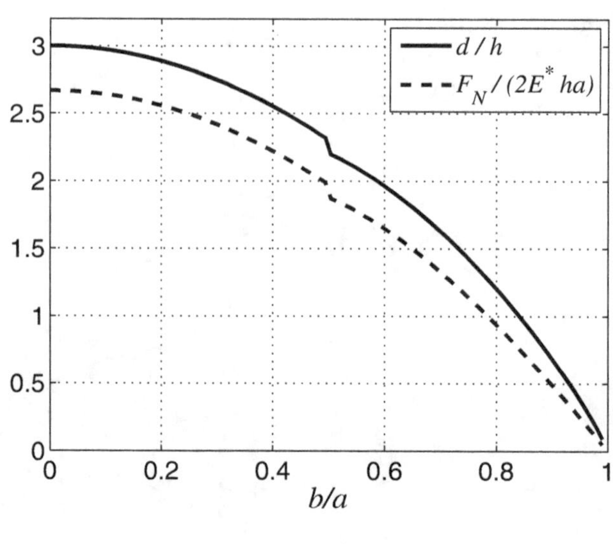

Fig. 10.10 Normalized indentation depth d/h and normalized normal force $F_N/(2E^*ha)$ as functions of $\varepsilon = b/a$ for the normal contact with a concave parabolic indenter. Left-hand side according to (10.21), and right-hand side according to (10.23)

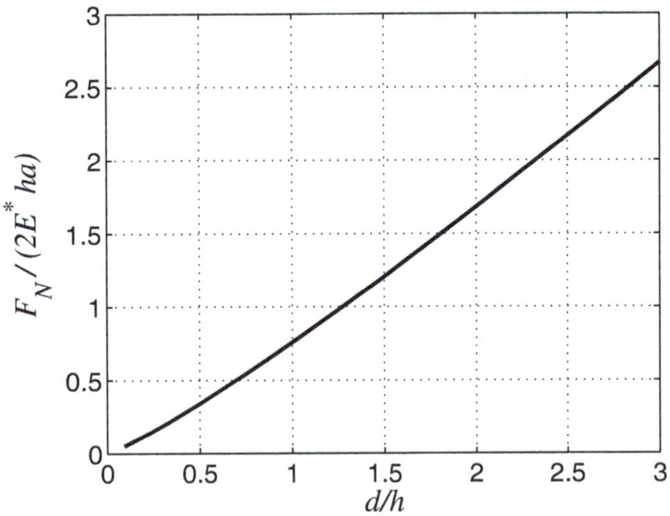

Fig. 10.11 Normalized normal force as a function of the normalized indentation depth for the normal contact of a concave parabolic indenter. Left-hand side according to (10.21), and right-hand side according to (10.23)

Note that (25) of Barber's publication (corresponding to (10.23)) contains a small printing error.

Figures 10.10 and 10.11 offer a visual representation of the relationships between the global contact quantities. For $\varepsilon < 0.5$, (10.21) were used; for all others, (10.23) were used. They achieve quite a good match at the transition point. The limits stated in this section for $b = 0$ are quite clear to see.

10.1.4 The Flat Cylindrical Punch with a Central Circular Recess

To fully understand the line of reasoning present in the literature, and due to the technical importance of this problem, this section is devoted to describing a contact problem that so far has completely evaded an analytical solution. It has even evaded one in the form of integrals, series expansions, or recursions as in the previous sections.

We consider the frictionless normal contact of a flat cylindrical punch of radius a, which features a central circular recess of radius b and depth e. Let the normal force F_N be sufficiently large to form another contact domain in the interior of the recess of radius c (for $c = 0$ it would result in the contact of the annular punch discussed in Sect. 10.1.1). A schematic diagram of the contact problem is given in Fig. 10.12.

This is a four-part boundary value problem, since in the zones $0 \leq r \leq c$ and $b \leq r \leq a$ the displacement are known. In the zones $c < r < b$ and $r > a$ the (vanishing) normal stress are the given values. The problem was considered by Barber (1983b), who reduced it to two coupled Fredholm equations using the method of complex potentials by Green and Collins (see, for example, the preceding publication by Barber 1983a). The equations are then solved numerically. The author provides graphical representations for the normal force and the indentation depth as functions of the contact radius.

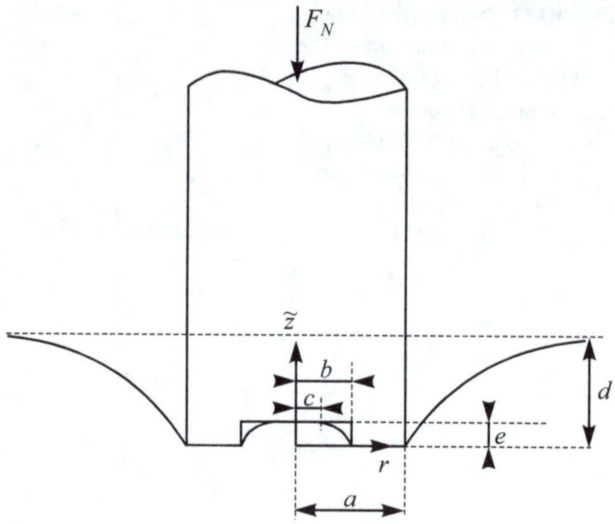

Fig. 10.12 Normal contact of a flat punch with a central recess and an elastic half-space

10.1.5 The Torus

We now consider the frictionless normal contact between a toroidal indenter and an elastic half-space. Let the torus of radii R_1 and R_2 be pressed by a normal force F_N into the elastic half-space to the depth d (see Fig. 10.13).

A ring-shaped contact area of the thickness $2h$ is formed. Argatov and Nazarov (1996, 1999) found the asymptotic solution of this problem for $h \ll R_2$. The relationships between F_N, d, and h are given by:

$$F_N \approx \frac{\pi^2 E^*}{2} \frac{R_2}{R_1} h^2,$$
$$d \approx \frac{h^2}{2R_1} \left(\ln \frac{R_2}{h} + 4 \ln 2 + \frac{1}{2} \right). \tag{10.25}$$

As a first-order approximation, the stress state under the torus (i.e., in the contact area) can be treated as two-dimensional and is given by:

$$\sigma_{zz}(r;h) \approx -\frac{E^* d}{2 \ln(16 R_2/h)} \frac{1}{\sqrt{h^2 - (r - R_2)^2}}. \tag{10.26}$$

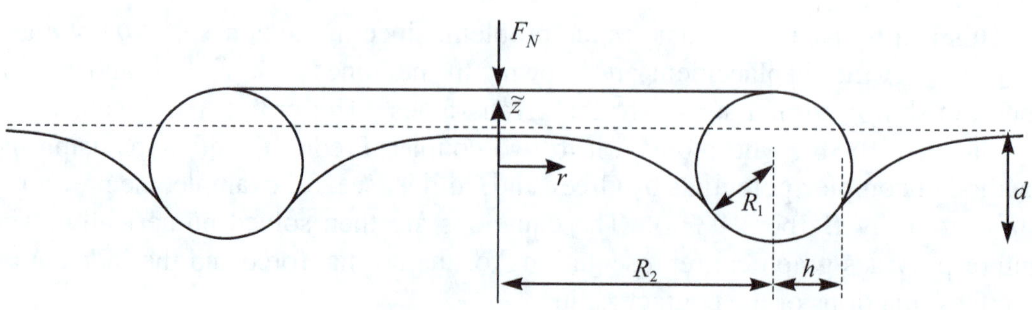

Fig. 10.13 Normal contact (cross-section) between a torus and an elastic half-space

10.1.6 The Toroidal Indenter with a Power-Law Profile

An analogy to the classic torus in the preceding section, the toroidal indenter with an arbitrary profile in the shape of a power-law could also be solved through an asymptotic solution, which was first published by Argatov et al. (2016). A cross-section of the contact problem is given in Fig. 10.14. Let the torus have the radius R and the resulting ring-shaped contact area the width $2h$. The indenter is pressed by a normal force F_N into the half-space to the depth d. The following results are only valid for thin rings; i.e., for $h \ll R$.

Consider an indenter having the rotationally symmetric profile

$$f(r) = c|r - R|^n, \quad n \in \mathbb{R}^+, \tag{10.27}$$

where c is a constant and n a positive real number. The relationship between h and d is given approximately by the expression:

$$d(h) \approx \frac{nch^n}{\kappa(n)}\left(\ln\frac{R}{h} + 4\ln 2 + \frac{1}{n}\right), \tag{10.28}$$

with the scaling factor from Chap. 2:

$$\kappa(n) := \sqrt{\pi}\frac{\Gamma(n/2+1)}{\Gamma\left[(n+1)/2\right]}. \tag{10.29}$$

Here, $\Gamma(\cdot)$ denotes the Gamma function

$$\Gamma(z) := \int_0^\infty t^{z-1}\exp(-t)\mathrm{d}t. \tag{10.30}$$

The normal force F_N is approximately equal to:

$$F_N(h) \approx \frac{n\pi^2 cRE^*}{\kappa(n)}h^n. \tag{10.31}$$

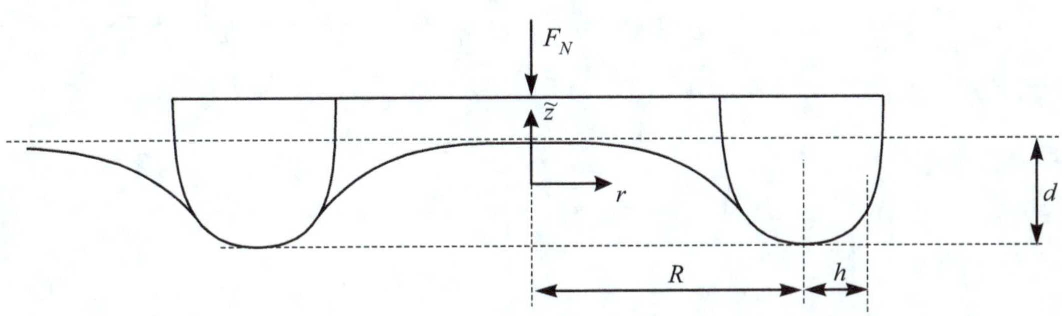

Fig. 10.14 Normal contact (cross-section) between a toroidal body and an elastic half-space

For $n = 2$ and $c = 1/(2R_1)$, we obtain the results of the preceding section. The stress distribution is, in first approximation, two-dimensional and (independent of n) identical to the term in (10.26).

10.1.7 The Indenter Which Generates a Constant Pressure on the Circular Ring

It poses no problem to determine the indenter shape which generates a constant pressure p_0 on the circular ring $b < r < a$. The pressure distribution is a superposition $p(r) = p_1(r) + p_2(r)$ of both distributions

$$\begin{aligned} p_1(r) &= p_0, \quad r < a \\ p_2(r) &= -p_0, \quad r < b. \end{aligned} \tag{10.32}$$

Chap. 2 (Sect. 2.5.6) provides the following displacements of the half-space under the influence of a constant pressure on a circular area of radius a:

$$\begin{aligned} w_1(r; a, p_0) &= \frac{4p_0 a}{\pi E^*} \mathrm{E}\left(\frac{r}{a}\right), \quad r \le a, \\ w_1(r; a, p_0) &= \frac{4p_0 r}{\pi E^*} \left[\mathrm{E}\left(\frac{a}{r}\right) - \left(1 - \frac{a^2}{r^2}\right) \mathrm{K}\left(\frac{a}{r}\right)\right], \quad r > a. \end{aligned} \tag{10.33}$$

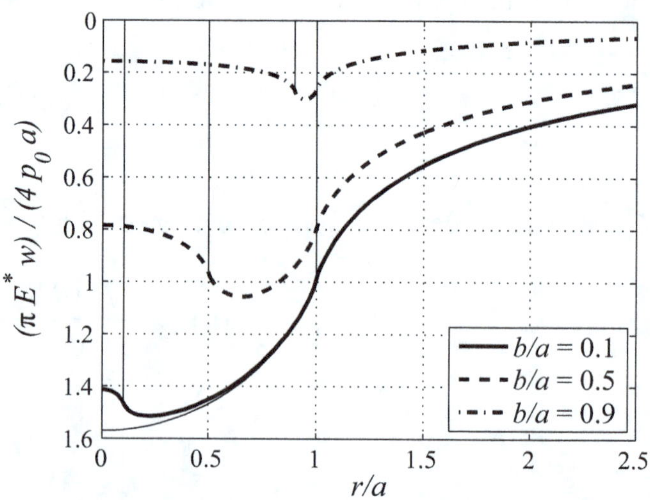

Fig. 10.15 Normalized curve of the half-space displacements under the influence of a constant pressure on a ring-shaped area $b < r < a$ for different ratios of the radii. The thin solid line corresponds to the case $b = 0$

It follows that the displacement under the influence of a constant pressure on a ring-shaped area is:

$$w(r;a,b,p_0) = \frac{4p_0}{\pi E^*}\begin{cases} a\mathrm{E}\left(\frac{r}{a}\right) - b\mathrm{E}\left(\frac{r}{b}\right), & r \le b \\ a\mathrm{E}\left(\frac{r}{a}\right) - r\left[\mathrm{E}\left(\frac{b}{r}\right) - \left(1-\frac{b^2}{r^2}\right)\mathrm{K}\left(\frac{b}{r}\right)\right], & b < r \le a \\ r\left[\mathrm{E}\left(\frac{a}{r}\right) - \left(1-\frac{a^2}{r^2}\right)\mathrm{K}\left(\frac{a}{r}\right) - \mathrm{E}\left(\frac{b}{r}\right) \right. & \\ \quad \left. + \left(1-\frac{b^2}{r^2}\right)\mathrm{K}\left(\frac{b}{r}\right)\right], & r > a. \end{cases} \tag{10.34}$$

These displacements are shown in a normalized representation in Fig. 10.15.

10.2 Frictionless Normal Contact with JKR Adhesion

Ring-shaped contacts also occur in systems which are sufficiently small or soft so that surface forces play a role. One example are hollow ring-shaped positioners in micro-assemblers; such structures are also typical for various biological systems.

As in the compact adhesive contacts, all adhesive models described in Chap. 3 can be applied to ring-shaped contacts, which was the theory of Johnson, Kendall, and Roberts (1971), Maugis (1992), and others. We can also distinguish whether a contact is frictionless or truly *adhesive*, i.e., sticking without tangential slip. We will restrict our consideration to frictionless normal contacts in the JKR approximation. This means that we will assume the range of the adhesive interactions to be small compared to all characteristic lengths of the system. In the JKR approximation, adhesion properties can be completely characterized by the effective surface energy (work of adhesion) per unit area, Δw. The notation Δw used in this chapter differs from the standard notation $\Delta\gamma$ used in all other parts of this book. This is to avoid confusion with the many γ used in the solutions of ring-shaped contacts. Let the half-space have, as always, the effective elasticity modulus E^*.

For contact areas in the form of very thin circular rings, Argatov et al. (2016) presented a very elegant solution for the toroidal indenter in the style of the MDR. It presents the relationships between the global quantities—normal force F_N, indentation depth d, and half of the contact width $h = (a-b)/2$ (with the contact radii b and a)—in the case of the adhesive normal contact. First, the variable

$$\delta = \frac{2h}{a+b} = \frac{a-b}{a+b} \tag{10.35}$$

is introduced, which (in the case of thin circular rings) represents a small parameter. If the adhesion is structured according to the framework of the JKR theory and the relationships of the indentation depth and the normal force of the non-adhesive contact, $d_{\text{n.a.}} = d_{\text{n.a.}}(\delta)$ and $F_{N,\text{n.a.}} = F_{N,\text{n.a.}}(\delta)$, are known, the relationships of the adhesive contact can be obtained through the addition of an appropriate "punch solution".

$$d(\delta) = d_{\text{n.a.}}(\delta) - 2\ln\left(\frac{16}{\delta}\right)\sqrt{\frac{(a+b)\Delta w}{\pi E^*}\delta},$$
$$F_N(\delta) = F_{N,\text{n.a.}}(\delta) - \sqrt{\pi^3(a+b)^3 E^* \Delta w \delta}. \tag{10.36}$$

The critical state, in which the contact loses its stability and detaches, results from the local maximums of these expressions as functions of δ.

In general though, the solutions are usually only available in the form of asymptotic expansions, which will be presented in the following sections.

10.2.1 The Hollow Flat Cylindrical Punch

Let a hollow flat cylindrical punch with the radii b (inner) and a (outer) be pressed into an elastic half-space. And let adhesion act in the ring-shaped contact area $b \le r \le a$ with the effective surface energy Δw. As in the previous sections, we introduce the ratio $\varepsilon = b/a$. The following solution was first presented by Willert et al. (2016).

In Sect. 10.1.1 the following relationship between the normal force $F_{N,\text{n.a.}}$ and the indentation depth d for the non-adhesive contact was derived:

$$F_{N,\text{n.a.}} = 2E^* d a \gamma(\varepsilon). \tag{10.37}$$

Here, the index "n.a." indicates the non-adhesive quantity. The complete expression of the function $\gamma_1(\varepsilon)$ can be referenced in (10.4), (10.9), and (10.10). It should be noted that the two analytical approximations result from series expansions at $\varepsilon = 0$, thus they are only accurate for small values of ε. In Fig. 10.2 a graphical representation of both approximations of the function is given. The elastic energy is then given by:

$$U_{\text{el}} = \int_0^d F_{N,\text{n.a.}}(\delta)\mathrm{d}\delta = E^* d^2 a \gamma(\varepsilon). \tag{10.38}$$

The surface energy, according to the JKR theory, is given by the expression:

$$U_{\text{adh}} = -\Delta w A_c = -\pi \Delta w a^2\left(1 - \frac{b^2}{a^2}\right) = -\pi \Delta w a^2 \gamma_2(\varepsilon), \tag{10.39}$$

with the contact area A_c. The total potential energy is then:

$$U_{\text{tot}} = E^* d^2 a \gamma(\varepsilon) - \pi \Delta w a^2 \gamma_2(\varepsilon). \tag{10.40}$$

The relation between the normal force and the indentation depth in the adhesive case is then given by:

$$F_N = \frac{\partial U_{\text{tot}}}{\partial d} = 2E^* da\gamma(\varepsilon) = F_{N,\text{n.a.}}. \tag{10.41}$$

The contact experiences a loss of its stability at:

$$d_c = -\frac{d_0}{\sqrt{\gamma(\varepsilon) - \varepsilon\gamma'(\varepsilon)}}, \tag{10.42}$$

(the prime denotes the derivative with respect to the argument). The corresponding critical normal force, which is also called the adhesive force, is given by:

$$F_c = -\frac{F_0\gamma(\varepsilon)}{\sqrt{\gamma(\varepsilon) - \varepsilon\gamma'(\varepsilon)}}. \tag{10.43}$$

Here the values of the full cylinder, according to Kendall (1971), were used:

$$d_0 := \sqrt{\frac{2\pi a \Delta w}{E^*}},\ F_0 := \sqrt{8\pi a^3 E^* \Delta w}. \tag{10.44}$$

Using the results of Collins for $\gamma(\varepsilon)$ (see (10.4)) yields the expression in the denominator:

$$\frac{\varepsilon}{\gamma(\varepsilon)}\frac{\mathrm{d}\gamma}{\mathrm{d}\varepsilon} = -\frac{\frac{4}{\pi^2}\varepsilon^3 + \frac{8}{3\pi^2}\varepsilon^5 + \frac{32}{9\pi^4}\varepsilon^6 + O(\varepsilon^7)}{1 - \frac{4}{3\pi^2}\varepsilon^3 - \frac{8}{15\pi^2}\varepsilon^5 - \frac{16}{27\pi^4}\varepsilon^6 + O(\varepsilon^7)}. \tag{10.45}$$

The Case $\varepsilon \to 1$

As noted in this section, Collins' solution is only usable for small values of ε. For the other limiting case $\varepsilon \to 1$, i.e., with $\delta \to 0$ (δ from (10.35)), the solution by Argatov et al. (2016) can be used.

The relation $F_N = F_{N,\text{n.a.}}$ remains valid of course. Independent of whether a force-controlled or displacement-controlled trial is being conducted, the contact loses its stability at:

$$d = d_c = -2\ln\left(\frac{16}{\delta}\right)\sqrt{\frac{(a+b)\Delta w}{\pi E^*}\delta},$$
$$F_N = F_c = -\sqrt{\pi^3(a+b)^3 E^* \Delta w \delta}. \tag{10.46}$$

The curves of the critical indentation depth and the adhesive force, both normalized to the values of the solid cylinder, are depicted in Fig. 10.16. For $\varepsilon > 0.85$, the results of (10.46) were used; for all others, (10.42) to (10.45) was used. The values at the transition point are clearly in good agreement. Additionally, the approximate solutions are displayed for which the numerical expression (10.10) was inserted into (10.42) and (10.43).

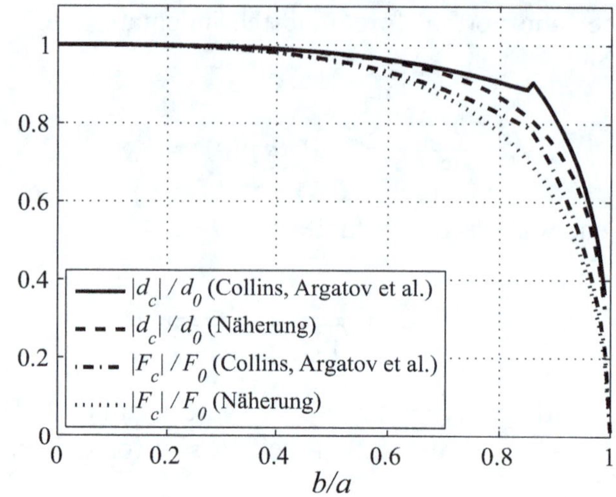

Fig. 10.16 Dependency of the critical indentation depth and adhesive force for the adhesive normal contact with a flat annular punch on $\varepsilon = b/a$. The curves are normalized to the values of the full cylinder ($b = 0$). For $\varepsilon < 0.85$, (10.42), (10.43), and (10.45) were used, and for $\varepsilon > 0.85$, the expressions from (10.46) were used. The smooth transition is clearly visible. The approximate solutions refer to (10.42) and (10.43) with the approximation (10.10)

10.2.2 The Toroidal Indenter with a Power-Law Profile

We consider the adhesive normal contact between a toroidal indenter and an elastic half-space. The indenter has the profile:

$$f(r) = c|r - R|^n, \quad n \in \mathbb{R}^+, \tag{10.47}$$

with a constant c, the radius of the torus R, and a positive real number n. Let the annular contact have the width $2h$ and let the value

$$\delta = \frac{h}{R} \tag{10.48}$$

be small. In this case we can use (10.36). Section 10.1.6 documents the derivation of the following equation for the non-adhesive relationships between the normal force F_N, indentation depth d, and the normalized contact width δ:

$$\begin{aligned} d_{\text{n.a.}}(\delta) &\approx \frac{nc\delta^n R^n}{\kappa(n)}\left[\ln\left(\frac{16}{\delta}\right) + \frac{1}{n}\right], \\ F_{N,\text{n.a.}}(\delta) &\approx \frac{n\pi^2 cE^*}{\kappa(n)} R^{1+n}\delta^n. \end{aligned} \tag{10.49}$$

Here, the index "n.a." indicates the non-adhesive quantities. The function $\kappa(n)$ can be gathered from (10.29). Using (10.36), the adhesive relationships are then given by:

$$d(\delta) \approx \frac{nc\delta^n R^n}{\kappa(n)} \left[\ln\left(\frac{16}{\delta}\right) + \frac{1}{n} \right] - 2\ln\left(\frac{16}{\delta}\right) \sqrt{\frac{2R\Delta w}{\pi E^*}\delta},$$
$$F_N(\delta) \approx \frac{n\pi^2 c E^*}{\kappa(n)} R^{1+n}\delta^n - \sqrt{8\pi^3 R^3 E^* \Delta w \delta}. \tag{10.50}$$

In a force-controlled trial, the contact experiences a loss of stability at

$$\delta_c^{\,2n-1} = \frac{2}{\pi}\left(\frac{\kappa(n)}{cn^2}\right)^2 R^{1-2n}\frac{\Delta w}{E^*}. \tag{10.51}$$

The corresponding adhesive force has the value

$$F_c = -R\left[\frac{(E^*)^{n-1}(\Delta w)^n \kappa(n)\pi^{3n-2}2^n}{n^{2n+1}c}\right]^{\frac{1}{2n-1}} (2n-1). \tag{10.52}$$

It is obvious that these expressions can be valid only for $n > 0,5$. For $n \to \infty$ we obtain the results of the hollow flat cylindrical punch from (10.46). Other special cases which we will briefly examine are:

The V-Shaped Toroidal Indenter with $n = 1$

In case of a V-shaped toriidal indenter it is $c = \tan\theta$, with the slope angle θ of the V-profile, $n = 1$ and $\kappa(n = 1) = \pi/2$. This yields:

$$\delta_c = \frac{\pi}{2\tan\theta}\frac{\Delta w}{RE^*},$$
$$F_c = -R\frac{\Delta w \pi^2}{\tan\theta}. \tag{10.53}$$

Thus the adhesive force is independent of the elasticity properties of the half-space.

The Classic Torus $n = 2$

With $c = 1/(2R_1)$, the radius of the torus R_1, $n = 2$ and $\kappa(n = 2) = 2$, we obtain:

$$\delta_c = \sqrt[3]{\frac{1}{\pi R_1}\frac{\Delta w}{R^3 E^*}},$$
$$F_c = -3R\sqrt[3]{\left(\frac{R_1 E^*(\Delta w)^2\pi^4}{2}\right)}. \tag{10.54}$$

10.3 Torsional Contact

The purely torsional contact problem between a rigid indenter and an elastic half-space is also referred to as the Reissner–Sagoci problem. The boundary conditions at the surface of the half-space at $z = 0$, in the axisymmetric case for a ring-shaped contact area with the radii b and $a > b$, are as follows:

$$\begin{aligned} u_{\varphi}(r, z=0) &= f(r), \quad b \le r \le a, \\ \sigma_{\varphi z}(r, z=0) &= 0, \quad r < b, r > a, \end{aligned} \tag{10.55}$$

with the rotational displacement and the tangential stresses. This consideration only deals with the pure torsion problem without slip, which means that all other stresses and displacements vanish.

10.3.1 The Hollow Flat Cylindrical Punch

In drilling applications, sometimes hollow or concave drill bits are used. This approximately corresponds to the contact problem described in this section. For the contact of a hollow flat cylindrical punch, the function $f(r)$ of the torsional displacement is given by:

$$f(r) = \varphi r, \tag{10.56}$$

with the twisting angle φ of the punch around its axis of symmetry. The contact problem was studied by Borodachev and Borodacheva (1966b), Shibuya (1976), and Gladwell and Gupta (1979).

Borodachev and Borodacheva (1966b) applied the same method which Collins (1962, 1963) utilized for the solution of the normal contact problem of a flat annular punch. They numerically solved the Fredholm equation that arose over the course of the solution process and obtained the following relationship between the torsional moment M_z and the torsion angle φ:

$$M_z = \frac{16}{3} G a^3 \varphi \gamma(\varepsilon), \tag{10.57}$$

once again introducing the shorthand notation $\varepsilon = b/a$. G is the shear modulus of the elastic half-space. A series expansion of the function $\gamma(\varepsilon)$ is only possible for small values of ε. The authors presented the approximate solution:

$$\gamma = 1 + 0.0094\varepsilon^4 - 0.1189\varepsilon^5 - 0.0792\varepsilon^7 - 0.0094\varepsilon^8 - 0.0645\varepsilon^9 + O(\varepsilon^{10}). \tag{10.58}$$

For the stresses in the contact area they provided the expression:

$$\begin{aligned} \sigma_{\varphi z}(r) \approx -\frac{4G\theta}{\pi} \Bigg\{ & \frac{r}{\sqrt{a^2 - r^2}} \left[1 + \frac{\varepsilon^5}{\pi^2} \left(\alpha_0 + \alpha_2 \frac{r^2}{a^2} + \alpha_4 \frac{r^4}{a^4} + \alpha_6 \frac{r^6}{a^6} \right) \right] \\ & + \frac{\varepsilon b^4}{\pi r^3 \sqrt{r^2 - b^2}} \left(\beta_0 + \beta_2 \frac{b^2}{r^2} + \beta_4 \frac{b^4}{r^4} + \beta_6 \frac{b^6}{r^6} \right) \Bigg\}, \end{aligned} \tag{10.59}$$

with the shorthand notation:

$$\alpha_0 = 0.0839 + 0.0913\varepsilon^2 + 0.0929\varepsilon^4 + 0.0007\varepsilon^5 + 0.1469\varepsilon^6,$$
$$\alpha_2 = 0.0531 + 0.0698\varepsilon^2 + 0.101\varepsilon^4,$$
$$\alpha_4 = 0.0434 + 0.0844\varepsilon^2, \quad \alpha_6 = 0.0579,$$
$$\beta_0 = 0.2896 + 0.0917\varepsilon^2 + 0.0562\varepsilon^4 + 0.0025\varepsilon^5 + 0.0625\varepsilon^6,$$
$$\beta_2 = 0.2083 + 0.075\varepsilon^2 + 0.05\varepsilon^4,$$
$$\beta_4 = 0.15 + 0.1\varepsilon^2, \quad \beta_6 = 0.2. \tag{10.60}$$

Gladwell and Gupta (1979) solved the contact problem in a very elegant manner through a combination of appropriate potentials. They gave the sought stresses and displacements inside and outside the contact area as:

$$u_\varphi(r) = \varphi a \begin{cases} \dfrac{r}{a} - 2b_1 \dfrac{r}{b}\sqrt{1 - \dfrac{r^2}{b^2}}, & 0 < r < b \\ \dfrac{r}{a}, & b \leq r \leq a \\ \dfrac{2r}{\pi a}\left[\arcsin\left(\dfrac{a}{r}\right) - \dfrac{a}{r}\sqrt{1 - \dfrac{a^2}{r^2}}\right] & \\ \quad - 6a_1 \dfrac{a^2}{r}\sqrt{1 - \dfrac{a^2}{r^2}}, & r > a, \end{cases},$$

$$\sigma_{\varphi z}(r) = G\varphi\left[\frac{4r}{\pi a} R\left(\frac{r}{a}\right) - \frac{9b_1 r}{\varepsilon b} U\left(\frac{b}{r}\right) - 9a_1 \frac{a^4}{r^4} U\left(\frac{r}{a}\right)\right],$$
$$b \leq r \leq a, \tag{10.61}$$

with the functions U and R and the shorthand notations defined as follows:

$$U(x) = \begin{cases} \arcsin x - \dfrac{x}{\sqrt{1 - x^2}}\left(1 - \dfrac{x^2}{3}\right), & x < 1 \\ \dfrac{\pi}{2}, & x \geq 1, \end{cases}$$

$$R(x) = \begin{cases} \dfrac{1}{\sqrt{1 - x^2}}, & x < 1 \\ 0, & x \geq 1, \end{cases}$$

$$b_1 = \frac{600\varepsilon^2}{675\pi^2 - 48\varepsilon^5}, \quad a_1 = \frac{4\varepsilon^3 b_1}{15\pi} = \frac{160\varepsilon^5}{675\pi^3 - 48\pi\varepsilon^5}, \tag{10.62}$$

which is in very good agreement with the results of Shibuya (1976), who converted this problem into a coupled system of infinitely many linear equations and solved these numerically.

If we then determine the torsional moment via integration of the stress distribution (this step was not performed in the publication by Gladwell and Gupta) and

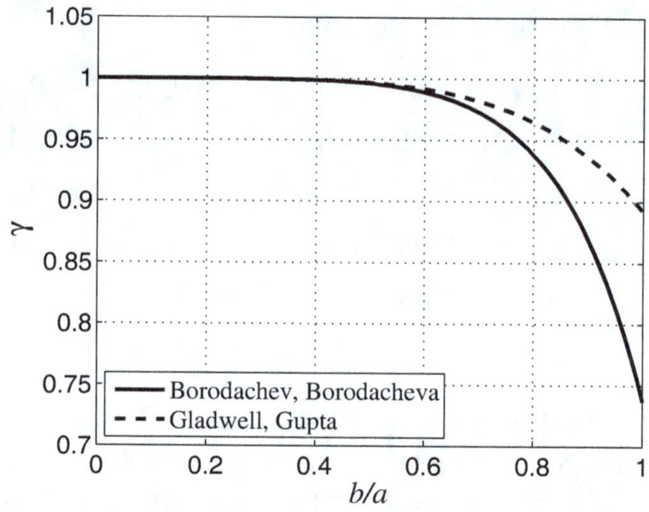

Fig. 10.17 Normalized torsional moment $\gamma = 3M_z/16G\varphi a^3$ as a function of the ratio of radii $\varepsilon = b/a$ according to (10.58) (Borodachev and Borodacheva 1966b) and (10.63) (Gladwell and Gupta 1979)

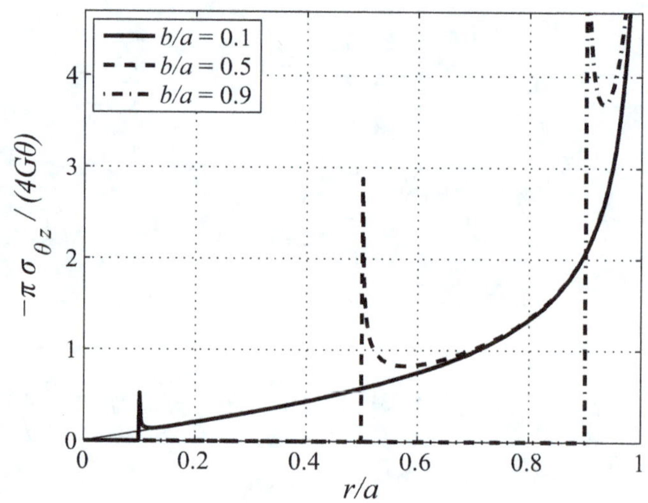

Fig. 10.18 Normalized torsional stresses according to Borodachev and Borodacheva (1966b) as a function of the radial coordinate for different values of b/a. The thin solid line represents the solution of the full cylinder

expand the expression for γ in Taylor series to powers of ε we obtain:

$$\gamma = 1 - \frac{16}{15\pi^2}\varepsilon^5 + O(\varepsilon^6). \tag{10.63}$$

Figure 10.17 shows the solutions for γ, according to Borodachev and Borodacheva (1966b) and Gladwell and Gupta (1979). For values of $\varepsilon < 0.6$ they are in very good agreement. We must keep in mind that both solutions are intended only for small values of ε. Figures 10.18 and 10.19 display the solution for the tangential stress distribution on the basis of (10.59) and (10.61). The stress singularities at the sharp edges of the indenter and the convergence with the solution for the flat cylindrical punch from Chap. 5 (see Sect. 5.1.1) are easily recognized.

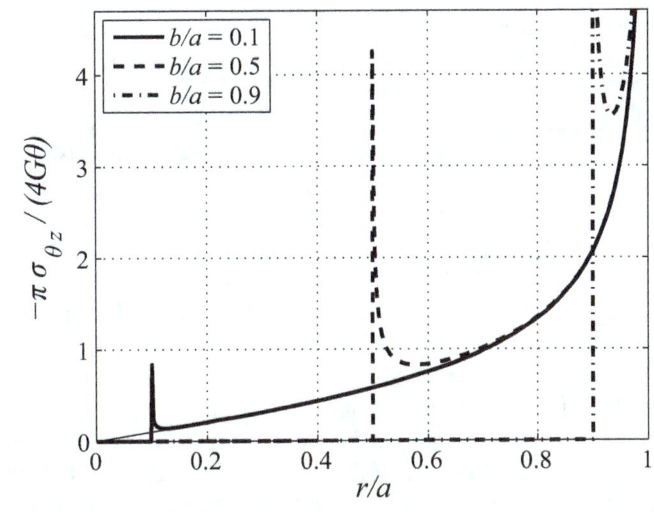

Fig. 10.19 Normalized torsional stresses according to Gladwell and Gupta (1979) as a function of the radial coordinate for different values of b/a. The thin solid line represents the solution of the full cylinder

References

Antipov, Y.A.: Analytic solution of mixed problems of mathematical physics with a change of boundary conditions over a ring. Mech. Solids **24**(3), 49–56 (1989)

Argatov, I.I., Nazarov, S.A.: The pressure of a narrow ring-shaped punch on an elastic half-space. J. Appl. Math. Mech. **60**(5), 799–812 (1996)

Argatov, I.I., Nazarov, S.A.: The contact problem for a narrow annular punch. Increasing contact zone. In: Studies in elasticity and plasticity, pp. 15–26. University Press, St. Petersburg (1999)

Argatov, I.I., Li, Q., Pohrt, R., Popov, V.L.: Johnson–Kendall–Roberts adhesive contact for a toroidal indenter. Proc. R. Soc. London Ser. A **472**, 20160218 (2016). https://doi.org/10.1098/rspa.20160218

Barber, J.R.: Determining the contact area in elastic indentation problems. J. Strain Anal. **9**(4), 230–232 (1974)

Barber, J.R.: Indentation of the semi-infinite elastic solid by a concave rigid punch. J. Elast. **6**(2), 149–159 (1976)

Barber, J.R.: The solution of elasticity problems for the half-space by the method of Green and Collins. Appl. Sci. Res. **40**(2), 135–157 (1983a)

Barber, J.R.: A four-part boundary value problem in elasticity: indentation by a discontinuously concave punch. Appl. Sci. Res. **40**(2), 159–167 (1983b)

Borodachev, N.M.: On the nature of the contact stress singularities under an annular Stamp. J. Appl. Math. Mech **40**(2), 347–352 (1976)

Borodachev, N.M., Borodacheva, F.N.: Penetration of an annular stamp into an elastic half-space. Mech. Solids **1**(4), 101–103 (1966a)

Borodachev, N.M., Borodacheva, F.N.: Twisting of an elastic half-space by the rotation of a ring-shaped punch. Mech. Solids **1**(1), 63–66 (1966b)

Collins, W.D.: On some triple series equations and their applications. Arch. Ration. Mech. Anal. **11**(1), 122–137 (1962)

Collins, W.D.: On the solution of some Axi-symmetric boundary value problems by means of integral equations. VIII. Potential problems for a circular annulus. Proc. Edinb. Math. Soc. Ser. 2 **13**(3), 235–246 (1963)

Gladwell, G.M.L., Gupta, O.P.: On the approximate solution of elastic contact problems for a circular annulus. J. Elast. **9**(4), 335–348 (1979)

Grinberg, G.A., Kuritsyn, V.N.: Diffraction of a plane electromagnetic wave by an ideally conducting plane ring and the electrostatic problem for such a ring. Soviet Phys. Tech. Phys. **6**, 743–749 (1962)

Gubenko, V.S., Mossakovskii, V.I.: Pressure of an axially-symmetric circular die on an elastic half-space. J. Appl. Math. Mech **24**(2), 477–486 (1960)

Johnson, K.L., Kendall, K., Roberts, A.D.: Surface Energy and the Contact of Elastic Solids. Proc. Royal Soc. Lond. Ser. A **324**, 301–313 (1971)

Kendall, K.: The adhesion and surface energy of elastic solids. J. Phys. D Appl. Phys. **4**(8), 1186–1195 (1971)

Maugis, D.: Adhesion of spheres: The JKR-DMT-transition using a Dugdale model. J. Colloid Interface Sci. **150**(1), 243–269 (1992)

Roitman, A.B., Shishkanova, S.F.: The solution of the annular punch problem with the aid of Recursion relations. Soviet Appl. Mech. **9**(7), 725–729 (1973)

Shibuya, T.: A mixed boundary value problem of an elastic half-space under torsion by a flat annular rigid stamp. Bull. JSME **19**(129), 233–238 (1976)

Shibuya, T.: Indentation of an elastic half-space by a concave rigid punch. Z. Angew. Math. Mech. **60**(9), 421–427 (1980)

Shibuya, T., Koizumi, T., Nakahara, I.: An elastic contact problem for a half-space indented by a flat annular rigid stamp. Int. J. Eng. Sci. **12**(9), 759–771 (1974)

Willert, E., Li, Q., Popov, V.L.: The JKR-adhesive normal contact problem of Axi-symmetric rigid punches with a flat annular shape or concave profiles. Facta Univ. Seri. Mech. Eng. **14**(3), 281–292 (2016)

11 Appendix

A significant number of contact problems concerning rotationally symmetric bodies can be solved via reduction to the non-adhesive frictionless normal contact problem. The problem of an adhesive normal contact is thereby reduced entirely to the solution of a non-adhesive normal contact. Likewise, the tangential contact problem in the Cattaneo–Mindlin approximation can be reduced to the frictionless normal contact problem using the principle of superposition by Cattaneo–Mindlin–Jäger–Ciavarella. And finally, contacts of viscoelastic bodies can be reduced to the corresponding elastic problem, i.e., the non-adhesive elastic normal contact, using the method by Lee und Radok.

This problem of the elastic, frictionless, non-adhesive normal contact of rotationally symmetric bodies can in turn be reduced via the MDR to the problem of the indentation by a rigid cylindrical indenter. If the solution for the contact between a rigid cylindrical punch of an arbitrary radius and a given elastic medium is known, then the solution for an arbitrarily shaped punch consists of a simple superposition of indentations by punches of varying radii.

The problem of an indentation by a flat cylindrical punch cannot be further reduced and thus requires an actual formal solution. In this appendix, we perform this basic step and describe the "construction of the building" of the MDR. In doing this, we follow the appendix of the book by Popov (2015). In Sect. 11.8 of this chapter an overview of the special functions used in this book is provided. In (the closing) Sect. 11.9, the historical solution of the contact problem for an arbitrary axisymmetric profile with a compact contact area according to Föppl (1941) and Schubert (1942) is given.

V.L. Popov, M. Heß, E. Willert, *Handbook of Contact Mechanics*,
https://doi.org/10.1007/978-3-662-58709-6_11

11.1 The Flat Punch Solution for Homogeneous Materials

This section demonstrates that the pressure distribution at the surface of an elastic half-space

$$p(r;a) = p_0\left(1-\frac{r^2}{a^2}\right)^{-1/2}, \quad r^2 = x^2+y^2,\ r<a \tag{11.1}$$

leads to a constant normal displacement of the circle $r < a$ and, therefore, corresponds to the indentation by a rigid circular punch. The basis for the calculation is the general relationship between an arbitrary pressure distribution $p(x', y')$ and the resulting surface deformation w given by the fundamental solution of Boussinesq:

$$w(x,y) = \frac{1}{\pi E^*}\iint p(x',y')\frac{\mathrm{d}x'\mathrm{d}y'}{r}, \quad r = \sqrt{(x-x')^2+(y-y')^2}, \tag{11.2}$$

with the effective elasticity modulus E^*. The integral (11.2) is calculated using the system of coordinates and notations shown in Fig. 11.1. Due to rotational symmetry of the stress distribution, the vertical displacement of a surface point is dependent solely on the distance r of this point from the origin O. Therefore, it is sufficient to determine the displacements of the points on the x-axis. In the following, we will calculate the vertical deformation at point A. For this, we must determine the displacement at point A caused by the stress in the "varying" point B, and subsequently integrate over all possible positions of point B within the load zone. Due to rotational symmetry, the stress at point B also depends solely on the distance t of the point to the origin O. For this distance we obtain $t^2 = r^2 + s^2 + 2rs\cos\varphi$. Therefore, the pressure distribution is:

$$\begin{aligned} p(s,\varphi) &= p_0\left(1-\frac{r^2+s^2+2rs\cos\varphi}{a^2}\right)^{-1/2} \\ &= p_0 a(a^2-r^2-s^2-2rs\cos\varphi)^{-1/2} \\ &= p_0 a(\alpha^2-2\beta s-s^2)^{-1/2}, \end{aligned} \tag{11.3}$$

where we have introduced $\alpha^2 = a^2 - r^2$ and $\beta = r\cos\varphi$.

For the z-component of the displacement of a point within the load zone we obtain:

$$w = \frac{1}{\pi E^*}p_0 a \int_0^{2\pi}\left(\int_0^{s_1}(\alpha^2-2\beta s-s^2)^{-1/2}\mathrm{d}s\right)\mathrm{d}\varphi. \tag{11.4}$$

Here, s_1 is the positive root of the equation $\alpha^2 - 2\beta s - s^2 = 0$. The integral over d$s$ is calculated to:

$$\int_0^{s_1}(\alpha^2-2\beta s-s^2)^{-1/2}\mathrm{d}s = \frac{\pi}{2} - \arctan(\beta/\alpha). \tag{11.5}$$

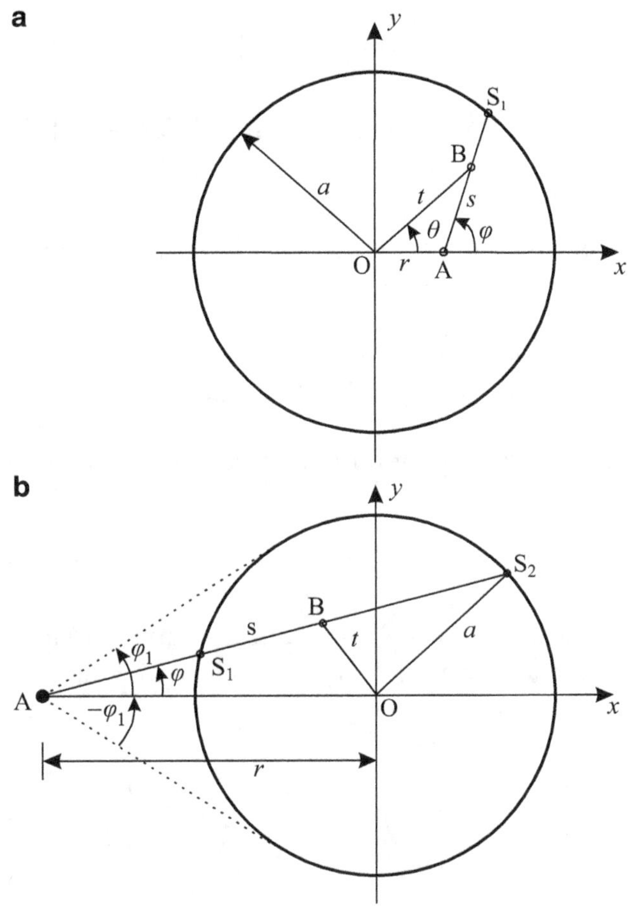

Fig. 11.1 Calculation of the vertical deformation caused by normal stresses acting on a circular area: (a) at a location within the load zone, (b) at a location outside of the load zone

It is apparent that $\arctan(\beta(\varphi)/\alpha) = -\arctan(\beta(\varphi + \pi)/\alpha)$. Thus, by integrating over φ, the term with arctan is eliminated. Therefore,

$$w(r;a) = \frac{1}{\pi E^*} p_0 a \int_0^{2\pi} \frac{\pi}{2} d\varphi = \frac{\pi p_0 a}{E^*} =: d = \text{const}, \quad r < a, \tag{11.6}$$

where the indentation depth d is introduced.

We now consider a point A outside the load zone (Fig. 11.1b). In this case:

$$p(s,\varphi) = p_0 a (\alpha^2 + 2\beta s - s^2)^{-1/2}. \tag{11.7}$$

The displacement is then given by the equation:

$$w = \frac{1}{\pi E^*} p_0 a \int_{-\varphi_1}^{\varphi_1} \left(\int_{s_1}^{s_2} (\alpha^2 + 2\beta s - s^2)^{-1/2} \, \mathrm{d}s \right) \mathrm{d}\varphi \tag{11.8}$$

where s_1 and s_2 are the roots of the equation:

$$\alpha^2 + 2\beta s - s^2 = 0. \tag{11.9}$$

Accordingly, it follows that:

$$\int_{s_1}^{s_2} \left(\alpha^2 + 2\beta s - s^2\right)^{-1/2} \mathrm{d}s = \pi. \tag{11.10}$$

The remaining integration in (11.8) now results trivially in $w = \frac{2}{E^*} p_0 a \varphi_1$ or, taking into account the obvious geometric relation from Fig. 11.1b, $\varphi_1 = \arcsin(a/r)$,

$$w(r;a) = \frac{2}{E^*} p_0 a \cdot \arcsin(a/r), \quad r \geq a, \tag{11.11}$$

which, accounting for (11.6), can also be written as:

$$w(r;a) = \frac{2}{\pi} d \cdot \arcsin(a/r), \quad r \geq a. \tag{11.12}$$

From the result (11.6) it follows directly that the assumed pressure distribution is generated by an indentation by a rigid cylindrical punch. The total force acting on the load zone equals:

$$F_N = \int_0^a p_0 \left(1 - \frac{r^2}{a^2}\right)^{-1/2} 2\pi r \mathrm{d}r = 2\pi p_0 a^2. \tag{11.13}$$

The contact stiffness is defined as the relation of force to displacement:

$$k_z = 2aE^*. \tag{11.14}$$

The pressure distribution (11.1) can also be represented in the following form with the consideration of (11.6):

$$p(r) = \frac{1}{\pi} \frac{E^* d}{\sqrt{a^2 - r^2}}. \tag{11.15}$$

11.2 Normal Contact of Axisymmetric Profiles with a Compact Contact Area

In this section we will present the solution for the normal contact problem for axisymmetric profiles with a compact contact area in its general form. For this, we will use the solution of the contact problem of a rigid flat cylindrical punch obtained in the preceding section. Simultaneously, a derivation of the fundamental equations of the MDR will be performed.

We will now consider the contact between a rigid indenter of the shape $\tilde{z} = f(r)$ and an elastic half-space characterized by the effective elasticity modulus E^*. Let the indentation depth under the effect of the normal force F_N be d and the contact radius a. For a given profile shape, any of these three quantities uniquely determines the other two. The indentation depth is a unique function of the contact radius, which is denoted by:

$$d = g(a). \tag{11.16}$$

Let us examine the process of the indentation from the first touch to the final indentation depth d, denoting the values of the force, the indentation depth, and the contact radius during the indentation by $\tilde{F}_N$, $\tilde{d}$ and $\tilde{a}$, respectively. The process can then be viewed as a change in the indentation depth from $\tilde{d} = 0$ to $\tilde{d} = d$, with the contact radius changing from $\tilde{a} = 0$ to $\tilde{a} = a$, and the contact force from $\tilde{F}_N = 0$ to $\tilde{F}_N = F_N$. The normal force at the end of the process can be written in the following form:

$$F_N = \int_0^{F_N} \mathrm{d}\tilde{F}_N = \int_0^a \frac{\mathrm{d}\tilde{F}_N}{\mathrm{d}\tilde{d}} \frac{\mathrm{d}\tilde{d}}{\mathrm{d}\tilde{a}} \mathrm{d}\tilde{a}. \tag{11.17}$$

Taking into account that the differential stiffness of a zone of radius $\tilde{a}$ is given by:

$$\frac{\mathrm{d}\tilde{F}_N}{\mathrm{d}\tilde{d}} = 2E^*\tilde{a} \tag{11.18}$$

(see (11.14)), and using the notation (11.16), we obtain:

$$F_N = 2E^* \int_0^a \tilde{a} \frac{\mathrm{d}g(\tilde{a})}{\mathrm{d}\tilde{a}} \mathrm{d}\tilde{a}. \tag{11.19}$$

Integration by parts now results in:

$$\begin{aligned} F_N &= 2E^* \left[a \cdot g(a) - \int_0^a g(\tilde{a}) \mathrm{d}\tilde{a} \right] = 2E^* \left[\int_0^a [g(a) - g(\tilde{a})] \mathrm{d}\tilde{a} \right] \\ &= 2E^* \left[\int_0^a [d - g(\tilde{a})] \mathrm{d}\tilde{a} \right]. \end{aligned} \tag{11.20}$$

We now turn our attention to calculating the pressure distribution in the contact area. An infinitesimal indentation of an area of radius $\tilde{a}$ generates the following contribution to the pressure distribution (see (11.15)):

$$\mathrm{d}p(r) = \frac{1}{\pi} \frac{E^*}{\sqrt{\tilde{a}^2 - r^2}} \mathrm{d}\tilde{d}, \quad \text{for } r < \tilde{a}. \tag{11.21}$$

The pressure distribution at the end of the indentation process equals the sum of the incremental pressure distributions

$$p(r) = \int_{d(r)}^{d} \frac{1}{\pi} \frac{E^*}{\sqrt{\tilde{a}^2 - r^2}} \mathrm{d}\tilde{d} = \int_{r}^{a} \frac{1}{\pi} \frac{E^*}{\sqrt{\tilde{a}^2 - r^2}} \frac{\mathrm{d}\tilde{d}}{\mathrm{d}\tilde{a}} \mathrm{d}\tilde{a}, \tag{11.22}$$

or under consideration of the notation (11.16)

$$p(r) = \frac{E^*}{\pi} \int_{r}^{a} \frac{1}{\sqrt{\tilde{a}^2 - r^2}} \frac{\mathrm{d}g(\tilde{a})}{\mathrm{d}\tilde{a}} \mathrm{d}\tilde{a}. \tag{11.23}$$

The function $g(a)$ from (11.16), therefore, uniquely defines both the normal force (11.20) and the pressure distribution (11.23). The normal contact problem is reduced to the determination of the function $g(a)$ (11.16).

The function $d = g(a)$ can be determined as follows. The infinitesimal surface displacement at the point $r = a$ for infinitesimal indentation by $\mathrm{d}\tilde{d}$ of a contact area of radius $\tilde{a} < a$ is, according to (11.12), equal to:

$$\mathrm{d}w(a) = \frac{2}{\pi} \arcsin\left(\frac{\tilde{a}}{a}\right) \mathrm{d}\tilde{d}. \tag{11.24}$$

The total vertical displacement at the end of the indentation process is, therefore, equal to:

$$w(a) = \frac{2}{\pi} \int_{0}^{d} \arcsin\left(\frac{\tilde{a}}{a}\right) \mathrm{d}\tilde{d} = \frac{2}{\pi} \int_{0}^{a} \arcsin\left(\frac{\tilde{a}}{a}\right) \frac{\mathrm{d}\tilde{d}}{\mathrm{d}\tilde{a}} \mathrm{d}\tilde{a}, \tag{11.25}$$

or with the notation (11.16), it is:

$$w(a) = \frac{2}{\pi} \int_{0}^{a} \arcsin\left(\frac{\tilde{a}}{a}\right) \frac{\mathrm{d}g(\tilde{a})}{\mathrm{d}\tilde{a}} \mathrm{d}\tilde{a}. \tag{11.26}$$

This vertical displacement, however, is obviously equal to $w(a) = d - f(a)$:

$$d - f(a) = \frac{2}{\pi} \int_{0}^{a} \arcsin\left(\frac{\tilde{a}}{a}\right) \frac{\mathrm{d}g\,(\tilde{a})}{\mathrm{d}\tilde{a}} \mathrm{d}\tilde{a}. \tag{11.27}$$

Integration by parts and consideration of (11.16) leads to the equation:

$$f(a) = \frac{2}{\pi} \int_{0}^{a} \frac{g(\tilde{a})}{\sqrt{a^2 - \tilde{a}^2}} \mathrm{d}\tilde{a}. \tag{11.28}$$

This is Abel's integral equation, which is solved with respect to $g(a)$ in the following way (Bracewell 1965):

$$g(a) = a \int_0^a \frac{f'(\tilde{a})}{\sqrt{a^2 - \tilde{a}^2}} d\tilde{a}. \tag{11.29}$$

With the determination of the function $g(a)$, the contact problem is completely solved.

It is easy to see that (11.16), (11.20), (11.23), (11.28), and (11.29) coincide exactly with the equations for the MDR (Chap. 2, (2.17)), which proves the validity of the MDR.

11.3 Adhesive Contact of Axisymmetric Profiles with a Compact Contact Area

The MDR described in Sect. 11.2 for the solution of non-adhesive contact of rotationally symmetric bodies can easily be generalized to adhesive contacts. The generalization is based on the fundamental idea of Johnson, Kendall, and Roberts that the contact with adhesion can be determined from *the contact without adhesion plus a rigid-body translation.* In other words, the configuration of the adhesive contact can be obtained by initially pressing the body *without consideration of the adhesion* to a certain contact radius a (Fig. 11.2a) and, subsequently, retracting to a certain critical height Δl while maintaining the constant contact area (Fig. 11.2b). Since both the indentation of any rotationally symmetric profile and the subsequent rigid body translation are mapped correctly by the MDR, this is valid also for the superposition of these two movements.

The still unknown critical length $\Delta l = \Delta l(a)$ can be determined using the principle of virtual work. According to this principle, the system is in equilibrium when the energy does not change for small variations of its generalized coordinates. Applied to the adhesive contact, it means that the change in the elastic energy for a small reduction of the contact radius from a to $a - \Delta x$ is equal to the change in

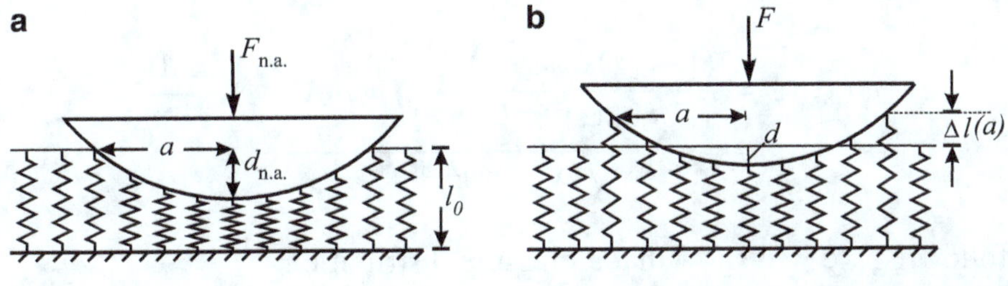

Fig. 11.2 Qualitative representation of the pressing and retraction process of a one-dimensional indenter with an elastic foundation, which exactly represents the characteristics of the adhesive contact between a rigid spherical punch and an elastic half-space. **a** Indentation without accounting for adhesion; **b** retraction with constant contact radius

the surface energy $2\pi a \Delta x \Delta \gamma$, where $\Delta \gamma$ is the separation work of the contacting surfaces per unit area. Since the MDR maps the relation of force to displacement exactly, the elastic energy is also reproduced exactly. The change in the elastic energy can, therefore, be calculated directly in the MDR model. Due to the detachment of each spring at the edge of the contact, the elastic energy is reduced by $E^* \Delta x \Delta l^2$. Balance of the changes in the elastic and the adhesive energy results in:

$$2\pi a \Delta x \Delta \gamma = E^* \Delta x \Delta l^2. \tag{11.30}$$

It follows that:

$$\Delta l = \sqrt{\frac{2\pi a \Delta \gamma}{E^*}}. \tag{11.31}$$

This criterion for the detachment of the edge springs in the adhesive MDR model was found by Heß (2011) and is known as the *rule of Heß*.

11.4 The Flat Punch Solution for FGMs

In Sect. 11.1, the solution of the contact problem between a rigid, cylindrical flat punch and an homogeneous elastic half-space was derived via the Boussinesq fundamental solution. Analogously, the corresponding solution can be found for an elastically inhomogeneous half-space with the variable elastic modulus

$$E(z) = E_0 \left(\frac{z}{c_0}\right)^k \quad \text{with } -1 < k < 1. \tag{11.32}$$

The fundamental solution in this case is given by:

$$w(r) = \frac{(1-\nu^2)\cos\left(\frac{k\pi}{2}\right) c_0^k}{h_N(k,\nu)\pi E_0} \frac{F_N}{r^{1+k}}, \tag{11.33}$$

with $h_N(k,\nu)$, which was introduced in (9.5) in Chap. 9. Using the superposition principle and (11.33) the normal displacement of the half-space surface can be determined for any pressure distribution $p(x', y')$:

$$w(x,y) = \frac{(1-\nu^2)\cos\left(\frac{k\pi}{2}\right) c_0^k}{h_N(k,\nu)\pi E_0} \iint p(x',y') \frac{\mathrm{d}x'\mathrm{d}y'}{r^{1+k}}$$
$$\text{with } r = \sqrt{(x-x')^2 + (y-y')^2}. \tag{11.34}$$

In the following we will show that a pressure distribution

$$p(r;a) = p_0 \left(1 - \frac{r^2}{a^2}\right)^{\frac{k-1}{2}}, \quad r^2 = x^2 + y^2,\ r < a \tag{11.35}$$

generates a constant normal displacement inside the circle with radius a. Utilizing the transformed variables depicted in Fig. 11.1a and by introducing the short-cuts $\alpha^2 = a^2 - r^2$ and $\beta = r\cos\varphi$, (11.35) can be written in the form:

$$p(s,\varphi) = \frac{p_0 a^{1-k}}{\left(\sqrt{\alpha^2 - 2\beta s - s^2}\right)^{1-k}}. \tag{11.36}$$

The surface normal displacement of a point within the circular area stressed by the pressure distribution (11.36) is determined by (11.34):

$$w(r;a) = \frac{(1-\nu^2)\cos\left(\frac{k\pi}{2}\right) c_0^k\, p_0 a^{1-k}}{h_N(k,\nu)\pi E_0} \int_0^{2\pi}\int_0^{s_1} \frac{1}{\left(\sqrt{\alpha^2 - 2\beta s - s^2}\right)^{1-k}} \frac{\mathrm{d}s}{s^k}\mathrm{d}\varphi, \tag{11.37}$$

where $s_1 = -\beta + \sqrt{\beta^2 + \alpha^2}$ is the positive root of the denominator under the integral in (11.37). The inner integral over s results in

$$\int_0^{s_1} \frac{1}{\left(\sqrt{\alpha^2 - 2\beta s - s^2}\right)^{1-k}} \frac{\mathrm{d}s}{s^k} = \frac{\pi}{2\cos\left(\frac{k\pi}{2}\right)} - \frac{\Gamma\left(1-\frac{k}{2}\right)\Gamma\left(\frac{1+k}{2}\right)}{\sqrt{\pi}} \frac{\beta}{\sqrt{\alpha^2+\beta^2}}\, {}_2\mathrm{F}_1\left(\frac{1}{2}, \frac{1+k}{2}; \frac{3}{2}; \frac{\beta^2}{\alpha^2+\beta^2}\right). \tag{11.38}$$

With the symmetry relation $\beta(\varphi + \pi) = -\beta(\varphi)$ it is clear that the second, hypergeometric term will not contribute after integrating over φ. Hence, the normal displacement inside the pressured ring is given by:

$$w(r;a) = \frac{(1-\nu^2) c_0^k \pi p_0 a^{1-k}}{h_N(k,\nu)E_0} =: d = \text{const}, \quad r < a. \tag{11.39}$$

As such, the pressure distribution (11.35) results from the indentation by a rigid flat cylindrical punch.

We now proceed to the calculation of the normal displacement of a point outside the loading zone. Taking into account the variables introduced in Fig. 11.1b, the pressure distribution (11.35) then takes the form:

$$p(s,\varphi) = \frac{p_0 a^{1-k}}{\left(\sqrt{\alpha^2 + 2\beta(\varphi)s - s^2}\right)^{1-k}}, \tag{11.40}$$

and the normal displacements can be determined from the integral:

$$w(r;a) = \frac{(1-\nu^2)\cos\left(\frac{k\pi}{2}\right) c_0^k p_0 a^{1-k}}{h_N(k,\nu)\pi E_0} \int_{-\varphi_1}^{\varphi_1} \int_{s_1}^{s_2} \frac{1}{\left(\sqrt{\alpha^2+2\beta s-s^2}\right)^{1-k}} \frac{\mathrm{d}s}{s^k} \mathrm{d}\varphi, \tag{11.41}$$

whereas $s_{1/2} = \beta \mp \sqrt{\beta^2+\alpha^2}$. The inner integral resolves to:

$$\int_{s_1}^{s_2} \frac{1}{\left(\sqrt{\alpha^2+2\beta s-s^2}\right)^{1-k}} \frac{\mathrm{d}s}{s^k} =$$

$$\frac{\sqrt{\pi}\Gamma\left(\frac{1+k}{2}\right)}{\Gamma\left(1+\frac{k}{2}\right)} \left(\frac{\alpha^2+\beta^2}{\beta^2}\right)^{k/2} {}_2\mathrm{F}_1\left(\frac{k}{2}, \frac{1+k}{2}; 1+\frac{k}{2}; \frac{\alpha^2+\beta^2}{\beta^2}\right). \tag{11.42}$$

The subsequent integration over φ after some transformations leads to the sought normal displacements outside the loaded area:

$$w(r;a) =$$

$$\frac{2(1-\nu^2)\cos\left(\frac{k\pi}{2}\right) c_0^k p_0 a^{1-k}}{h_N(k,\nu)E_0(1+k)} \left(\frac{a}{r}\right)^{k+1} {}_2\mathrm{F}_1\left(\frac{1+k}{2}, \frac{1+k}{2}; \frac{3+k}{2}; \frac{a^2}{r^2}\right), \tag{11.43}$$

which (taking into account (11.39)) can be written in the form:

$$w(r;a) = \frac{2d\cos\left(\frac{k\pi}{2}\right)}{\pi(1+k)} \left(\frac{a}{r}\right)^{k+1} {}_2\mathrm{F}_1\left(\frac{1+k}{2}, \frac{1+k}{2}; \frac{3+k}{2}; \frac{a^2}{r^2}\right). \tag{11.44}$$

Integration of the pressure distribution over the contact area will give the total normal force:

$$F_N(a) = \int_0^a p_0 \left(1-\frac{r^2}{a^2}\right)^{\frac{k-1}{2}} 2\pi r \mathrm{d}r = \frac{2\pi p_0 a^2}{k+1}. \tag{11.45}$$

The normal contact stiffness results from (11.39) and (11.45):

$$k_z := \frac{\mathrm{d}F_N}{\mathrm{d}d} = \frac{2h_N(k,\nu)E_0}{(1-\nu^2)c_0^k(1+k)} a^{1+k}. \tag{11.46}$$

Application of (11.39) also allows for the reformulation of the pressure distribution in terms of the indentation depth:

$$p(r;a) = \frac{h_N(k,\nu)E_0 d}{\pi(1-\nu^2)c_0^k(a^2-r^2)^{\frac{1-k}{2}}}. \tag{11.47}$$

11.5 Normal Contact of Axisymmetric Profiles with a Compact Contact Area for FGMs

Starting from the contact solution for the indentation of a power-law graded elastic half-space by a rigid, flat-ended cylindrical punch, we derive the solution of an arbitrarily shaped, axisymmetric contact problem with a compact contact area. Analogous to the derivation for elastically homogeneous materials from Sect. 11.2, we first assume that the indentation depth can be written as an unambiguous function of the contact radius:

$$d = g(a). \tag{11.48}$$

In the following, formulas for the normal force, pressure distribution, and the normal displacements of the surface are derived, all of which depend only on the function g. For this purpose we will make use of an idea by Mossakovskii (1963), according to which the solution of the general axisymmetric contact problem results from a superposition of (differential) flat-ended punch solutions. As a final step the determination of the function g will be shown.

For the calculation of the normal force, we assume the incremental contact stiffness according to (11.46) which, taking (11.48) into account, will lead to:

$$\mathrm{d}\tilde{F}_N = \frac{2h_N(k,\nu)E_0}{(1-\nu^2)c_0^k(1+k)}\tilde{a}^{1+k}\mathrm{d}\tilde{d} = \frac{2h_N(k,\nu)E_0}{(1-\nu^2)c_0^k(1+k)}\tilde{a}^{1+k}\frac{\mathrm{d}g(\tilde{a})}{\mathrm{d}\tilde{a}}\mathrm{d}\tilde{a}. \tag{11.49}$$

The indentation process can be viewed as a change in the indentation depth from $\tilde{d} = 0$ to $\tilde{d} = d$, with the contact radius changing from $\tilde{a} = 0$ to $\tilde{a} = a$ and the contact force from $\tilde{F}_N = 0$ to $\tilde{F}_N = F_N$. The normal force at the end of the process can be calculated from (11.49) by integration:

$$F_N(a) = \frac{2h_N(k,\nu)E_0}{(1-\nu^2)c_0^k(1+k)}\int_0^a \tilde{a}^{1+k}\frac{\mathrm{d}g(\tilde{a})}{\mathrm{d}\tilde{a}}\mathrm{d}\tilde{a}. \tag{11.50}$$

Integration by parts results in:

$$\begin{aligned} F_N(a) &= \frac{2h_N(k,\nu)E_0}{(1-\nu^2)c_0^k}\int_0^a \tilde{a}^k\,[g(a)-g(\tilde{a})]\,\mathrm{d}\tilde{a} \\ &= \int_{-a}^{a} c_N(\tilde{a})\,[d-g(\tilde{a})]\,\mathrm{d}\tilde{a}, \end{aligned} \tag{11.51}$$

wherein $c_N(\tilde{a})$ denotes the foundation modulus defined by (9.4).

The calculation of the pressure distribution is done in the same way. An infinitesimal indentation of an area of radius $\tilde{a}$ generates the following contribution to the pressure distribution (see (11.47)):

$$\begin{aligned} \mathrm{d}p(r;\tilde{a}) &= \frac{h_N(k,\nu)E_0}{\pi(1-\nu^2)c_0^k(\tilde{a}^2-r^2)^{\frac{1-k}{2}}}\mathrm{d}\tilde{d} \\ &= \frac{h_N(k,\nu)E_0}{\pi(1-\nu^2)c_0^k(\tilde{a}^2-r^2)^{\frac{1-k}{2}}}\frac{\mathrm{d}g(\tilde{a})}{\mathrm{d}\tilde{a}}\mathrm{d}\tilde{a} \quad \text{for } r<\tilde{a}. \end{aligned} \tag{11.52}$$

The pressure distribution at the end of the indentation process equals the sum of the incremental pressure distributions:

$$p(r) = \frac{h_N(k,\nu)E_0}{\pi(1-\nu^2)c_0^k}\int_r^a \frac{\mathrm{d}g(\tilde{a})}{\mathrm{d}\tilde{a}}\frac{\mathrm{d}\tilde{a}}{(\tilde{a}^2-r^2)^{\frac{1-k}{2}}} = -\frac{c_N(c_0)}{\pi c_0^k}\int_r^a \frac{w'_{1D}(\tilde{a})}{(\tilde{a}^2-r^2)^{\frac{1-k}{2}}}\mathrm{d}\tilde{a}, \tag{11.53}$$

wherein w_{1D} denotes the displacement of the Winkler foundation introduced by (9.9).

For the calculation of the normal displacements outside the loaded area, we first need the displacement proportion caused by an incremental indentation with the radius $\tilde{a}$. Thereby, (11.44) will provide the relation

$$\mathrm{d}w(r;\tilde{a}) = \begin{cases} \mathrm{d}\tilde{d} & \text{for } 0 \le r < \tilde{a} \\ \dfrac{2\cos\left(\frac{k\pi}{2}\right)}{\pi(1+k)}\left(\dfrac{\tilde{a}}{r}\right)^{1+k} & \\ \quad \cdot {}_2F_1\left(\dfrac{1+k}{2}, \dfrac{1+k}{2}; \dfrac{3+k}{2}; \dfrac{\tilde{a}^2}{r^2}\right)\mathrm{d}\tilde{d} & \text{for } r > \tilde{a}. \end{cases} \tag{11.54}$$

Summation over all incremental contributions $0 \le \tilde{a} \le a$, accounting for the condition $r > a$, gives:

$$w(r;a) = \\ \frac{2\cos\left(\frac{k\pi}{2}\right)}{\pi(1+k)}\int_0^a \left(\frac{\tilde{a}}{r}\right)^{1+k} {}_2F_1\left(\frac{1+k}{2}, \frac{1+k}{2}; \frac{3+k}{2}; \frac{\tilde{a}^2}{r^2}\right)\frac{\mathrm{d}g(\tilde{a})}{\mathrm{d}\tilde{a}}\mathrm{d}\tilde{a}, \tag{11.55}$$

which (after integrating by parts) results in the determining relation for the displacements outside the contact area:

$$\begin{aligned} w(r;a) &= \frac{2\cos\left(\frac{k\pi}{2}\right)}{\pi}\int_0^a \frac{\tilde{a}^k[g(a)-g(\tilde{a})]}{(r^2-\tilde{a}^2)^{\frac{1+k}{2}}}\mathrm{d}\tilde{a} \\ &= \frac{2\cos\left(\frac{k\pi}{2}\right)}{\pi}\int_0^a \frac{\tilde{a}^k w_{1D}(\tilde{a})}{(r^2-\tilde{a}^2)^{\frac{1+k}{2}}}\mathrm{d}\tilde{a} \quad \text{for } r>a. \end{aligned} \tag{11.56}$$

Equations (11.51), (11.53), and (11.56) reproduce the MDR relations (9.11) und (9.12) for the normal contact of power-law graded elastic bodies. They only depend on the yet not determined function $g(\tilde{a})$. To give an expression for this function we first calculate the normal displacements inside the contact area based on an incremental formulation and (11.54). We obtain:

$$\begin{aligned} w(r;a) = &\int_r^a \frac{\mathrm{d}g(\tilde{a})}{\mathrm{d}\tilde{a}}\mathrm{d}\tilde{a} \\ &+ \frac{2\cos\left(\frac{k\pi}{2}\right)}{\pi(1+k)} \int_0^r \left(\frac{\tilde{a}}{r}\right)^{1+k} {}_2\mathrm{F}_1\left(\frac{1+k}{2}, \frac{1+k}{2}; \frac{3+k}{2}; \frac{\tilde{a}^2}{r^2}\right) \frac{\mathrm{d}g(\tilde{a})}{\mathrm{d}\tilde{a}}\mathrm{d}\tilde{a} \\ &\text{for } r < a. \end{aligned} \tag{11.57}$$

Integration by parts results in:

$$w(r;a) = g(a) - \frac{2\cos\left(\frac{k\pi}{2}\right)}{\pi} \int_0^r \frac{\tilde{a}^k g(\tilde{a})}{(r^2-\tilde{a}^2)^{\frac{1+k}{2}}}\mathrm{d}\tilde{a} \quad \text{for } 0 < r < a, \tag{11.58}$$

which of course must equal the normal displacements due to the indenting profile:

$$w(r;a) = d - f(r) \quad \text{for } 0 < r < a. \tag{11.59}$$

Hence, the functions f and g are connected via the Abel transform

$$f(r) = \frac{2\cos\left(\frac{k\pi}{2}\right)}{\pi} \int_0^r \frac{\tilde{a}^k g(\tilde{a})}{(r^2-\tilde{a}^2)^{\frac{1+k}{2}}}\mathrm{d}\tilde{a}. \tag{11.60}$$

Inverse transform of this equation yields a determining relation for the function $g(\tilde{a})$:

$$g(\tilde{a}) = |\tilde{a}|^{1-k} \int_0^{|\tilde{a}|} \frac{f'(r)}{(\tilde{a}^2-r^2)^{\frac{1-k}{2}}}\mathrm{d}r, \tag{11.61}$$

which naturally corresponds to (9.2) in Chap. 9.

11.6 Adhesive Contact of Axisymmetric Profiles with a Compact Contact Area for FGMs

The calculation of adhesive contacts with FGMs is performed completely analogously to the case of adhesive contact of homogeneous materials. Only the Winkler foundation is redefined by (9.3):

$$\Delta k_z = c_N(x) \cdot \Delta x \tag{11.62}$$

with $c_N(x)$ defined by (9.4). The energy balance now takes on the form:

$$2\pi a \Delta x \Delta \gamma = c_N(a) \Delta x \Delta l^2. \tag{11.63}$$

It follows that:

$$\Delta l = \sqrt{\frac{2\pi a \Delta \gamma}{c_N(a)}}. \tag{11.64}$$

For a rigid indenter,

$$c_N(x) = h_N(k, \nu) E^* \left(\frac{|x|}{c_0} \right)^k \tag{11.65}$$

and (11.64) can be written in the following form:

$$\Delta l = \sqrt{\frac{2\pi a \Delta \gamma c_0^k a^{1-k}}{E^* h_N(k, \nu)}} \tag{11.66}$$

which justifies (9.45).

11.7 Tangential Contact of Axisymmetric Profiles with a Compact Contact Area

There exists a very close analogy between normal and tangential contact problems. The normal force and pressure distribution for an indentation d of a flat cylindrical indenter of radius a are given by the equations:

$$\begin{aligned} F_N &= 2E^* a d, \\ p(r) &= \frac{1}{\pi} \frac{E^* d}{\sqrt{a^2 - r^2}}, \end{aligned} \tag{11.67}$$

(see Sect. 11.1). A tangential displacement $u^{(0)}$ leads to the tangential force and stress distribution given by Johnson (1985):

$$\begin{aligned} F_x &= 2G^* a u^{(0)}, \\ \tau(r) &= \frac{1}{\pi} \frac{G^* u^{(0)}}{\sqrt{a^2 - r^2}}, \end{aligned} \tag{11.68}$$

which differs from those for the normal contact only in the notation. Now we consider the simultaneous impression of an axisymmetric profile $z = f(r)$ in the normal and tangential directions and characterize these movements through the normal and tangential displacements as functions of the contact radius:

$$d = g(a), \quad u^{(0)} = h(a). \tag{11.69}$$

The force and stress during the indentation starting at initial contact are then:

$$F_N = 2E^* \int_0^a \tilde{a} \frac{\mathrm{d}g(\tilde{a})}{\mathrm{d}\tilde{a}} \mathrm{d}\tilde{a}, \quad p(r) = \frac{E^*}{\pi} \int_r^a \frac{1}{\sqrt{\tilde{a}^2 - r^2}} \frac{\mathrm{d}g(\tilde{a})}{\mathrm{d}\tilde{a}} \mathrm{d}\tilde{a} \tag{11.70}$$

and

$$F_x = 2G^* \int_0^a \tilde{a} \frac{\mathrm{d}h(\tilde{a})}{\mathrm{d}\tilde{a}} \mathrm{d}\tilde{a}, \quad \tau(r) = \frac{G^*}{\pi} \int_r^a \frac{1}{\sqrt{\tilde{a}^2 - r^2}} \frac{\mathrm{d}h(\tilde{a})}{\mathrm{d}\tilde{a}} \mathrm{d}\tilde{a}. \tag{11.71}$$

We now consider the following two-step process. The punch is initially pressed in a purely normal motion until the contact radius c is reached, and then further pressed in such a way that it moves simultaneously normally and tangentially until contact radius a is reached, whereby

$$\mathrm{d}h = \lambda \cdot \mathrm{d}g. \tag{11.72}$$

The normal force and the pressure distribution at the end of this process is still given by (11.70), while the tangential force and stress distribution obviously result in:

$$F_x = 2G^* \int_c^a \tilde{a} \frac{\mathrm{d}h(\tilde{a})}{\mathrm{d}\tilde{a}} \mathrm{d}\tilde{a} = 2G^* \lambda \int_c^a \tilde{a} \frac{\mathrm{d}g(\tilde{a})}{\mathrm{d}\tilde{a}} \mathrm{d}\tilde{a} \tag{11.73}$$

and

$$\tau(r) = \begin{cases} \dfrac{G^*}{\pi} \lambda \displaystyle\int_c^a \frac{1}{\sqrt{\tilde{a}^2 - r^2}} \frac{\mathrm{d}g(\tilde{a})}{\mathrm{d}\tilde{a}} \mathrm{d}\tilde{a}, & \text{for } r < c, \\ \dfrac{G^*}{\pi} \lambda \displaystyle\int_r^a \frac{1}{\sqrt{\tilde{a}^2 - r^2}} \frac{\mathrm{d}g(\tilde{a})}{\mathrm{d}\tilde{a}} \mathrm{d}\tilde{a}, & \text{for } c < r < a. \end{cases} \tag{11.74}$$

In the area $c < r < a$, the normal and tangential stress distribution have the same form:

$$\tau(r) = \lambda \frac{G^*}{E^*} p(r). \tag{11.75}$$

If we set

$$\lambda \frac{G^*}{E^*} = \mu \tag{11.76}$$

then the examined contact will have the following properties:

$$\begin{aligned} u(r) &= u^{(0)} = \text{const}, && \text{for } r < c, \\ \tau(r) &= \mu p(r), && \text{for } c < r < a. \end{aligned} \tag{11.77}$$

These conditions correspond exactly to the stick and slip conditions in a tangential contact with the coefficient of friction μ. Therefore, the force (11.73) and the stress distribution (11.74) under consideration of (11.75) solve the tangential contact problem:

$$F_x = 2\mu E^* \int_c^a \tilde{a}\frac{\mathrm{d}g(\tilde{a})}{\mathrm{d}\tilde{a}}\mathrm{d}\tilde{a},$$

$$\tau(r) = \begin{cases} \mu\frac{E^*}{\pi}\left(\int_r^a \frac{1}{\sqrt{\tilde{a}^2 - r^2}}\frac{\mathrm{d}g(\tilde{a})}{\mathrm{d}\tilde{a}}\mathrm{d}\tilde{a} \right. \\ \qquad \left. - \int_r^c \frac{1}{\sqrt{\tilde{a}^2 - r^2}}\frac{\mathrm{d}g(\tilde{a})}{\mathrm{d}\tilde{a}}\mathrm{d}\tilde{a}\right), & \text{for } r < c, \\ \mu\frac{E^*}{\pi}\int_r^a \frac{1}{\sqrt{\tilde{a}^2 - r^2}}\frac{\mathrm{d}g(\tilde{a})}{\mathrm{d}\tilde{a}}\mathrm{d}\tilde{a}, & \text{for } c < r < a, \end{cases} \tag{11.78}$$

or

$$\tau(r) = \mu\begin{cases} p(r;a) - p(r;c), & \text{for } r < c, \\ p(r;a), & \text{for } c < r < a, \end{cases} \tag{11.79}$$

where we denote the normal pressure distribution at the contact radii a and c by $p(r;a)$ and $p(r;c)$, respectively. The tangential displacement in the contact is obtained by integration of (11.72):

$$u^{(0)} = \mu\frac{E^*}{G^*}[g(a) - g(c)]. \tag{11.80}$$

It is easy to see that (11.78) and (11.80) coincide with (4.39), thereby proving the validity of the MDR for tangential contacts.

11.8 Definitions of Special Functions Used in this Book

At certain points in this book, non-elementary functions have been utilized. For ease of reference, we have provided a short summary of their definitions and significant properties.

11.8.1 Elliptical Integrals

Several definite integrals that cannot be solved in the form of elementary functions are themselves defined as (non-elementary) functions. The so-called elliptical integrals belong to this class of functions. Let k be an elliptical modulus with $0 \leq k^2 \leq 1$ and $0 \leq \theta \leq \pi/2$. The incomplete elliptical integrals of the first and second kind are defined as:

$$\mathrm{F}(\theta, k) := \int_0^\theta \frac{\mathrm{d}\varphi}{\sqrt{1-k^2\sin^2\varphi}},$$

$$\mathrm{E}(\theta, k) := \int_0^\theta \sqrt{1-k^2\sin^2\varphi}\mathrm{d}\varphi. \tag{11.81}$$

For the case of $\theta = \pi/2$, this results in the so-called complete elliptical integrals:

$$\mathrm{K}(k) := \int_0^{\pi/2} \frac{\mathrm{d}\varphi}{\sqrt{1-k^2\sin^2\varphi}} = \mathrm{F}\left(\frac{\pi}{2}, k\right),$$

$$\mathrm{E}(k) := \int_0^{\pi/2} \sqrt{1-k^2\sin^2\varphi}\mathrm{d}\varphi = \mathrm{E}\left(\frac{\pi}{2}, k\right). \tag{11.82}$$

It should be noted that several mathematical databases provide the elliptical integrals as a function of the modulus $m = k^2$.

11.8.2 The Gamma Function

Euler's Gamma function can be defined in an integral form by the expression:

$$\Gamma(z) := \int_0^\infty t^{z-1}\exp(-t)\mathrm{d}t, \quad \mathrm{Re}\{z\} > 0. \tag{11.83}$$

It is easy to prove the recursion property

$$\Gamma(z) = (z-1)\Gamma(z-1). \tag{11.84}$$

The Gamma function is, therefore, a generalization of the faculty function for arbitrary complex arguments, since (11.84) immediately returns the following relationship for positive integer arguments of the function:

$$\Gamma(n) = (n-1)!, \quad n \in \mathbb{N}^*. \tag{11.85}$$

The complete Gamma function defined in (11.83) can be generalized to the "lower" or "upper" incomplete Gamma function by restricting the integration limits,

$$\Gamma_o(z,a) := \int_a^\infty t^{z-1}\exp(-t)\mathrm{d}t, \quad \mathrm{Re}\{z\} > 0,\ a \in \mathbb{R}^+$$

$$\Gamma_u(z,a) := \int_0^a t^{z-1}\exp(-t)\mathrm{d}t, \quad \mathrm{Re}\{z\} > 0,\ a \in \mathbb{R}^+ \tag{11.86}$$

with the obvious property

$$\Gamma_o(z,a) + \Gamma_u(z,a) = \Gamma(z). \tag{11.87}$$

11.8.3 The Beta Function

The complete Beta function relates to the Gamma function according to the definition:

$$\mathrm{B}(x,y) := \frac{\Gamma(x)\Gamma(y)}{\Gamma(x,y)}, \quad \mathrm{Re}\{x,y\} > 0. \tag{11.88}$$

The integral definition can also be shown in this representation:

$$\mathrm{B}(x,y) = \int_0^1 u^{x-1}(1-u)^{y-1}\mathrm{d}u, \quad \mathrm{Re}\{x,y\} > 0. \tag{11.89}$$

The Beta function is clearly symmetric, with $\mathrm{B}(x,y) = \mathrm{B}(y,x)$. From (11.85) and (11.88) it is apparent that the binomial coefficient

$$\begin{aligned}\binom{n}{k} &:= \frac{n!}{k!(n-k)!} = \frac{\Gamma(n+1)}{\Gamma(k+1)\Gamma(n-k+1)} \\ &= \frac{1}{(n+1)\mathrm{B}(k+1,n-k+1)}, \quad n,k \in \mathbb{N},\ n \geq k\end{aligned} \tag{11.90}$$

can be expressed by the Beta function. From the definition of the Beta function also follows the recursion property

$$\mathrm{B}(x,y) = \mathrm{B}(x+1,y) + \mathrm{B}(x,y+1). \tag{11.91}$$

By restricting the integration limits in (11.89), the complete Beta function can be generalized to the "lower" or "upper" incomplete Beta function. Usually the upper limit is set, resulting in the definition

$$\mathrm{B}(z;x,y) = \int_0^z u^{x-1}(1-u)^{y-1}\mathrm{d}u, \quad \mathrm{Re}\{x,y\} > 0,\ z \in [0;1]. \tag{11.92}$$

11.8.4 The Hypergeometric Function

Many different hypergeometric functions exist. In this book, only the Gaussian hypergeometric function

$$ {}_2\mathrm{F}_1(a,b;c;z) := \sum_{n=0}^{\infty} \frac{\Gamma(a+n)\Gamma(b+n)\Gamma(c)}{\Gamma(a)\Gamma(b)\Gamma(c+n)} \frac{z^n}{n!}. \tag{11.93} $$

is used. It is the solution of the hypergeometric differential equation

$$ z(1-z)y''(z) + [c-(a+b+1)z]y'(z) - aby(z) = 0. \tag{11.94} $$

The derivative of this function is given by:

$$ \frac{\mathrm{d}}{\mathrm{d}z}\left[{}_2\mathrm{F}_1(a,b;c;z)\right] = \frac{ab}{c}\, {}_2\mathrm{F}_1(a+1,b+1;c+1;z). \tag{11.95} $$

Certain special cases of the hypergeometric function can be expressed by elementary functions, such as:

$$ {}_2\mathrm{F}_1(1,1;1;z) = \frac{1}{1-z}. \tag{11.96} $$

11.8.5 The Struve H-Function

The Struve H-function is a Bessel-type function. It can be expanded to the power series

$$ \mathrm{H}_n(z) := \sum_{k=0}^{\infty} \frac{(-1)^k}{\Gamma\left(k+\frac{3}{2}\right)\Gamma\left(k+n+\frac{3}{2}\right)} \left(\frac{z}{2}\right)^{2k+n+1}, \quad z \in \mathbb{C} \tag{11.97} $$

and is the solution of the inhomogeneous Bessel differential equation

$$ y''(z)z^2 + y'(z)z + (z^2 - n^2)y(z) = \frac{4}{\sqrt{\pi}\,\Gamma(n+1/2)} \left(\frac{z}{2}\right)^{n+1}. \tag{11.98} $$

Additionally, the following differentiation property of the Struve H-function can be demonstrated:

$$ \frac{\mathrm{d}}{\mathrm{d}z}\left[\mathrm{H}_n(z)\right] = \mathrm{H}_{n-1}(z) - \frac{n}{z}\mathrm{H}_n(z). \tag{11.99} $$

11.9 Solutions to Axisymmetric Contact Problems According to Föppl and Schubert

In this section, we will present the derivation of the solution for the contact problem of an arbitrary axially symmetric indenter, which was published by Föppl (1941) and Schubert (1942). This is the first known publication to perform the derivation of the equations that lay the foundation for this book. Over the course of the history of contact mechanics, they were derived multiple times using different approaches. For instance, they were later found by Galin (1946) as well (likely independently). They reached a high degree of international recognition through the paper of Sneddon (1965)—one of the most cited publications in the history of contact mechanics. Yet Sneddon merely offered a different derivation approach to obtain already known solutions, including those of Galin (whom he cites). And much later, new interpretations and derivations of the same equations were published multiple times. Some of these were quite useful since they provided a new perspective on the issues and facilitated different generalizations and developments. This includes the interpretations by Jäger (1995), who treated the indentation of a curved body as the superposition of infinitesimal indentations of flat cylindrical punches (although even this idea was not original and was previously used by Mossakovskii). The MDR is also based on the equations of Föppl–Schubert–Galin–Sneddon–Jäger, however, it offers an intuitive physical–mnemonic interpretation, which can be directly generalized to many other contact problems.

Although 77 years have passed since the publications of Föppl and Schubert, this first historic derivation still surprisingly remains the most direct and simple of all. It is valuable to understand this derivation both for historic and didactic reasons. The presentation of the derivation of Föppl and Schubert very closely follows the original publication, however, modified notations are used to highlight the direct connection to the equations of the MDR.

We consider an axially symmetric pressure distribution $p(r)$ in a circle of radius a (see Fig. 11.3). We calculate the displacement at the "point of observation" A caused by an infinitesimally small force in the "source point" B and then integrate over all source points. The location of the source point is parametrized by the linear coordinate s and the angle φ. The vertical displacement of point A by the force $\mathrm{d}F_N = p(\rho)s\mathrm{d}s\mathrm{d}\varphi$ in point B is given by the fundamental solution (2.2):

$$\mathrm{d}w(r) = \frac{p(\rho)s\mathrm{d}s\mathrm{d}\varphi}{\pi E^* s} = \frac{1}{\pi E^*}p(\rho)\mathrm{d}s\mathrm{d}\varphi. \tag{11.100}$$

The total vertical displacement caused by the entire pressure distribution results from the integration of

$$w(r) = \frac{1}{\pi}\int_0^{\pi}\left(\frac{1}{E^*}\int_{s_1}^{s_2} p(\rho)\mathrm{d}s\right)\mathrm{d}\varphi. \tag{11.101}$$

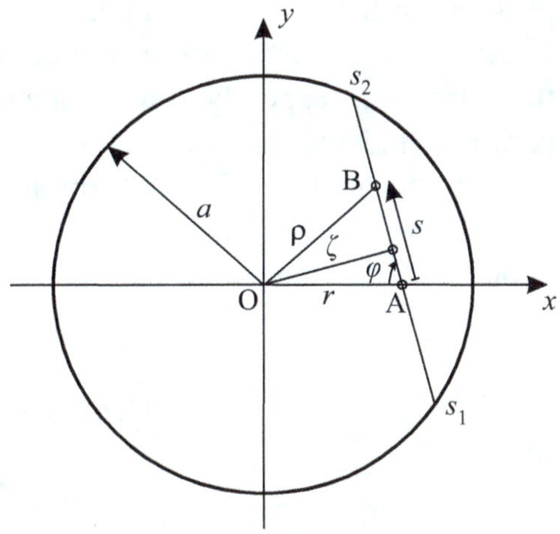

Fig. 11.3 Schematic diagram to display the notation proposed by Föppl (1941)

In his paper, Föppl (1941) proposed replacing the parametrization s and φ with new variables ρ and ξ, which uniquely define the location of point B too, and relate to s and φ according to the following equations:

$$s = \sqrt{\rho^2 - \xi^2} + \sqrt{r^2 - \xi^2}, \quad \xi \leq \rho \leq a,$$
$$\varphi = \arcsin\left(\frac{\xi}{r}\right), \quad 0 \leq \xi \leq r. \tag{11.102}$$

Accordingly, the derivations are:

$$\frac{\partial s}{\partial \rho} = \frac{\rho}{\sqrt{\rho^2 - \xi^2}},$$
$$\frac{\partial \varphi}{\partial \xi} = \frac{1}{\sqrt{r^2 - \xi^2}}. \tag{11.103}$$

Denoting the expression in parenthesis in (11.101) by $w_{1D}(\xi)$:

$$w_{1D}(\xi) = \frac{1}{E^*}\int_{s_1}^{s_2} p(\rho)\mathrm{d}s = \frac{2}{E^*}\int_{\xi}^{a} \frac{p(\rho)\rho\mathrm{d}\rho}{\sqrt{\rho^2 - \xi^2}}. \tag{11.104}$$

For the vertical displacement (11.101), we then obtain:

$$w(r) = \frac{2}{\pi}\int_0^{\pi/2} w_{1D}(\xi)\mathrm{d}\varphi = \frac{2}{\pi}\int_0^{r} \frac{w_{1D}(\xi)\mathrm{d}\xi}{\sqrt{r^2 - \xi^2}}. \tag{11.105}$$

Both (11.104) and (11.105) are identical to (2.16) and (2.14) of the MDR. These were already derived in the paper by Föppl (1941), as previously described in this chapter, and enable the calculation of the displacement field resulting from a known pressure distribution. In his paper, Föppl examined the pressure distributions $(1 - r^2/a^2)^{-1/2}$, $(1 - r^2/a^2)^{1/2}$ and a constant pressure distribution, demonstrating that the former corresponds to a constant displacement and the latter corresponds to a parabolic indenter.

The contribution of his doctoral candidate, Schubert (1942), was the inversion of the integral equations (11.104) and (11.105). Since these are Abel transforms, Schubert found the solutions

$$w_{1D}(\xi) = \xi \int_0^{\xi} \frac{w'(\rho)}{\sqrt{\xi^2 - \rho^2}} \mathrm{d}\rho \tag{11.106}$$

and

$$p(\rho) = -\frac{E^*}{\pi} \int_{\rho}^{a} \frac{w'_{1D}(\xi)}{\sqrt{\xi^2 - \rho^2}} \mathrm{d}\xi, \tag{11.107}$$

which are identical to (2.6) and (2.13) of the MDR.

Equations (11.106) and (11.107) completely solve the contact problem: with three-dimensional form $w(\rho)$ and using (11.106), one can calculate the auxiliary function $w_{1D}(\xi)$, which then determines the pressure distribution by (11.107). Schubert used this approach to solve the contact problems of the flat punch, the cone, and the power-law profiles of second, fourth, and sixth-order, the concave power-law profiles of the second and fourth-order, and the cylindrical indenter with rounded edges.

The publications of Föppl and Schubert of course did not contain the MDR interpretations of their equations, which requires a couple of additional steps. In the interpretation of the MDR, $w_{1D}(\xi)$ is the vertical displacement in the equivalent MDR model. The necessary property for the transition to the MDR interpretation is the relation $w_{1D}(\xi = 0) = w(r = 0)$ (which guarantees that indentation depth of the three-dimensional profile is also the indentation depth of the equivalent MDR profile). It follows from (11.105), when the limit $w_{1D}(\xi = 0)$ is substituted for $w_{1D}(\xi)$ in the limit case $r \to 0$ and the identity

$$\frac{2}{\pi} \int_0^{r} \frac{\mathrm{d}\xi}{\sqrt{r^2 - \xi^2}} \equiv 1. \tag{11.108}$$

is taken into account. Then

$$\begin{aligned} w(r = 0) = \lim_{r \to 0}[w(r)] &= w_{1D}(\xi = 0) \cdot \frac{2}{\pi} \int_0^{r} \frac{\mathrm{d}\xi}{\sqrt{r^2 - \xi^2}} \\ &= w_{1D}(\xi = 0). \end{aligned} \tag{11.109}$$

It follows trivially from (11.104) that the contact radius is determined by the equation $w_{1D}(a) = 0$. The equation determining the force (2.11) follows from (11.107):

$$F_N = 2\pi \int_0^a p(r) r \mathrm{d}r = -2E^* \int_0^a \left(\int_\rho^a \frac{w_{1D}'(\xi)}{\sqrt{\xi^2 - r^2}} \mathrm{d}\xi \right) r \mathrm{d}r$$

$$= -2E^* \int_0^a w_{1D}'(\xi) \left(\int_0^\xi \frac{r \mathrm{d}r}{\sqrt{\xi^2 - r^2}} \right) \mathrm{d}\xi$$

$$= -2E^* \int_0^a \xi w_{1D}'(\xi) \mathrm{d}\xi = 2E^* \int_0^a w_{1D}(\xi) \mathrm{d}\xi. \quad (11.110)$$

Thus, all fundamental equations of the MDR are determined and the only task remaining is to "put them into words".

In closing, it should be noted that the publication by Schubert also contained the complete solution of the plane contact problem, which he applied to the following profiles: flat punch with symmetric load, flat punch with axisymmetric load, wedge profiles, parabolically rounded wedge profile, power-law profiles of the second, fourth, and sixth-order, concave power-law profiles of the second and fourth-order, and the flat punch with rounded edges.

It is most regrettable that this excellent work, which in itself nearly represents a small "handbook of contact mechanics", remained nearly unknown for a long time and has only just been "rediscovered" in recent years.

References

Bracewell, R.: The Fourier Transform and Its Applications. McGraw-Hill, New York (1965)

Föppl, L.: Elastische Beanspruchung des Erdbodens unter Fundamenten. Forsch. Geb. Ingenieurwes. A **12**(1), 31–39 (1941)

Galin, L.A.: Three-dimensional contact problems of the theory of elasticity for punches with a circular planform. Prikladnaya Matematika I Mekhanika **10**, 425–448 (1946)

Heß, M.: Über die exakte Abbildung ausgewählter dreidimensionaler Kontakte auf Systeme mit niedrigerer räumlicher Dimension. Cuvillier, Göttingen (2011)

Jäger, J.: Axi-symmetric bodies of equal material in contact under torsion or shift. Arch. Appl. Mech. **65**, 478–487 (1995)

Johnson, K.L.: Contact mechanics. Cambridge University Press, Cambridge (1985)

Mossakovskii, V.I.: Compression of elastic bodies under conditions of adhesion (axi-symmetric case). J. Appl. Math. Mech. **27**(3), 630–643 (1963)

Popov, V.L.: Kontaktmechanik und Reibung. Von der Nanotribologie bis zur Erdbebendynamik, 3rd edn. Springer, Heidelberg (2015). ISBN 978-3-662-45974-4. 397 p.

Schubert, G.: Zur Frage der Druckverteilung unter elastisch gelagerten Tragwerken. Ing. Arch. **13**(3), 132–147 (1942)

Sneddon, I.N.: The relation between load and penetration in the axi-symmetric Boussinesq problem for a punch of arbitrary profile. Int. J. Eng. Sci. **3**(1), 47–57 (1965)

Index

V.L. Popov, M. Heß, E. Willert, *Handbook of Contact Mechanics*,
https://doi.org/10.1007/978-3-662-58709-6

www.ingramcontent.com/pod-product-compliance
Lightning Source LLC
LaVergne TN
LVHW060056220926
848484LV00008BA/1953